Algebra

SERIES ON UNIVERSITY MATHEMATICS

Forthcoming

Vol. 1: Differential Geometry
S S Chern

Vol. 2: Lie Group
W Y Hsiang

Vol. 3: Classical Geometries
W Y Hsiang

Vol. 4: Foundation of Linear Algebra
W Y Hsiang & T T Moh

SERIES ON UNIVERSITY MATHEMATICS VOL. 5

ALGEBRA

T T Moh
Department of Mathematics
Purdue University
USA

World Scientific
Singapore • New Jersey • London • Hong Kong

Published by

World Scientific Publishing Co. Pte. Ltd
5 Toh Tuck Link, Singapore 596224
USA offic 27 Warren Street, Suite 401-402, Hackensack, NJ 07601
UK offic 57 Shelton Street, Covent Garden, London WC2H 9HE

British Library Cataloguing-in-Publication Data
A catalogue record for this book is available from the British Library.

Series on University Mathematics — Vol. 5
ALGEBRA

ISBN-13 978-981-02-1195-0
ISBN-10 981-02-1195-3
ISBN-13 978-981-02-1196-7 (pbk)
ISBN-10 981-02-1196-1 (pbk)

PREFACE

Algebra is generous : she often gives more than is asked of her.

D'Alembert

The present book comes from the first part of the lecture notes I used for a first-year graduate algebra course at the University of Minnesota, Purdue University, and Peking University. The Chinese versions of these notes were published by The Peking University Press in 1986, and by Linking Publishing Co of Taiwan in 1987.

The aim of this book is not only to give the student quick access to the basic knowledge of algebra, either for future advancement in the field of algebra, or for general background information, but also to show that algebra is truly a master key or a 'skeleton key' to many mathematical problems. As one knows, the teeth of an ordinary key prevent it from opening all but one door, whereas the skeleton key keeps only the essential parts, allowing it to unlock many doors.

Sometimes I like to think that 'fashion' is a space-traveler, while 'wisdom' is a time-traveler. Frequently, the time-traveler only touches a small circle among the elite. Most people think that Mathematics is dry and difficult. Most mathematicians feel the same way towards algebra. How unfortunate! When Heisenberg presented his quantum theory, he had to re-invent matrix theory. Mathematicians, and algebraists especially, should present the subject more interestingly to attract the attention of the student and the concerned reader.

I wish to present this book as an attempt to help the student to re-establish the contacts between algebra and other branches of mathematics and sciences. I prefer the intuitive approaches to algebra, and have included many examples and exercises to illustrate its power. I hope that the present book fulfills these goals.

To teach a core course for one semester, the materials of §6, Chapter I, §7, Chapter II, §4, Chapter III, §3 -§7, Chapter IV, part of §3 and §8 -§9, Chapter V may be omitted.

We wish to thank Jem Corcoran for proof-reading.

T.T.Moh
W. Lafayette, 1992

CONTENTS

Algebra

CHAPTER I

Set Theory and Number Theory

§1 Set Theory

We shall assume the elementary concepts of set[1] theory in this book as the *union*, the *intersection*, the *inclusion* and the *mapping*. The set theory symbols used in this book will be listed in the appendix I.

Definition 1.1. *Let* S *and* T *be sets,* ρ: $\mathbf{S} \to \mathbf{T}$ *be a map from the set* S *to the set* T. *If* $\rho(s_1) = \rho(s_2)$ *implies* $s_1 = s_2$ *for any two elements* $s_1, s_2 \in \mathbf{S}$, *then we say the map* ρ *is 1–1 or* ***injective***. *If for any given* $t \in \mathbf{T}$, *there is a* $s \in \mathbf{S}$ *with* $\rho(s) = t$, *then we say the map* ρ *is onto or* ***surjective***. *If* ρ *is injective and surjective, then we say that* ρ *is* ***bijective***. ∎

One of the most important concepts in the set theory is the *cardinal number*. We have the following definition,

Definition 1.2. *Let* S *and* T *be sets. If there is a bijective map* ρ: $\mathbf{S} \to \mathbf{T}$, *then we say that* S *and* T *have the same* ***cardinality***. ∎

Discussion

(1) If the cardinalities of the set **S** and the set of integers $\{1, 2, \cdots, n\}$ are the same, then we say the set **S** is a *finite set*, and the cardinality of **S** is n. Otherwise, the set **S** is said to be an *infinite set*. Furthermore, if the cardinalities of the set **S** and the set of all positive integers $\{1, 2, \cdots, n, \cdots\}$ are the same, then we say that the set **S** is a *countably infinite set*. If a set **S** is either finite or countably infinite, then we say the set **S** is *countable*. If the set **S** is not countable, then we say the set **S** is *uncountable*.

[1]Due to German Mathematician Cantor 1874.

Certainly the set of non-negative integers $\{0, 1, 2, 3, 4, \cdots\}$ is countable. Furthermore the set of all integers $\mathbf{Z}$ can be listed as $\{0, -1, 1, -2, 2, \cdots, -i, i, \cdots\}$, hence is also countable.

(2) ***Pigeon hole principle***: Given two finite sets $\mathbf{S}$ and $\mathbf{T}$ with the same cardinality. Let ρ be a map $\mathbf{S} \to \mathbf{T}$. Then ρ is injective $\Leftrightarrow$ surjective $\Leftrightarrow$ bijective.

One may imagine that the set $\mathbf{S}$ is a finite set of pigeons and the set $\mathbf{T}$ is the set of the pigeon holes with the same cardinality. Then the above principle says that if every pigeon gets into a different hole, then every hole is occupied. On the other hand, if every hole is occupied, then every pigeon must get into a different hole.

(3) In fact, the pigeon hole principle is true only for the finite cardinalities. If a set $\mathbf{S}$ is infinite, then it follows from the set theory (using some form of the *axiom of choice*) that there is a countably infinite subset $\mathbf{R}$. It is easy to use the argument of *Hilbert's hotel* as follows to show that the pigeon hole principle is false: let there be a hotel with countably infinite many rooms $\{r_1, r_2, \cdots\}$ filled with guests $\{g_1, g_2, \cdots\}$. Suppose that there appear countably infinite many new guests $\{n_1, n_2, \cdots\}$. A simple way of management is to ask the old guest g_m to move to the room r_{2m}, and assign the new guest n_m to the room r_{2m-1}. It is clear that all guests, old or new, will each have a room. Imitating this example, the reader will have no trouble to set a mapping ρ: $\mathbf{R} \to \mathbf{R}$ which is injective while not surjective. Furthermore, we may extend the map ρ to $\mathbf{S}$ by defining $\rho(s) = s$ for all $s \notin \mathbf{R}$. Then it is easy to see the extension of ρ is injective while not surjective. On the other hand, we may require the first two guests g_1, g_2 to stay in the first room r_1, while the guest g_m stays in the room r_{m-1} for all $m \geq 3$. Then we construct a map which is surjective while not injective. ∎

Theorem 1.1. *Let the sets* $\mathbf{S}_i$ *be countable sets for* $i = 1, 2, \cdots$. *Then the union set* $\mathbf{S} = \bigcup_{i=1}^{\infty} \mathbf{S}_i$ *is a countable set.*

Proof. Let the set $\mathbf{S}_i = \{a_{i1}, a_{i2}, \cdots, a_{ij}, \cdots\}$. We shall use the following *triangle counting* to form a sequence,

$$
\begin{array}{l}
a_{11}, a_{12}, a_{13}, \cdots, a_{1j}, \cdots \\
\quad \nearrow \quad \nearrow \quad \nearrow \qquad\quad \nearrow \\
a_{21}, a_{22}, a_{23}, \cdots, a_{2j}, \cdots \\
\quad \nearrow \quad \nearrow \quad \nearrow \qquad\quad \nearrow \\
a_{31}, a_{32}, a_{33}, \cdots, a_{3j}, \cdots \\
\cdots\cdots\cdots\cdots\cdots\cdots\cdots \\
\cdots\cdots\cdots\cdots\cdots\cdots\cdots \\
\quad \nearrow \quad \nearrow \quad \nearrow \qquad\quad \nearrow \\
a_{i1}, a_{i2}, a_{i3}, \cdots, a_{ij}, \cdots \\
\cdots\cdots\cdots\cdots\cdots\cdots\cdots
\end{array}
$$

In other words, let us define a sequence $\{c_1, c_2, \cdots, c_k, \cdots\}$ with $c_k = a_{ij}$ where $k = ((i+j-1)(i+j-2)/2) + j$. In this sequence let us delete all c_k which equals to c_n for some $n < k$. Then it is obvious that the set of the deleted sequence is the union set $\mathbf{S}$. Thus we establish that the union set $\mathbf{S}$ is countable. ∎

Corollary. *The set of rational numbers* $\mathbf{Q}$ *is countable.*

Proof: We always have $\mathbf{Q}=\bigcup_{i=1}^{\infty}\frac{1}{i}\mathbf{Z}$, while $\frac{1}{i}\mathbf{Z}$ is a countable set for each i. Our Corollary follows from the preceding Proposition. ∎

Theorem 1.2. *The set of real numbers* $\mathbf{R}$ *is uncountable.*

Proof: Suppose that the set of real numbers $\mathbf{R}$ is countable, i.e., $\mathbf{R}=\{r_1,r_2,\cdots,r_i,\cdots\}$. We will use the following ***diagonal counting*** to deduce a contradiction.

Let r_i be expressed as the decimal number $r_i = a_i.b_{i1}b_{i2}\cdots b_{ij}\cdots$, where a_i is the integer part of r_i and $0 \le b_{ij} \le 9$. We have the following diagram,

$$\begin{array}{l} r_1=a_1.b_{11}b_{12}\cdots b_{1j}\cdots \\ r_2=a_2.b_{21}b_{22}\cdots b_{2j}\cdots \\ r_3=a_3.b_{31}b_{32}\cdots b_{3j}\cdots \\ \cdots\cdots\cdots\cdots\cdots \\ \cdots\cdots\cdots\cdots\cdots \\ r_i=a_i.b_{i1}b_{i2}\cdots b_{ij}\cdots \\ \cdots\cdots\cdots\cdots\cdots \end{array}$$

Although there are some ambiguities about the decimal expressions of the real numbers, i.e., $2.340000\cdots = 2.339999\cdots$, however, it will not interfere with the following arguments. Let the real number $r = 0.c_1c_2\cdots c_i\cdots$ be defined $c_i = 5$ if $b_{ii} \neq 5$, otherwise $c_i = 4$. Then r is a real number which is not in the list $\{r_1,r_2,\cdots,r_i,\cdots\}$. Thus we establish that the set of real numbers $\mathbf{R}$ is not countable. ∎

In Algebra the processes of taking the "***direct product***" and "***quotient***" are commonly used. Their definitions are given as follows,

Definition 1.3. *Let* $\{\mathbf{S}_i\}$, $i \in \mathbf{I}$, *be a collection of sets* $\mathbf{S}_i$. *Then the direct product of the set* $\{\mathbf{S}_i\}$, $\prod_{i\in I}\mathbf{S}_i$, *is defined to be the set* $\{(s_i)_{i\in I} : s_i \in \mathbf{S}_i\}$, *i.e., the set of all maps* $s : I \to \bigcup_{i\in I}\mathbf{S}_i$ *with* $s(i) = s_i \in \mathbf{S}_i$. *The set* $\mathbf{I}$ *will be called as the **index set** of the direct product. If the index set* $\mathbf{I}$ *is the set of positive integers* $\mathbf{Z}_+$, *sometimes we write the elements of the direct product as the sequences* $\{s_1, s_2, \cdots\}$. ∎

Definition 1.4. *Let* $\mathbf{T}$ *and* $\mathbf{I}$ *be sets, and let* $\mathbf{T}_i$, $i \in \mathbf{I}$, *be subsets of* $\mathbf{T}$ *such that* $\mathbf{T}$ *is a **disjoint union** of* $\mathbf{T}_i$, *i.e.,* $\mathbf{T}=\bigcup_{i\in I}\mathbf{T}_i$ *and* $\mathbf{T}_i\bigcap\mathbf{T}_j = \emptyset$ *for* $i \neq j$, *then the set of the subsets* $\{\mathbf{T}_i\}$ *is called a **quotient set** of* $\mathbf{T}$. ∎

Discussion

(1) Let $\mathbf{T}$ be the set of the people of a country. According to the law, the set $\mathbf{T}$ may be separated into the subsets: $\mathbf{T}_1$ = the set of all juveniles and $\mathbf{T}_2$ = the set of all adults. Then $\{\mathbf{T}_1, \mathbf{T}_2\}$ is a quotient set of $\mathbf{T}$ with two elements.

(2) Another way to discuss the quotient set is through the concept of ***equivalence relations*** which is defined as follows. A relation "$\sim$" is said to be an equivalence relation if and only if the following conditions are satisfied,

(r) Reflexion: $t \sim t$ for all $t \in \mathbf{T}$.

(s) Symmetry: $t \sim s \Longrightarrow s \sim t$.
(t) Transition: $t \sim s, s \sim r \Longrightarrow t \sim r$.

(3) Let $\{\mathbf{T}_i\}$ be a quotient set of $\mathbf{T}$. Then we may define a relation "$\approx$" as follows,

$$t \approx s \Longleftrightarrow t, s \in \text{ the same } \mathbf{T}_j$$

Then it is easy to see that the relation $\approx$ is indeed an equivalence relation.

(4) Suppose that we are given an equivalence relation $\sim$. Let the subset $\mathbf{T}_t$ be defined as $\mathbf{T}_t = \{s : s \sim t\}$. Then it is easy to see that $\mathbf{T}$ is a disjoint union of $\{\mathbf{T}_t\}$. Hence $\{\mathbf{T}_t\}$ is a quotient set of $\mathbf{T}$. ∎

To sum up the above discussions (3) & (4), we have the following new definition,

Definition 1.4*. *Let "$\sim$" be an equivalence relation on a set* $\mathbf{T}$. *Let* $\mathbf{T}_t = \{s : s \sim t\}$. *The subset* $\mathbf{T}_t$ *is called an equivalence subset of* $\mathbf{T}$. *Then* $\{\mathbf{T}_t\}$ *is a called the* ***quotient set*** *of* $\mathbf{T}$ *with respect to the relation "$\sim$".* ∎

Let us now introduce one of the axioms, ***Mathematical induction***, for the set of positive integers $\mathbf{Z}_+$ as follows, for a detailed discussion, the reader is referred to Appendix II.

Mathematical induction: Let $\{P(i)\}_{i \in Z_+}$ be a set of statements indexed by the set of positive integers $\mathbf{Z}_+$. If we can verify that

(a) the statement $P(1)$ is true.
(b) for $n \geq 2$, the truth of $P(\ell)$ for all $\ell \leq n$ implies the truth of $P(n)$.

then $P(n)$ is true for all positive integers n. ∎

Discussion

(1) Mathematical induction is an axiom satisfied by the set of positive integers $\mathbf{Z}_+$. Although it can not be deduced from a small set of axioms, we may understand the rationale of it; it follows from (1) above, we know that $P(1)$ is true. Furthermore, let $n = 2$ in (2), we may conclude that $P(2)$ is true. Then let $n = 3$ in (2), we conclude that $P(3)$ is true. Recursively, we conclude that $P(n)$ is true for all n. ∎

We shall use ***Zorn's lemma*** very often in Algebra. It is known that Zorn's lemma is equivalence to the ***axiom of choice*** and the ***well ordering principle***. Although it is an axiom of the set theory which can not be proved, it is important to understand Zorn's lemma. For this purpose, let us introduce the concepts of the ***partial ordering*** and the ***total ordering***. Sometime a total ordering is simply called an ***ordering***.

Definition 1.5. *Given a set* $\mathbf{T}$ *and a relation* $\geq$ *on* $\mathbf{T}$. *If the relation* $\geq$ *satisfies the following conditions, then it is called a* ***partial ordering***,

(a) $t_1 \geq t_1$ *for all* $t_1 \in \mathbf{T}$.
(b) $t_1 \geq t_2$ *and* $t_2 \geq t_3 \Longrightarrow t_1 \geq t_3$.
(c) $t_1 \geq t_2$ *and* $t_2 \geq t_1 \Longrightarrow t_1 = t_2$. ∎

Definition 1.6. *Given a set* **T** *and a partial ordering* $\geq$ *on* **T**. *If* $\geq$ *satisfies the following condition (d), then* $\geq$ *is called a* ***total ordering***,

(d) *for any* $t_1, t_2 \in$ **T**, *we always have either* $t_1 \geq t_2$ *or* $t_2 \geq t_1$.

∎

Definition 1.7. *Let* **T** *be a set, and* $\geq$ *be a partial ordering on* **T**. *Let* **S** *be a subset of* **T**. *If an element* $t \in$ **T** *satisfies* $t \geq s$ *for all* $s \in$ **S**, *then the element* t *is said to be an* ***upper bound*** *of* **S**. *If for a given element* $t \in$ **T**, *the relation* $t_1 \geq t$ *implies* $t_1 = t$, *then the element* t *is said to be a* ***maximal element*** *of* **T**. ∎

Definition 1.8. *Let* **T** *be a set, and* $\geq$ *be a partial ordering on* **T**. *Let* **S** *be a subset of* **T**. *If the restriction of* $\geq$ *to* **S** *is a total ordering of* **S**, *then* **S** *is said to be a* ***chain***. ∎

Now we may state Zorn's lemma as follows,

Zorn's lemma. *Let* **T** *be a non-empty set and a partial ordering* $\geq$ *on* **T**. *If every chain of* **T** *has an upper bound, then there is a maximal element in* **T**. ∎

Discussion

(1) Since Zorn's lemma is in fact an axiom for set theory, it can not be deduced from a simpler system of axioms.

(2) Zorn's lemma is a tool which helps us to simplify some proofs. Moreover, some results can only be deduced from Zorn's lemma. For instance, let us establish that in any bounded domain **D** of the real plane, there is always a maximal open disc. We shall use Zorn's lemma. Let **T** be the set of all open discs in the domain **D**. Let the usual set theoretic inclusion $\supseteq$ be the partial ordering. Then it is easy to establish that (1) the set **T** is non-empty. (2) let $\{\mathbf{D}_i : i \in \mathbf{I}\}$ be a chain, then $\bigcup_{i\in I}\mathbf{D}_i$ is obvious an open disc and hence an element in **T**, thus an upper bound of the chain. It follows from Zorn's lemma that there is a maximal element in **T** which is what we want. ∎

Exercises

(1) Let **Q**[x] be the set of all polynomials with rational coefficients. Prove that **Q**[x] is a countable set.

(2) Given any set **T**. Prove that the set theoretic inclusion $\supseteq$ is a partial ordering on the set of all subsets of **T**.

(3) Prove that the usual ordering $\geq$ is a partial ordering, and in fact a total ordering, for the set of all integers **Z**.

(4) Prove that the partial ordering in the above problem (2) satisfies the conditions oı Zorn's lemma. Prove that the partial ordering in problem (3) does not satisfy the conditions of Zorn's lemma.

(5) Use the Mathematical induction to prove $1^2+2^2+\cdots+n^2=(n(n+1)(2n+1)/3!)$.

(6) Let $\rho : S \rightarrow T$ be a map from the set S to the set T. Prove that ρ is injective if and only if one of the following two conditions are satisfied;
  (i) There exists a map $\tau : T \rightarrow S$, such that $\tau\rho =$ the identity map on S.
  (ii) For any set U, and any two maps $\tau_1, \tau_2 : U \rightarrow S$, the relation $\rho\tau_1 = \rho\tau_2$ implies $\tau_1 = \tau_2$

(7) Let $\rho : S \rightarrow T$ be a map from the set S to the set T. Prove that ρ is surjective if and only if one of the following two conditions are satisfied;
  (i) There exists a map $\tau : T \rightarrow S$, such that $\rho\tau =$ the identity map on T.
  (ii) For any set U, and any two maps $\tau_1, \tau_2 : T \rightarrow U$. the relation $\tau_1\rho = \tau_2\rho$ implies $\tau_1 = \tau_2$

---

§2 Unique Factorization Theorem

The set of the natural numbers $\{1, 2, 3, \cdots\}$ will be called the set of the positive integers, denoted by $\mathbf{Z}_+$. The set of the integers $\{\cdots, -3, -2, -1, 0, 1, 2, 3, \cdots\}$ will be denoted by $\mathbf{Z}$.

Mathematics originates from the natural numbers $\mathbf{Z}_+$. One way is to start with the ***Peano's axioms*** of the natural numbers and then introduce the four arithmetical operations, $+, -, \div, \times$. We can prove ***the commutative laws***, ***the associative laws*** and ***the distributive laws*** thereafter in a logical manner. Using the natural numbers thus built, we may then construct the set of rational numbers $\mathbf{Q}$, the set of real numbers $\mathbf{R}$ and the set of complex numbers $\mathbf{C}$. For our readers, this logic method will be tedious and unnecessary. A portion of the necessary ingredients of those logic arguments is attached in an appendix (cf Appendix II Peano's axioms). We will assume that the reader is familiar with the arithmetical operations of $\mathbf{Z}, Q, R$ and $\mathbf{C}$.

One of the most important operations in the theory of integers is the *long division algorithm*. This operation had been known to many ancient civilizations. In modern mathematics, it is known as the *Euclidean algorithm*[2,3]. Let us introduce the following concept,

Definition 1.9. *Let a be a real number. Let $[a]$ be the largest integer which is less then or equal to a.*

∎

---

[2] Euclid: Greek Mathematician lived at Alexandria, Egypt 306 B.C..

[3] The term 'algorithm' is a corruption of Persian algebraist al-Khwārizmi, 9th century.

Discussion

(1) The existence of $[a]$ is intuitively obvious, while equivalence to one of the fundamental properties of the real numbers, the *Archimedean property*[4], which proclaims that for any two real numbers n and $d > 0$, there exists a natural number q with $q \cdot d > n$.

(2) For instance, $[3.1] = 3$, $[-3.2] = -4$ and $[5] = 5$. ∎

Theorem 1.3. (Euclidean algorithm). *Let d be a positive real number and n an arbitrary real number. Then there must be an integer q and a real number r, such that*

$$n = q \cdot d + r, \qquad 0 \leq r < d$$

Proof: Let $q = [n/d]$, $r = n - q \cdot d$. Then we have

$$\begin{aligned} q &\leq n/d < q+1 \\ q \cdot d &\leq n < q \cdot d + d \\ 0 &\leq n - q \cdot d = r < d \end{aligned}$$

∎

Corollary 1. *In the above theorem, the numbers q and r are uniquely determined by n and d.*

Proof: Let q' and r' be another pair of real numbers with

$$n = q' \cdot d + r'(= q \cdot d + r), \qquad 0 \leq r' < d$$

Then we have

$$(q - q') \cdot d = r' - r$$

We may assume that $q - q' \geq 0$. Then we get

$$0 \leq (q - q') \cdot d = r' - r \leq r' < d$$

Therefore we conclude

$$q = q', \qquad r = r'$$

∎

[4] Archimedes: Greek Mathematician and Scientist 287-212 B.C..

Example 1. *Using Euclidean algorithm, we may define the continuous fraction of any real number. Let us take an example; From* $\pi = 3.14159265358979238 46\cdots$, *we get*

$$\pi = \frac{\pi}{1} = 3 + \frac{0.14159265358979238 46\cdots}{1}$$
$$= 3 + \cfrac{1}{\cfrac{1}{0.14159265358979238 46\cdots}}$$
$$= 3 + \cfrac{1}{7 + \cfrac{0.00885142787144737 07\cdots}{0.14159265358979238 46\cdots}}$$
$$= 3 + \cfrac{1}{7 + \cfrac{1}{\cfrac{0.14159265358979238 46\cdots}{0.00885142787144737 07\cdots}}}$$
$$= 3 + \cfrac{1}{7 + \cfrac{1}{15 + \cfrac{0.00882123551808317 69\cdots}{0.00885142787144737 07\cdots}}}$$
$$= 3 + \cfrac{1}{7 + \cfrac{1}{15 + \cfrac{1}{1 + \cfrac{0.008851427871447\cdots}{0.000030192353364\cdots}}}}$$
$$= 3 + \cfrac{1}{7 + \cfrac{1}{15 + \cfrac{1}{1 + \cfrac{1}{292 + \cdots}}}}$$

Let us discard the decimal parts and only keep the integer parts in the above, and call the resulting rational numbers the partial continuous fractions. Then we get the partial continuous fractions of π *as follows,*

$$3, \quad 3 + \frac{1}{7}, \quad 3 + \cfrac{1}{7 + \cfrac{1}{15}}, \quad 3 + \cfrac{1}{7 + \cfrac{1}{15 + \cfrac{1}{1}}}, \quad 3 + \cfrac{1}{7 + \cfrac{1}{15 + \cfrac{1}{1 + \cfrac{1}{292}}}}$$

The above rational numbers are 3, 22/7, 333/106, 355/113, 103993/33102. The first approximation, 3, was known to most ancient civilizations. The second approximation, 22/7, was due to Archimedes (250 B.C.) and is still used in high schools today. The third one was not very significant. The fourth one, 355/113 = 3.1415929203⋯, was very close to the true value of π, was discovered by Tsu Chhung-Chih (470 A.D.) in China and independently by Vieta (1593 A.D.) in France.

It is generally known in number theory that the partial continuous fractions are the best rational approximations with restrictions on the sizes of the denominators. ∎

Another application of Euclidean algorithm is the unique factorization property of the integers $\mathbf{Z}$. For this purpose, let us introduce,

Definition 1.10. *Let a, b, c be integers. If $a = b \cdot c$. then we say that a is a **multiple** of b and b is a **divisor** of a, in symbol, $b \mid a$. If $b \mid a_1, b \mid a_2, \cdots, b \mid a_n$, then we say that b is a **common divisor** of $a_1, a_2, \cdots, a_n$. The greatest one among the common divisors of $a_1, a_2, \cdots, a_n$ will be called the **greatest common divisor**, in symbol g.c.d., of $a_1, a_2, \cdots, a_n$. If $a_1 \mid b, a_2 \mid b, \cdots a_n \mid b$, then we say that b is a **common multiple** of $a_1, a_2, \cdots, a_n$. The smallest non-negative integer which is a common multiple of $a_1, a_2, \cdots, a_n$ will be called the **least common multiple**, in symbol ℓ.c.m. of $a_1, a_2, \cdots, a_n$.* ∎

Theorem 1.4. *Suppose that one of a_1, a_2 is non-zero. Then the greatest common divisor of a_1, a_2 is the smallest positive integer in the set $\mathbf{S}=\{b_1 \cdot a_1 + b_2 \cdot a_2 : b_i \in Z\}$. We will use (a_1, a_2) to denote the greatest common divisor of a_1, a_2.*

Proof: Let the smallest positive integer be $d = c_1 \cdot a_1 + c_2 \cdot a_2$. Applying Euclidean algorithm to the pair d, a_1, there exist q_1 and r_1 with

$$a_1 = q_1 \cdot d + r_1, \qquad 0 \le r_1 < d$$

If $r_1 \neq 0$, then we get

$$r_1 = a_1 - q_1 \cdot d = (1 - c_1 \cdot q_1)a_1 + (-c_2 \cdot q_1)a_2 \in \mathbf{S}$$

Note that then r_1 is a positive element in $\mathbf{S}$ which is less then d. A contradiction! We conclude that $r_1 = 0$, i.e.,

$$d \mid a_1$$

Similarly, we can prove

$$d \mid a_2$$

Namely, d is a common divisor of a_1 and a_2. Let d' be another common divisor of a_1, a_2. Then we have

$$d' \mid a_1, \quad d' \mid a_2 \Longrightarrow d' \mid c_1 \cdot a_1 + c_2 \cdot a_2 = d$$

Since d is a positive integer, then we have $d \ge d'$. Therefore it follows from the definition that d is the greatest common divisor of a_1, a_2. ∎

In the following we will establish the ***unique factorization theorem*** for $\mathbf{Z}$, which is sometimes called the ***fundamental theorem of Arithmetics***. For this purpose, we will introduce the concepts of the ***irreducible numbers*** and the ***prime numbers***. Note that in $\mathbf{Z}$ the only invertible elements with respect to multiplication are $1, -1$, i.e., if n and n^{-1} are both integers, then n must be 1 or -1.

Definition 1.11. *Let $a \neq 0, 1, -1$ be an integer. Then a is said to be an irreducible number, if in any factorization of $a = b \cdot c$, for some integers b and c, implies either $b = \pm 1$ or $c = \pm 1$. If for $f, g \in \mathbf{Z}, a \mid f \cdot g \Longrightarrow a \mid f$ or $a \mid g$, then a is called a prime number.* ∎

Lemma. *In $\mathbf{Z}$, a number a is an irreducible number if and only if it is a prime number.*

Proof: ($\Longrightarrow$) Let a be irreducible. We may assume that a is positive, otherwise replace it by $-a$. Suppose that we have

$$a \mid f \cdot g, \qquad f, g \in \mathbf{Z}$$

Suppose a is not a divisor of f. Since a is irreducible, i.e., the only positive divisors of a are 1 and a, then the greatest common divisor of a and f must be 1. It follows from Theorem 1.2. that

$$1 = c_1 \cdot a + c_2 \cdot f$$

Multiplying by g on both sides of the above equation, we get

$$g = g \cdot c_1 \cdot a + c_2 \cdot (f \cdot g)$$

Therefore we have $a \mid g$ and a is a prime number.

($\Longleftarrow$) Let a be a prime number and $a = b \cdot c$. Then we have $a \mid b \cdot c$ which implies $a \mid b$ or $a \mid c$, say $a \mid b, b = a \cdot d$. Trivially, we have $a = a \cdot (c \cdot d)$ and $1 = c \cdot d$. Therefore c is multiplicatively invertible and must be 1 or -1. We conclude that a is irreducible. ∎

Discussion

(1) For the general *rings* (cf Chapter III), the concepts of irreducible elements and prime elements are different. The coincidence of these two concepts establishes the Unique Factorization Theorem (see below).

(2) An expression $a = \prod_i p_i$ with all p_i prime numbers will be called *a prime decomposition of a*. ∎

Theorem 1.5. (Unique Factorization Theorem). *Let $a > 1$ be any positive integer. Then a has a prime decomposition $a = \prod_i p_i$. Moreover, all prime decompositions of a are identical up to a reordering of p_i.*

Proof: We shall prove the existence and then the uniqueness of the prime decomposition. The present theorem is void for $a = 1$. Let us start with $a = 2$. Since 2 is a prime number, then the equation $2 = 2$ is a prime decomposition of 2. Let any positive integer $a > 2$ be given. Using Mathematical induction, we assume that any positive integer less than a has a prime decomposition. If a is a prime number, then $a = a$ is a prime decomposition of a. Suppose that a is not prime. Then it follows from the preceding lemma that a is reducible. Therefore there are positive integers b, c with

$$a = b \cdot c \qquad a > b > 1, \quad a > c > 1$$

By the assumption of Mathematical induction, both b and c have prime decompositions. Their product is a prime decomposition of a. Thus we establish that any positive $a > 1$ has a prime decomposition.

Let us prove the uniqueness of the prime decomposition. Let $a = \prod_j q_j$ be another prime decomposition of a. Then we have

$$p_1 \mid a = q_1(\prod_{j>1} q_j)$$

which implies

$$p_1 \mid q_1 \text{ or } p_1 \mid (\prod_{j>1} q_j)$$

If $p_1 \nmid q_1$, then we have $p_1 \mid q_2(\prod_{j>2} q_j)$, namely

$$p_1 \mid q_2 \text{ or } p_1 \mid (\prod_{j>2} q_j)$$

Step by step, there must be a q_s with

$$p_1 \mid q_s$$

While q_s is a prime number which is irreducible by the preceding lemma. Thus we have

$$p_1 = q_s$$

Now let us consider $a/p_1 = a/q_s = \prod_{i>1} p_i = \prod_{j \neq s} q_j$. If this number is 1, then we are done. If it is bigger than 1, then by Mathematical induction, the two product expressions $\prod_{i>1} p_i = \prod_{j \neq s}$ are just a reordering of each other. We are done again. ∎

Example 2. *Let p_i be prime numbers. Then $p_1, p_2, \cdots, p_n$ are not divisors of $\prod_1^n p_i + 1$. Otherwise, we will have for some integer a the following*

$$a \cdot p_j = \prod_1^n p_i + 1$$

$$(a - \prod_{i \neq j} p_i) p_j = 1$$

namely $p_j \mid 1$. A contradiction! From the above, we may easily deduce that there are infinitely many prime numbers; otherwise, let $\{p_i\}$ be all possible prime numbers, then $\prod_i p_i + 1$ can not have any p_i as a divisor, and any divisor of $\prod_i p_i + 1$ will increase the list of prime numbers. ∎

Exercises

(1) Use the Euclidean algorithm to prove that every positive integer has a decimal expansion.

(2) Find the greatest common divisor of 273 and 1729.

(3) Write 1030301 as a product of prime numbers.

(4) Find the first five partial continuous fractions of e.

(5) In practice (for instance, the coding theory) it is not easy to find the prime decomposition of an integer n. Find the prime decomposition of 12345678910111213141516.

(6) **n!-expansion:** Measure any real number r by an integer, then measure the remainder by $1/2!$, further measure the remainder by $1/3!$, and then by $1/4!$, etc. In other words, for any real number $0 \leq r < 1$, we have the following expansion,

$$r = \frac{a_2}{2!} + \frac{a_3}{3!} + \frac{a_4}{4!} + \cdots + \frac{a_n}{n!} + \cdots, \qquad 0 \leq a_n < n$$

where all a_n are integers. Prove the existence of the expansion and that a real number r is rational if and only if the $n!$-expansion of it is finite.

(7) **Egypt numeral system:** Show that any positive rational number a can be written as

$$a = \sum_{finite} \frac{1}{n_i} \qquad \text{where all } n_i \text{ are distinct positive integers}$$

(8) Let n be a positive integer and r a real number. Prove

(i) $[[nr]/n] = [r]$.

(ii) $[r] + [r + \frac{1}{n}] + \cdots + [r + \frac{n-1}{n}] = [nr]$.

(9) Let r, s be two real numbers. Prove

$$[2r] + [2s] \geq [r] + [r + s] + [s]$$

(10) Let d, m be the greatest common divisor and the least common multiple of two positive integers a and b respectively. Show that

$$d \cdot m = a \cdot b$$

(11) Let n be a positive integer and p a prime number. Find the highest power of p which is a factor of $n!$.

§3 Congruence

Let us define the following basic relation between integers,

Definition 1.12. *Let $m \neq 0$ be an integer. Two integers a, b are said to be **congruent to each other** mod m if the following condition is satisfied,*

$$m \mid a - b$$

in symbol,

$$a \equiv b \ (\bmod m)$$

*On the other hand if $m \nmid a - b$, then we say a, b are **not congruent** to each other mod m, in symbol,*

$$a \not\equiv b \ (\bmod m)$$

∎

Discussion

(1) It is easy to see that the congruence relation is an equivalence relation:

(r) Reflexion: $a \equiv a \ (mod\ m)$.
(s) Symmetry: $a \equiv b \ (mod\ m) \Longrightarrow b \equiv a \ (mod\ m)$
(t) Transition: $a \equiv b \ (mod\ m), b \equiv c \ (mod\ m) \Longrightarrow m \mid a - b,\ m \mid b - c \Longrightarrow m \mid (a - b) + (b - c) = a - c \Longrightarrow a \equiv c \ (mod\ m)$.

(2) It then follows from Definition 1.4*. that the congruence relation will divide the set of integers into disjoint equivalence classes, which will be called the ***residue classes*** mod m.

∎

Definition 1.13. *Let us use $[a]_m$ to denote the residue class $\{b : b \equiv a \ (mod\ m)\}$. Let $\mathbf{Z}_m$ be the set of all residue classes mod m. Let b be an integer satisfying $0 \leq b < |m|$ and $b \in [a]_m$. Then b is said to be the **principal residue** of $[a]_m$.* ∎

Lemma. *We always have $\mathbf{Z}_m = \{[0]_m, [1]_m, \cdots, [|m| - 1]_m\}$*

Proof: For any $a \in \mathbf{Z}$, we always have, by Euclidean algorithm, two integers q and r with

$$a = q \cdot |m| + r \qquad 0 \leq r < |m|$$

Therefore we have $[a]_m = [r]_m$. Moreover, for any pair i, j with $0 \leq i < j < |m|$, we always have $[i]_m \neq [j]_m$. ∎

In the view of algebra, one important property of $\mathbf{Z}_m$ is that the three operations, $+, -, \cdot$, of Arithmetic can be carried out smoothly. We will define the addition $+$ for two elements $[a]_m$ and $[b]_m$ as $[a + b]_m$. The logic difficulty is that a definition of this form might depend on the ***representatives*** a, b of the classes $[a]_m, [b]_m$. We have to show that the result is independent of the representatives a, b to establish that the definition of addition is ***well-defined***. We have the following theorem,

Theorem 1.6. *Given* $[a]_m = [i]_m$ *and* $[b]_m = [j]_m$, *then we have the following,*
$[a+b]_m = [i+j]_m$,
$[a-b]_m = [i-j]_m$,
$[a\cdot b]_m = [i\cdot j]_m$,
Therefore we may define

$$[a]_m + [b]_m = [a+b]_m, \quad [a]_m - [b]_m = [a-b]_m, \quad [a]_m \cdot [b]_m = [a\cdot b]_m$$

Moreover, the above operations, $+, -, \cdot$, *satisfy the following laws,*

(1) *Associative law:*
$([a]_m + [b]_m) + [c]_m = [a+b+c]_m = [a]_m + ([b]_m + [c]_m)$
$([a]_m \cdot [b]_m)\cdot [c]_m = [a\cdot b\cdot c]_m = [a]_m \cdot ([b]_m \cdot [c]_m)$

(2) *Commutative law:*
$[a]_m + [b]_m = [a+b]_m = [b]_m + [a]_m$
$[a]_m \cdot [b]_m = [a\cdot b]_m = [b]_m \cdot [a]_m$

(3) *Distributive law:*
$[a]_m \cdot ([b]_m + [c]_m) = [a]_m \cdot [b]_m + [a]_m \cdot [c]_m$

(4) *Existence of units:*
$[0]_m$ *is the additive unit;* $[0]_m + [a]_m = [a]_m$
$[1]_m$ *is the multiplicative unit;* $[1]_m \cdot [a]_m = [a]_m$

(5) *Existence of additive inverse:*
$[a]_m + [-a]_m = [0]_m$

Proof: Let us give a detailed proof for the *well-defined* properties of the three operations; $+, -, \cdot$. As given $[a]_m = [i]_m, [b]_m = [j]_m$, we have,

$$m \mid a - i, \qquad m \mid b - j$$

Then we have

$$m \mid (a-i) + (b-j) = (a+b) - (i+j)$$

namely

$$a + b \equiv i + j \bmod m$$
$$[a+b]_m = [i+j]_m$$

Similarly, we may prove $[a-b]_m = [i-j]_m$. Moreover, we have

$$m \mid b\cdot(a-i) + i\cdot(b-j) = a\cdot b - i\cdot j$$
$$a\cdot b \equiv i\cdot j \bmod m$$
$$[a\cdot b]_m = [i\cdot j]_m$$

We have thus established that the three operations, $+, -, \cdot$, are well-defined. The proofs for the five laws listed in the theorem will be left to the reader as an exercise. ∎

Example 3.

(a) *Let $m = 2$. Then $\mathbf{Z}_2 = \{[0]_2, [1]_2\}$, $[0]_2$ = the set of even numbers, $[1]_2$ = the set of odd numbers. In plain words, the meanings of the equations $[0]_m + [0]_m = [0]_m, [0]_m + [1]_m = [1]_m, [1]_m + [1]_m = [0]_m$ are even + even= even, even + odd= odd, odd + odd= even.*

(b) *Let $m = 3$. It is easy to prove that $3 \mid (10-1), [10]_3 = [1]_3$, and $[10^n]_3 = [10]_3^n = [1]_3^n = [1]_3$. Therefore for any integer $n_s \cdots n_1 n_0$ in the decimal expansion, we have,*

$$[n_s \cdots n_1 n_0]_3 = [n_s + \cdots + n_1 + n_0]_3$$

For instance, $[741]_3 = [7+4+1]_3 = [12]_3 = [1+2]_3 = [0]_3$, i.e., $3 \mid 741$.

(c) *Similarly, let $m = 11$. Then we get $[10]_{11} = [-1]_{11}$ and $[10^n]_{11} = [10]_{11}^n = [-1]_{11}^n = [(-1)^n]_{11}$. Therefore for any integer $n_s \cdots n_1 n_0$ in the decimal expansion, we have,*

$$[n_s \cdots n_1 n_0]_{11} = [(-1)^s n_s + \cdots - n_1 + n_0]_{11}$$

For instance, $[5678]_{11} = [-5+6-7+8]_{11} = [2]_{11}$, i.e., $5678 \equiv 2 \bmod (11)$. ∎

Example 4. *Let us consider a **round robin** tournament, i.e., a contest with every team playing against all other teams. Let us use the congruence to arrange a schedule for the tournament. If there is an odd number of teams, then in every round one team will be idle. In this case, let us create a fictitious team to make the total number even with the understanding that a team playing against the fictitious team shall be idle. In other words, let us assume there is always an even number of teams; $t_1, \cdots, t_{2n}$.*

Let $m = 2n-1$ and $1 \le i \le m$. And at the a-th round, let the team t_i play with t_j if the following congruence relation determines a j with $1 \le j \le m$ and $j \ne i$,

$$j \equiv a - i \ (\bmod m)$$

Otherwise, the number j satisfies the above congruence equation with $1 \le j \le m$ must be $j = i$. In this case, we ask the team t_i play with t_{2n}.

To show that the above schedule is meaningful, we have to prove the following

(i) *if team t_i plays with team t_j then team t_j plays with team t_i.*

(ii) *for different rounds the opponents of team t_i are all distinct.*

Proof of (i): For a fixed a, let us prove that there is a unique i satisfying

$$i \equiv a - i \ (\bmod m)$$

Let i' be another solution of the above congruence equation, i.e.,

$$i' \equiv a - i' \ (\bmod m), \qquad 1 \le i' \le m$$

The difference of the above two equations will be

$$2 \cdot (i - i') \equiv 0 \ (\bmod m)$$

Since m is odd, then we have $i \equiv i' \pmod{m}$ and $i = i'$. Moreover, $i \equiv a - j \pmod{m} \Longleftrightarrow i + j \equiv a \pmod{m} \Longleftrightarrow j \equiv a - i \pmod{m}$. Thus the pairs $i \neq j$ which satisfy the above equations will be paired. Since m is odd and there are odd number of teams t_i with $1 \leq i \leq m$, then after the pairing there will be exactly one team t_k left with $k \equiv a - k \pmod{m}$. This team will play against team t_{2n}.

Proof of (ii): It is left to the reader. ∎

Definition 1.14. *The reduced residue classes mod m, in symbol $\mathbf{Z}_m^\times$, is the set $\{[a]_m : a, m$ coprime $\} \subseteq \mathbf{Z}_m$. The cardinality of the set $\mathbf{Z}_m^\times$ is called $\phi(m)$ where ϕ is called the Euler ϕ function.*[5]

Theorem 1.7. (Euler Theorem). *Let a and m be coprime. Then we have*

$$a^{\phi(m)} \equiv 1 \pmod{m}$$

Proof: Let us use a to define a map a: $\mathbf{Z}_m^\times \to \mathbf{Z}_m^\times$ by the following formula,

$$a([b]_m) = [ab]_m$$

It is easy to see that the above definition is well-defined. Let us show that the map is injective,

$$\begin{aligned} & a([b]_m = a([c]_m) \\ \Longrightarrow & [ab]_m = [ac]_m \\ \Longrightarrow & [a(b-c)]_m = [0]_m \\ \Longrightarrow & m \mid a(b-c), \qquad a, m \text{ coprime} \\ \Longrightarrow & m \mid b - c \\ \Longrightarrow & [b]_m = [c]_m \end{aligned}$$

It then follows from the pigeon hole principle that the map a is bijective. Therefore we have

$$\begin{aligned} & \prod_{[b_i]\in \mathbf{Z}_m^\times} [ab_i] = \prod_{[b_i]\in \mathbf{Z}_m^\times} [b_i] \\ \Longrightarrow & \prod_{[b_i]\in \mathbf{Z}_m^\times} (ab_i) = \prod_{[b_i]\in \mathbf{Z}_m^\times} (b_i) \pmod{m} \\ \Longrightarrow & (a^{\phi(m)} - 1) \prod_{[b_i]\in \mathbf{Z}_m^\times} (b_i) = 0 \pmod{m}, \qquad b_i, m \text{ coprime} \\ \Longrightarrow & m \mid (a^{\phi(m)} - 1) \prod_{[b_i]\in \mathbf{Z}_m^\times} (b_i), \qquad b_i, m \text{ coprime} \\ \Longrightarrow & m \mid (a^{\phi(m)} - 1) \\ \Longrightarrow & a^{\phi(m)} \equiv 1 \pmod{m} \end{aligned}$$

∎

[5] Swiss Mathematician 1707-1783.

It is easy to deduce the following theorem due to Fermat[6].

Theorem 1.8. (Fermat's Little Theorem). *Let p be a prime number with $p \nmid a$. Then we always have*

$$a^{p-1} \equiv 1 \pmod p$$

Proof: In the preceding Euler Theorem, let $m = p$. Then we have $\phi(p) = p - 1$. ∎

Corollary. *Let p be a prime number. Then we always have*

$$a^p \equiv a \pmod p$$

∎

Example 5. *We may use Fermat's Little Theorem to produce a code (due to Rivest-Schamir-Adleman) which is very hard to crack as follows;*

(i) *Any words, hence any message, can be represented by numbers as in the commonly used telegram coding. A long message can be cut into ones of the standard size. Therefore we may deal with messages of positive integers which are less than a certain integer n.*

(ii) *Let us take a prime number $p > n$ and another integer q which is comparable in size with p and coprime to $p-1$.*

(iii) *Let $m = p \cdot q$. Publicize the numbers m and n as the keys to send messages to the receiver while guarding the secret of the above factorization of m.*

(iv) *Direct any sender with the original message $a < n$ delivering a coded message b which is determined by*

$$b \equiv a^m \pmod m \qquad 1 \le b < m$$

(v) *Knowing that the factorization $m = p \cdot q$ and q is coprime to $p-1$, the receiver can find two integers c, d with*

$$c \cdot q - d(p-1) = 1$$
$$c \cdot q = 1 + d(p-1)$$

Upon receiving the coded message b, the receiver does the following computation using Fermat's Little Theorem,

$$b^c \equiv (a^m)^c \equiv (a^{pq})^c \equiv (a^{cq})^p \equiv (a^{1+d(p-1)})^p$$
$$\equiv a^p (a^{p-1})^{dp} \equiv a^p 1^p \equiv a^p = a \bmod m$$

Therefore the receiver recovers the original message a.

(vi) *A third person without knowing the factorization $m = p \cdot q$ cannot find the integer c and hence cannot apply Fermat's Little Theorem in the above way to recover the original message a.*

---

[6] Fermat: French mathematician, 1601-1675.

Certainly, a third person who can find the prime decomposition $m = \prod_i p_i$ is able to test all pairs $p = p_j, q = \prod_{i \neq j} p_i$ to see if q is coprime to $p-1$ first and then using Fermat's Little Theorem to see if the a' thus produced is meaningful. However, the practical problem of finding the prime decomposition of an integer m is time-consuming. At present, it is relatively fast for a high speed computer to (a) find a large prime p and a suitable q, (b) find c, d as required in (v), (c) find $a = b^c \pmod{m}$, while very slow to find the prime factorization of m. Those facts make the above code hard to crack by a third person.

∎

Lemma. *Let p be a prime number and i an integer with $1 \leq i \leq p-1$. Then there exists a unique j with $1 \leq j \leq p-1$ and $[i]_p[j]_p = [1]_p$. The unique $[j]_p$ is called the* **multiplicative inverse** *of $[i]_p$ and written as $[i]_p^{-1}$.*

Proof: Since i and p are coprime, then we have a, b such that

$$a \cdot i + b \cdot p = 1$$

namely

$$[a]_p \cdot [i]_p \equiv [1]_p$$

Let j be the principal residue of a mod p. Then j has the required property. Let j' be another number with $1 \leq j' \leq p-1$ and $[i]_p[j']_p = [1]_p$. Then we have

$$\begin{aligned} &[j]_p \cdot [i]_p - [j']_p \cdot [i]_p = ([j]_p - [j']_p) \cdot [i]_p = [(j - j') \cdot i]_p = [0]_p \\ &p \mid (j - j') \cdot i \end{aligned}$$

Since i is coprime to p, then we have

$$\begin{aligned} &p \mid j - j' \\ &j = j' \end{aligned}$$

∎

Theorem 1.9. (Wilson's Theorem)[7]. *Let p be a prime number. Then we have*

$$(p-1)! \equiv -1 \ (\bmod\ p)$$

Proof: Let i be an integer with $1 \leq i \leq p-1$ and $[i]_p^{-1} = [i]_p$. Then we have

$$\begin{aligned} &[1]_p = [i]_p[i]_p^{-1} = [i]_p^2 \\ &[i^2 - 1]_p = 0 \\ &p \mid i^2 - 1 = (i-1)(i+1) \\ &p \mid i - 1 \text{ or } p \mid i + 1 \end{aligned}$$

[7]Wilson: English Mathematician 1741-1793.

Thus we conclude $i = 1$ or $p-1$. Any other j will be paired with $[j]_p^{-1}$ to give a product .. Thus we conclude

$$\prod_{i=1}^{p-1} i \equiv 1 \cdot (p-1) \bmod p = -1 \bmod p$$

∎

Exercises

(1) Given any n points in the space, let us color the lines connecting the n points such that all lines meeting any one of the given points are of different colors. Show that n colors suffice.

(2) Prove that $3 \mid xyz$ for any integers x, y, z satisfying $x^2 + y^2 = z^2$.

(3) Let $p = 2, 3, 5$. Compute $\binom{n}{i}$ mod p for $1 \le n \le 10$.

(4) Find $\phi(p^n)$ for a prime number p.

(5) Prove $\phi(ab) = \phi(a)\phi(b)$ if a, b are coprime.

(6) Use problems (4) & (5) to compute $\phi(m)$ for any m.

(7) Prove that $(1+x)^p = 1 + x^p \pmod p$ for any integer x.

(8) Let a, b be two integers with greatest common divisor d. Let r be the product of all *distinct* prime factors of d. Prove that

$$\frac{\phi(ab)}{\phi(a)\phi(b)} = \frac{r}{\phi(r)}$$

(9) Let p be an odd prime number. Show that,

$$(\frac{p-1}{2}!)^2 \equiv (-1)^{\frac{p-1}{2}} \pmod p$$

(10) Let m be a positive integer, $a, b \in \mathbf{Z}$ and d be the greatest common divisor of m and a with $d \mid b$. Prove that there are exactly d solutions for the following congruence equation in the variable x,

$$a \cdot x + b \equiv 0 \pmod m \qquad 0 \le x < m$$

§4 Chinese Remainder Theorem

The Chinese remainder theorem can be traced back to the ancient Chinese Mathematical book *Sun Tzu Suan Ching*[8]. The following problem was mentioned in it: "*We have an unknown number of objects. There are two objects left if they are counted by threes. There are three objects left if they are counted by fives. There are two objects left if they are counted by sevens. How many objects are there?*" In our terminology of congruence, the problem is to find x satisfying

$$\begin{aligned} x &\equiv 2 \pmod 3 \\ x &\equiv 3 \pmod 5 \\ x &\equiv 2 \pmod 7 \end{aligned}$$

The answer of this problem was given in the book to be 23. The method of solving this type of equations was generalized by later mathematicians, especially Monk I-Hsing and Li Shun-Feng (both 8th century A.D.). The final form of this theorem was accomplished by Chhin Chiu-Shao (1247 A.D.). The method was generally known as *"finding one for the great remainders"* (Ta yen chiu i) in Chinese. The meaning of it can be best described by the following method in solving the original congruence equations; Let x_1, x_2, x_3 be solutions for the following auxiliary congruence equations,

$$\begin{array}{lll} x_1 \equiv 1 \pmod 3 & x_2 \equiv 0 \pmod 3 & x_3 \equiv 0 \pmod 3 \\ x_1 \equiv 0 \pmod 5 & x_2 \equiv 1 \pmod 5 & x_3 \equiv 0 \pmod 5 \\ x_1 \equiv 0 \pmod 7 & x_2 \equiv 0 \pmod 7 & x_3 \equiv 1 \pmod 7 \end{array}$$

To find x_1, x_2, x_3 is to solve special systems of congruence equations with one remainders 1 and all others zero. For instance, for the above equations we may take $x_1 = 70, x_2 = 21, x_3 = 15$. Let $x = 2 \cdot x_1 + 3 \cdot x_2 + 2 \cdot x_3 = 2 \cdot 70 + 3 \cdot 21 + 2 \cdot 15 = 233$. Then x clearly satisfies the original congruence equations. The preceding special congruence equations explain the term "*find one for the great remainders*". Moreover, the principal residue of $233 \pmod{3 \cdot 5 \cdot 7} = 23$, therefore 23 is the smallest positive solution.

Theorem 1.10. (Chinese Remainder Theorem). *Let integers $m_1, m_2, \cdots, m_n$ be pairwise coprime. Let $m = \prod_{i=1}^{n} m_i$. Then we have*

(i) *For any $1 \le j \le n$, the following system of congruence equations has solutions,*
$$x_j \equiv 1 \pmod{m_j},\ x_j \equiv 0 \pmod{m_i}\ \forall\, i \neq j$$
(ii) *Let $x = \sum a_j x_j$. Then x satisfies the following system of congruence equations,*
$$x \equiv a_j \pmod{m_j}\ \forall\, j.$$

Moreover, $[x]_m$ is the set of all solutions of the above system of congruence equations, its principal residue is the smallest non-negative solution.

[8] date uncertain, not later than 4th century A.D..

Proof: (i) It is known that $m_1, m_2, \cdots, m_n$ are pairwise coprime. Therefore, m_j and $\prod_{i\neq j} m_i$ are coprime. It follows from Theorem 1.4. that there are numbers a, b with

$$a \cdot m_j + b \cdot \prod_{i\neq j} m_i = 1$$

Let x_j be defined by

$$x_j = 1 - a \cdot m_j = b \cdot \prod_{i\neq j} m_i$$

Then x_j satisfies the required congruence equations.

(ii) Let $x = \sum_j a_j \cdot x_j$. Then we have

$$[x]_{m_i} = \left[\sum_j a_j \cdot x_j\right]_{m_i} = \sum_j [a_j]_{m_i} [x_j]_{m_i} = [a_i]_{m_i}$$

$$x \equiv a_i \pmod{m_i} \quad \forall\, i$$

(iii) Let y be another solution of the same system of congruence equations, i.e.,

$$y \equiv a_i \pmod{m_i} \quad \forall\, i$$

Then we have

$$x - y \equiv 0 \pmod{m_i} \quad \forall\, i$$

$$m_i \mid x - y \quad \forall\, i$$

While assuming that $m_1, m_2, \cdots, m_n$ are pairwise coprime, we conclude

$$m = \prod_i m_i \mid x - y$$

$$x \equiv y \pmod{m}$$

Hence we have $y \in [x]_m$. Conversely, it is easy to show that any number in the set $[x]_m$ is a solution of our system of congruence equations. ∎

In fact, we have another abstract version of the Chinese Remainder Theorem as follows,

Theorem 1.11. *Let integers $m_1, m_2, \cdots, m_n$ be pairwise coprime. Let $m = \prod_{i=1}^n m_i$. Let a map φ, from $\mathbf{Z}_m$ to the direct product $\prod_i \mathbf{Z}_{m_i}$, be defined as*

$$\varphi([a]_m) = ([a]_{m_1}, [a]_{m_2}, \cdots, [a]_{m_n})$$

Then φ is bijective.

Proof: It is easy to see that the cardinality m of $\mathbf{Z}_m$ equals the cardinality $(\prod_i m_i)$ of $\prod_i \mathbf{Z}_{m_i}$. Using the pigeon hole principle it suffices to show that φ is surjective

Let $([a]_{m_1}, [a]_{m_2}, \cdots, [a]_{m_n})$ be any element in $\prod_i \mathbf{Z}_{m_i}$. It follows from the Chinese remainder theorem that the following system of congruence equations have a common solution x,

$$x \equiv a_i \bmod m_i \quad \forall\, i = 1, 2, \cdots, n$$
$$[x]_{m_i} = [a]_{m_i}$$

Then we have

$$\varphi([x]_m) = ([x]_{m_1}, [x]_{m_2}, \cdots, [x]_{m_n})$$
$$= ([a]_{m_1}, [a]_{m_2}, \cdots, [a]_{m_n})$$

Therefore we conclude that φ is surjective, hence bijective. ∎

Discussion: It is easy to deduce Theorem 1.10. from Theorem 1.11. Assume Theorem 1.11., the essential part of the proof of Theorem 1.10. is to solve the following system of equations,

$$x \equiv a_i \pmod{m_i} \qquad \forall\, i$$

It follows from Theorem 1.11. that φ is surjective, i.e., there is an element a with

$$\varphi([a]_m) = ([a]_{m_1}, [a]_{m_2}, \cdots, [a]_{m_n})$$
$$= ([a_1]_{m_1}, [a_2]_{m_2}, \cdots, [a_n]_{m_n})$$
$$a = a_i \pmod{m_i} \qquad \forall\, i$$

Then $x = a$ is a solution. ∎

Exercises

(1) What is the minimal positive number with the following properties? If counted by sevens, there is one left. If counted by eights, there are two left. If counted by nines, there are three left.[9]

(2) **Lagrange Interpolation:**[10] Let us consider the Chinese Remainder Theorem for the *polynomial ring with real coefficient* $\mathbf{R}[x]$. Prove that for real numbers a_i and b_i with $b_i \neq b_j$ if $i \neq j$, the polynomial $f_j(x)$ defined as,

$$f_j(x) = \frac{\prod_{i \neq j}(x - b_i)}{\prod_{i \neq j}(b_j - b_i)}$$

satisfies the following system of congruence relations

$$f_j(x) \equiv 1 \pmod{(x_j - b_j)} \qquad f_j(x) \equiv 0 \pmod{(x_i - b_i)} \quad \forall\, i \neq j$$

[9] A problem in a book of Yang Hui (1275 A.D.).
[10] Lagrange: French Mathematician 1736-1813.

Moreover, the polynomial $f(x)$ defined as

$$f(x) = \sum_j a_j f_j(x) = \sum_{j=1}^{n} \frac{a_j}{\prod_{i\neq j}(b_j - b_i)} \frac{\prod_i (x - b_i)}{(x - b_j)}$$

satisfies the following system of congruence relations

$$f(x) \equiv a_i \pmod{(x_i - b_i)} \quad \forall i \quad i.e, \quad f(b_i) = a_i$$

(3) Find a real polynomial $f(x)$ of the smallest degree satisfying

$$f(x) \equiv 1 \pmod{(x)}$$
$$f(x) \equiv 2 \pmod{(x-1)}$$
$$f(x) \equiv 3 \pmod{(x-2)}$$

(4) Let m_1, m_2 be positive integers with the greatest common divisor d and the least common multiple m. Let a_1, a_2 be integers. Suppose that $d \mid (a_1 - a_2)$. Show that the following congruence equations have a unique solution $0 \le x < m$

$$x \equiv a_1 \pmod{m_1}$$
$$x \equiv a_2 \pmod{m_2}$$

---

§5 Complex Integers

Let $\mathbf{Z}[i]=\mathbf{Z}+\mathbf{Z}i=\{a+bi : a, b \in \mathbf{Z}\}$ where $i = \sqrt{-1}$. The set $\mathbf{Z}[i]$ is called the set of *complex integers* or the set of *Gaussian integers.*[11] Geometrically, the set of complex integers can be understood as the set of lattice points in the complex plane. It is easy to see that we may form the three arithmetic operations, $+, -, \cdot$, in the set of complex integers.

We will establish the Unique Factorization Theorem for the set of complex integers $\mathbf{Z}[i]$. The proof is similar to the one for the set of integers $\mathbf{Z}$. The complex integers is a special case of *rings of algebraic integers* which are important topics in *algebraic number theory.*

Definition 1.15. *Given any complex integer $a + bi \in \mathbf{Z}[i]$, let the norm, $N(a + bi)$, of it be defined as*

$$N(a + bi) = a^2 + b^2$$

∎

Discussion

(1) It is not hard to see the the norm is the square of the length from the origin to the point $a + bi$. ∎

One important property of the complex integers is that the *unique factorization theorem* holds for it. We will adopt the line of argument for the integers for our present situation. We shall prove the Euclidean algorithm (cf. §2 Theorem 1.3.) for the complex integers first.

[11] Gauss: German mathematician 1777-1855.

Theorem 1.12. (Euclidean Algorithm). *Let $\delta = d_1 + d_2 i$ be any non-zero complex integer. Let $\alpha = a_1 + a_2 i$ be any complex integer. Then there are complex integers $\beta = b_1 + b_2 i$ and $\gamma = r_1 + r_2 i$ such that*

$$\alpha = \beta \cdot \delta + \gamma, \qquad 0 \le N(\gamma) < N(\delta).$$

Proof: (1) Geometric proof: Clearly $\delta i = -d_2 + d_1 i$ is perpendicular to δ and the set $L = \{(c_1 + c_2 i)\delta : c_1 + c_2 i \in \mathbf{Z}[i]\}$ forms a lattice in the complex plane as in the following diagram,

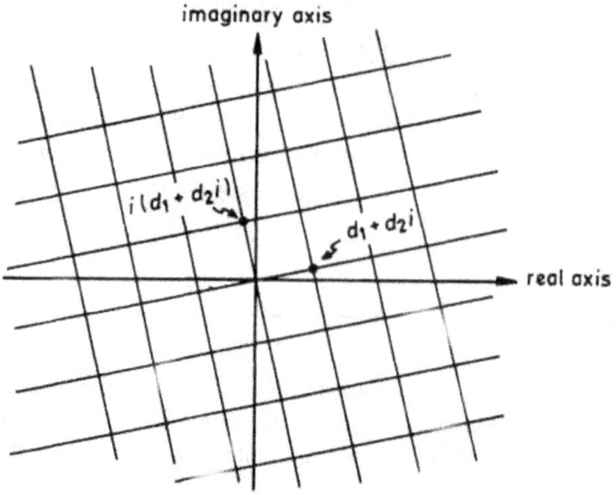

The grids of the new lattice are squares with sides of length $\sqrt{N(\delta)}$. Certainly the number $\alpha = a_1 + a_2 i$ must be in one of the grids. Therefore, one of the vertices, $\beta\delta = (b_1 + b_2 i)\delta$, must have a distance shorter than the side to the point α, i.e.,

$$\sqrt{N(\alpha - \beta\delta)} < \sqrt{N(\delta)}$$
$$N(\alpha - \beta\delta) < N(\delta)$$

Let $\gamma = \alpha - \beta\delta$. Then our theorem is proved.

(2) Algebraic proof: We may translate this proof into algebraic terms without mentioning the preceding diagram. In the following computation,

$$\frac{\alpha}{\delta} = \frac{a_1 + a_2 i}{d_1 + d_2 i} = \frac{(a_1 + a_2 i)(d_1 - d_2 i)}{(d_1 + d_2 i)(d_1 - d_2 i)}$$
$$= \frac{a_1 d_1 + a_2 d_2}{d_1^2 + d_2^2} + \frac{a_2 d_1 - a_1 d_2}{d_1^2 + d_2^2} i$$

et b_1 be the integer closest to $\frac{a_1d_1+a_2d_2}{d_1^2+d_2^2}$, b_2 be the integer closest to $\frac{a_2d_1-a_1d_2}{d_1^2+d_2^2}$, $\beta = b_1 + b_2 i$ and $\gamma = \alpha - \beta\delta$. Furthermore, let ϵ_1 and ϵ_2 be defined as follows,

$$\epsilon_1 = \frac{a_1d_1+a_2d_2}{d_1^2+d_2^2} - b_1,$$
$$\epsilon_2 = \frac{a_2d_1-a_1d_2}{d_1^2+d_2^2} - b_2$$

Then we have the following

$$\begin{aligned}
|\epsilon_1| &= |\frac{a_1d_1+a_2d_2}{d_1^2+d_2^2} - b_1| < \frac{1}{2},\\
|\epsilon_2| &= |\frac{a_2d_1-a_1d_2}{d_1^2+d_2^2} - b_2| < \frac{1}{2},\\
\gamma &= \alpha - \beta\delta = \delta(\frac{\alpha}{\delta - \beta})\\
&= (d_1 + d_2 i)(\epsilon_1 + \epsilon_2 i)\\
&= (d_1\epsilon_1 - d_2\epsilon_2) + (d_2\epsilon_1 + d_1\epsilon_2)i\\
N(\gamma) &= (d_1\epsilon_1 - d_2\epsilon_2)^2 + (d_2\epsilon_1 + d_1\epsilon_2)^2\\
&= d_1^2\epsilon_1^2 + d_1^2\epsilon_2^2 + d_2^2\epsilon_1^2 + d_2^2\epsilon_2^2\\
&< d_1^2 + d_2^2 = N(\delta)
\end{aligned}$$

∎

Discussion

(1) From the geometric proof of the preceding theorem, it is easy to see that for some pairs α, δ, there are four possible values for β and hence four values for γ. Therefore, in the Euclidean Algorithm, we do not require the *uniqueness* for β, γ. ∎

Theorem 1.13. *We always have* $N(\alpha\beta) = N(\alpha)N(\beta)$.

Proof: Let $\alpha = a_1 + a_2 i$ and $\beta = b_1 + b_2 i$. Then we have the following,

$$\begin{aligned}
N(\alpha\beta) &= N((a_1 + a_2 i)(b_1 + b_2 i)) = N((a_1b_1 - a_2b_2) + (a_1b_2 + a_2b_1)i)\\
&= (a_1b_1 - a_2b_2)^2 + (a_1b_2 + a_2b_1)^2 = a_1^2b_1^2 + a_2^2b_2^2 + a_1^2b_2^1 + a_2^2b_1^2\\
&= (a_1^2 + a_2^2)(b_1^2 + b_2^2) = N(a_1 + a_2 i)N(b_1 + b_2 i)\\
&= N(\alpha)N(\beta)
\end{aligned}$$

∎

The only integers with a multiplicative inverse are $1, -1$. For the complex integers, there are more multiplicatively invertible elements, see the following theorem.

Theorem 1.14. *A necessary and sufficient condition for a complex integer $\alpha = a_1 + a_2 i$ to be multiplicatively invertible, i.e., there is a complex integer $\beta = b_1 + b_2 i$ with $\alpha\beta = 1$, is $N(\alpha) = a_1^2 + a_2^2 = 1$. Therefore it must be one of $\pm 1, \pm i$.*

Proof. ($\Longrightarrow$) Note that since $\alpha, \beta \neq 0$, then $N(\alpha), N(\beta)$ are positive integers. From the following computation,

$$N(\alpha\beta) = N(\alpha)N(\beta) = 1$$

we must have $N(\alpha) = 1$.

($\Longleftarrow$) Since we have

$$(a_1 + a_2 i)(a_1 - a_2 i) = a_1^2 + a_2^2 = N(a_1 + a_2 i) = N(\alpha) = 1$$

then $a_1 - a_2 i$ is the multiplicative inverse of α. ∎

Definition 1.16. *Let α, β, γ be complex integers. If $\alpha = \beta \cdot \gamma$. then we say that α is a multiple of β and β is a divisor of α, in symbol, $\beta \mid \alpha$. If $\beta \mid \alpha_1, \beta \mid \alpha_2, \cdots, \beta \mid \alpha_n$, then we say that β is a common divisor of $\alpha_1, \alpha_2, \cdots, \alpha_n$. If β is a common divisor of $\alpha_1, \alpha_2, \cdots, \alpha_n$ and is a multiple of any other common divisor β' of $\alpha_1, \alpha_2, \cdots, \alpha_n$, then we say that β is a greatest common divisor of $\alpha_1, \alpha_2, \cdots, \alpha_n$.* ∎

The following theorem will provide a criterion of a greatest common divisor of a finite set of elements $\alpha_1, \alpha_2, \cdots, \alpha_n$.

Theorem 1.15. *Given any finite set of complex integers $\alpha_1, \alpha_2, \cdots, \alpha_n$, let the set $(\alpha_1, \alpha_2, \cdots, \alpha_n) = \{\sum_i \beta_i \alpha_i : \beta_i \in \mathbf{Z}[i]\}$. Then we have*

1. *There exists a complex integer δ with $(\alpha_1, \alpha_2, \cdots, \alpha_n) = (\delta) = \{\gamma\delta : \gamma \in \mathbf{Z}[i]\}$.*
2. *If $(\alpha_1, \alpha_2, \cdots, \alpha_n) = (\delta)$ with $\delta \neq 0$, then δ is a greatest common divisor of $\alpha_1, \alpha_2, \cdots, \alpha_n$.*
3. *If $(\delta) = (\delta') \neq 0$, then $\delta = \epsilon\delta$ where $\epsilon = \pm 1$ or $\pm i$.*

Proof. (1) If $n = 1$, then we take $\delta = \alpha_1$. Let us consider the case $n = 2$. If $\alpha_1 = \alpha_2 = 0$, then we take $\delta = 0$. Otherwise let δ_1 be an element in (α_1, α_2) with the smallest norm $N(\delta_1)$. According to the Euclidean algorithm, there are β_1, γ_1 with

$$\alpha_1 = \beta_1\delta_1 + \gamma_1, \qquad N(\gamma_1) < N(\delta_1)$$

It is obvious that

$$\gamma_1 = \alpha_1 - \beta_1\delta_1 \in (\alpha_1, \alpha_2)$$

It follows from the minimal property of the norm of δ_1 that $N(\gamma_1) = 0, \gamma_1 = 0$. Therefore, we have $\alpha_1 = \beta_1\delta_1$, i.e., $\delta_1 \mid \alpha_1$. Similarly we can prove that $\alpha_2 = \beta_2\delta_1, \delta_1 \mid \alpha_2$ for some suitable β_2. We conclude that

$$(\alpha_1, \alpha_2) \subseteq (\delta_1)$$

On the other hand, we have $(\alpha_1, \alpha_2) \supseteq (\delta_1)$. Therefore, we have $(\alpha_1, \alpha_2) = (\delta_1)$. It is routine to see

$$(\alpha_1, \alpha_2, \alpha_3, \cdots, \alpha_n) = (\delta_1, \alpha_3, \cdots, \alpha_n)$$

It is then a trivial exercise to use Mathematical induction to show that there is a δ with $(\alpha_1, \alpha_2, \cdots, \alpha_n) = (\delta)$.

(2) It follows from the definition that

$$\alpha_i \in (\alpha_1, \alpha_2, \cdots, \alpha_n) = (\delta) = \{\beta\delta : \beta \in \mathbf{Z}[i]\}$$

Therefore δ is a common divisor of $\alpha_1, \alpha_2, \cdots, \alpha_n$.Let $\delta = \sum_i \beta_i \alpha_i$ and δ' be another common divisor of $\alpha_1, \alpha_2, \cdots, \alpha_n$, i.e.,

$$\delta' \mid \alpha_i, \qquad \alpha_i = \gamma_i \delta' \qquad \forall \ i$$

Then we have

$$\delta = \sum_i \beta_i \alpha_i = \sum_i \beta_i \gamma_i \delta'$$

In other words, we have $\delta' \mid \delta$.

(3) Suppose that we have $(\delta) = (\delta')$. Then we have for suitable ϵ, ϵ' the following,

$$\delta = \epsilon\delta', \qquad \delta' = \epsilon'\delta$$

namely,

$$\delta = \epsilon\epsilon'\delta$$
$$1 = \epsilon\epsilon'$$

It follows from Theorem 1.14. that $\epsilon = \pm 1$ or $\pm i$. ∎

Similar to the integers, we will introduce the concepts of *irreducible* complex integers and *prime* complex integers. Due to the existences of many multiplicatively invertible elements, we will define the term *associativity* as follows,

Definition 1.17. *Let $\alpha \neq 0, \pm 1, \pm i$ be a complex integer, i.e., α is neither zero nor invertible. If in any factorization of $\alpha = \beta \cdot \gamma$, we must have β or $\gamma = \pm 1$ or $\pm i$, then α is said to be an **irreducible complex number**. If $\alpha \mid \beta \cdot \gamma \Longrightarrow \alpha \mid \beta$ or $\alpha \mid \gamma$, then α is called a **prime complex number**. Given any two complex integers α, α', if $\alpha = \epsilon\alpha'$ where ϵ is a multiplicatively invertible element, i.e., $\epsilon = \pm 1$ or $\pm i$, then α, α' are said to be **associated** to each other, in symbol, $\alpha \sim \alpha'$.* ∎

Discussion

(1) Suppose that α, α' are associated to each other, $\alpha \sim \alpha'$. Then α is irreducible $\Longleftrightarrow \alpha'$ is irreducible and α is prime $\Longleftrightarrow \alpha'$ is prime. ∎

Lemma. *In $\mathbb{Z}[i]$, a complex integer α is an irreducible complex integer if and only if it is a prime complex integer.*

Proof: ($\Longrightarrow$) Let α be irreducible and $\alpha \mid \beta\gamma$. Let $(\alpha, \beta) = (\delta)$ (cf Theorem 1.15.). Then for some suitable complex integers ϵ_1, ϵ_2, we have

$$\alpha = \epsilon_1 \delta, \qquad \beta = \epsilon_2 \delta.$$

Since α is irreducible, we must have either ϵ_1 is invertible or δ is invertible. If ϵ_1 is invertible, then we have,

$$\delta = \alpha\epsilon_1^{-1}, \qquad \beta = \epsilon_2\epsilon_1^{-1}\alpha, \qquad \alpha \mid \beta$$

Thus α satisfies the requirements for a prime complex integer. Suppose that ϵ_1 is not invertible. Then δ must be invertible. Therefore for some suitable complex integers ϵ_3, ϵ_4 we have

$$\delta = \epsilon_3\alpha + \epsilon_4\beta$$
$$1 = (\delta^{-1}\epsilon_3)\alpha + (\delta^{-1}\epsilon_4)\beta$$

Multiplying by γ on both sides of the above equation, we get

$$\gamma = (\delta^{-1}\epsilon_3)\gamma\alpha + (\delta^{-1}\epsilon_4)\gamma\beta$$

Therefore we have $\alpha \mid \gamma$ and α is prime.

($\Longleftarrow$) Let α be prime and $\alpha = \beta \cdot \gamma$. Then it implies $\alpha \mid \beta$ or $\alpha \mid \gamma$, say $\alpha \mid \beta, \beta = \alpha \cdot \epsilon$. Trivially, we have $\alpha = \alpha \cdot (\gamma \cdot \epsilon)$ and $1 = \gamma \cdot \epsilon$. Therefore γ is multiplicatively invertible and must be ± 1 or $\pm i$. We conclude that a is irreducible. ∎

Now we may state and prove the Unique Factorization Theorem for the complex integers $\mathbb{Z}[i]$ as follows,

Theorem 1.16. *Let α be any complex integer with $N(\alpha) > 1$. Then*

(1) *α has a **prime decomposition**, i.e., there are prime complex integers $\beta_1, \beta_2, \cdots, \beta_n$ such that $\alpha = \prod_i \beta_i$.*

(2) *Let $\alpha = \prod_{i=1}^n \beta_i = \prod_{j=1}^m \gamma_j$ with all β_i, γ_j prime complex integers. Then there is a reordering of $\gamma_1, \gamma_2, \cdots, \gamma_m$ such that β_i is associated to γ_i. Especially we have $n = m$.*

Proof: (1) If α is irreducible, then it is prime. therefore the equation $\alpha = \alpha$ is a prime decomposition of α. Suppose that α is reducible and $\alpha = \delta_1\delta_2$ with δ_1, δ_2 not invertible. Let us apply Mathematical induction to the norm $N(\alpha)$. Since we have

$$N(\alpha) = N(\delta_1)N(\delta_2) \qquad N(\delta_1) > 1,\ N(\delta_2) > 1$$
$$N(\delta_1) < N(\alpha), \qquad N(\delta_2) < N(\alpha)$$

Using Mathematical induction, we assume that any complex integer with norm less than $N(\alpha)$ has a prime decomposition. Therefore both δ_1 and δ_2 have prime decompositions. Their product is a prime decomposition of α.

(2) Let $\alpha = \prod_j \gamma_j$ be another prime decomposition of α. Then we have

$$\beta_1 \mid \alpha = \gamma_1(\prod_{j>1} \gamma_j)$$

which implies

$$\beta_1 \mid \gamma_1 \text{ or } \beta_1 \mid (\prod_{j>1} \gamma_j)$$

If $\beta_1 \nmid \gamma_1$, then we have $\beta_1 \mid \gamma_2(\prod_{j>2} \gamma_j)$, namely

$$\beta_1 \mid \gamma_2 \text{ or } \beta_1 \mid (\prod_{j>2} \gamma_j)$$

Step by step, there must be a γ_s with

$$\beta_1 \mid \gamma_s$$

After reordering $\gamma_1, \gamma_2, \cdots, \gamma_m$, we may assume that $s = 1$. The complex integer γ_s is prime, therefore irreducible by the preceding lemma. Thus for some invertible integer ϵ, we have

$$\gamma_1 = \epsilon\beta_1$$

Now let us consider $\alpha/\beta_1 = \prod_{i>1} \beta_i = \epsilon \prod_{j>1} \gamma_j$. If this number n is 1, then we are done. If it is bigger than 1, then by Mathematical induction on the total number $n + m$, the prime factors of the two product expressions $\prod_{i>1} \beta_i = \epsilon \prod_{j\neq s} \gamma_j$ are associated to each other after a reordering of $\gamma_2, \cdots, \gamma_m$. We are done again. ∎

Now we want to find all prime complex integers. Since $\mathbb{Z} \subset \mathbb{Z}[i]$, then we may use $\mathbb{Z}$ as a reference set. Let us assume that we know all prime integers in $\mathbb{Z}$. We have the following lemma,

Lemma. *Let α be a complex integer. Then we have*

(1) *The integer $N(\alpha)$ is prime $\Longrightarrow$ α is a complex prime integer.*

(2) *The complex integer α is prime $\Longrightarrow$ its complex conjugate $\overline{\alpha}$ is prime.*

Proof: (1) It follows from the definition.

(2) Let $\alpha = a_1 + a_2 i$. Then $\overline{\alpha} = a_1 - a_2 i$. If $\overline{\alpha} = (b_1 + b_2 i)(c_1 + c_2 i)$ is reducible, then we may assume that $N(b_1 + b_2 i) > 1, N(c_1 + c_2 i) > 1$. It is trivial to see $\alpha = \overline{\overline{\alpha}}$ is reducible too. ∎

Let us study the prime decomposition as a complex integer of a prime integer $p \in \mathbf{Z}$. Let us consider the residue of p (mod 4). We have

$$p \pmod 4 = \begin{cases} 1 \\ 2 \\ 3 \end{cases}$$

Note that if $p \pmod 4 = 2$, then $p = 2$. In this case, we have the following expression,

$$2 = -i(1+i)^2, \qquad N(1+i) = 2$$

It follows from our preceding lemma that $1+i$ is a prime complex integer, and 2 is associated to the square of it. We shall say that 2 is *ramified* in the extension $\mathbf{Z} \in \mathbf{Z}[i]$.

Let us consider a prime integer p with residue $p \equiv 3 \pmod 4$. In the complex integers $\mathbf{Z}[i]$, let the following be a prime decomposition of p,

(1) $$p = (a_1 + a_2 i) \cdots (v_1 + v_2 i)$$

Therefore we have the following relation for integers

$$N(a_1 + a_2 i) = a_1^2 + a_2^2 \mid p^2$$

Clearly, we have $a_1^2 + a_2^2 = p$ or p^2, one of a_1, a_2 must be even and the other one must be odd. From the following computation,

$$a_1^2 + a_2^2 = (1+2n)^2 + (2m)^2 = 1 + 4(n + n^2 + m^2) \equiv 1 \pmod 4$$

we conclude that $a_1^2 + a_2^2 \neq p$. Hence we have

$$N(a_1 + a_2 i) = (a_1 + a_2 i)(a_1 - a_2 i) = a_1^2 + a_2^2 = p^2 = N(p)$$

Let us look at the equation (1). After applying the norm to both sides of it, we know that the other factors must be units. Therefore p is a prime complex integer, and its prime decomposition is

$$p = p$$

We say those prime integers p are *inertia* in the extension $\mathbf{Z} \subset \mathbf{Z}[i]$.

Finally, let us study the case p with residue $p \equiv 1 \pmod 4$. We have the following interesting theorem,

Theorem 1.17. (Fermat). *Given an odd prime integer p. Then $p \equiv 1 \pmod 4 \iff$ there are integers a_1, a_2 such that $p = a_1^2 + a_2^2$.*

Proof: ($\Longleftarrow$) Since p is odd, then we must have that one of a_1, a_2 is odd with the other one even. Therefore we have

$$a_1^2 + a_2^2 = (1+2n)^2 + (2m)^2 = 1 + 4(n + n^2 + m^2) \equiv 1 \pmod 4$$

($\Longrightarrow$) (i) We claim that there is an integer b with $b^2 \equiv -1 \pmod p$. Let $p = 1 + 4\ell$. It follows from Theorem 1.9. that

$$(p-1)! = 1 \cdot 2 \cdot 3 \cdots (p-3)(p-2)(p-1) \equiv -1 \pmod p \tag{2}$$

Clearly we have the following pairing

$$i \equiv -(p-i) \pmod p$$

After the above pairing, we may rewrite the equation (2) as

$$(\frac{p-1}{2})!(-1)^{2\ell}(\frac{p-1}{2})! \equiv -1 \pmod p$$

Let $b = (\frac{p-1}{2})!$. Then we have $b^2 \equiv -1 \pmod p$.

(ii) We shall dismiss the possibility that p is irreducible in $\mathbf{Z}[i]$. Let us assume that p is irreducible in $\mathbf{Z}[i]$. Let δ be determined to satisfy $(p, b+i) = (\delta)$. Note that δ is a factor of p which is assumed to be irreducible. Therefore δ is either invertible or associated with p. We shall deduce contradictions in either case. Let us assume that δ is invertible. Then we have for some suitable complex integers α, β the following

$$\alpha p + \beta(b+i) = \delta$$
$$\delta^{-1}\alpha p + \delta^{-1}\beta(b+i) = 1 \tag{3}$$

Let $\delta^{-1}\alpha = a_1 + a_2 i$ and $\delta^{-1}\beta = b_1 + b_2 i$ and substitute into the equation (3). Then we have

$$\begin{aligned} &a_1 p + a_2 p i + (b_1 b - b_2) + (b_1 + b b_2) i \\ &= (a_1 p + b_1 b - b_2) + (a_2 p + b_1 + b b_2) i \\ &= 1 \end{aligned}$$

namely

$$a_1 p + b_1 b - b_2 = 1, \qquad a_2 p + b_1 + b b_2 = 0$$

Let us factor out p for the above equations. Then we get,

$$b_1 b - b_2 \equiv 1 \pmod p, \qquad b_1 \equiv -b b_2 \pmod p$$

The above implies

$$-b b_2 b - b_2 = -b_2(b^2+1) \equiv 1 \pmod p$$

which contradicts the equation of (i),

$$b^2 + 1 \equiv 0 \pmod p$$

Therefore we conclude that δ cannot be irreducible. In other words, δ is associated with p. Therefore there is an invertible complex integer ϵ with

$$\delta = \epsilon p, \qquad \delta \mid b+i, \qquad \epsilon p \mid b+i$$

namely

$$p \mid \epsilon^{-1}(b+i), \qquad p \nmid \epsilon^{-1}, \qquad p \mid (b+i)$$

Then for some suitable integers c_1, c_2, we have

$$\begin{aligned} b+i &= p(c_1 + c_2 i) \\ b &= pc_1 \\ 1 &= pc_2 \end{aligned}$$

The last equation is impossible and we have a contradiction.

(iii) From (ii) we conclude that p must be reducible in $\mathbf{Z}[i]$. Let $p = (a_1 + a_2 i)(b_1 + b_2 i)$ with $N(a_1 + a_2 i) > 1, N(b_1 + b_2 i) > 1$. Then we have

$$p^2 = N(p) = N(a_1 + a_2 i)N(b_1 + b_2 i) = (a_1^2 + a_2^2)(b_1^2 + b_2^2)$$

Therefore we have

$$p = a_1^2 + a_2^2$$

∎

Let us come back to our discussion about a prime integer p with $p \equiv 1 \pmod 4$. It follows from the above theorem that

$$p = (a_1 + a_2 i)(a_1 - a_2 i) = N(a_1 + a_2 i)$$

It follows from our lemma that both $a_1 + a_2 i$ and $a_1 - a_2 i$ are prime complex integers. Therefore the above is a prime decomposition of p. Those p are said to be *decomposed* in the extension $\mathbf{Z} \subset \mathbf{Z}[i]$.

The above process constructs some prime complex integers. We claim that *any prime complex integer must be associated with one of the above.* Let $\alpha = a_1 + a_2 i$ be any prime complex integer. Let p be a prime *integer* factor of $N(\alpha)$. We have

(i) If $p = 2$, then $2 = (1+i)(1-i) \mid (a_1 + a_2 i)(a_1 - a_2 i) = \alpha\bar{\alpha}$. Therefore, either $(1+i)$ or $(1-i)$ is associated with α.

(ii) If $p \equiv 3 \pmod 4$, then $p \mid \alpha\bar{\alpha}$. Therefore either p is associated with α or $\bar{p} = p$ is associated with α.

(iii) If $p \equiv 1 \pmod 4$, then $p = d_1^2 + d_2^2 = (d_1 + d_2 i)(d_1 - d_2 i) \mid \alpha\bar{\alpha}$. Therefore either $(d_1 + d_2 i)$ is associated with α or $(d_1 - d_2 i)$ is associated with α.

Therefore our claim is established.

∎

xercises

(1) Find a prime decomposition of $4 + 7i$.
(2) Find a prime decomposition of $3 + 4i$.
(3) Find a prime decomposition of $8 + 11i$.
(4) Let $\alpha = 5 - 13i, \delta = -2 + 3i$. Find $\beta, \gamma \in \mathbb{Z}[i]$ such that

$$\alpha = \beta\delta + \gamma, \qquad 0 \leq N(\gamma) < N(\delta)$$

(5) Find the primes among the following list,

$$4 + 5i, \quad 4 - 5i, \quad 7 + i, \quad -2 - 3i, \quad 5 + 9i$$

(6) Use the Euclidean Algorithm to find a greatest common divisor of $7 + i$ and $5 + 9i$.
(7) Let $\alpha \neq 0 \in \mathbb{Z}[i]$. Show that α is a prime $\Longleftrightarrow$ there does not exist a $\beta \neq \pm 1, \pm i$ with $(\alpha) \subset (\beta)$.
(8) Let $\alpha, \beta, \gamma, \delta \in \mathbb{Z}[i]$ with $(\alpha, \beta) = \mathbb{Z}[i]$. Show that there exists a $x \in \mathbb{Z}[i]$ such that

$$x - \gamma \in (\alpha), \qquad x - \delta \in (\beta)$$

6 Real Numbers and p-adic Numbers

In the treatment of similar triangles, the ancient Greek first treated two triangles with ides proportional to an integer ratio, which is elementary, and then generalized the method o a rational ratio. For an irrational ratio, they used the approximation method. The nodern treatments of real numbers are due to Cauchy[12] and Dedekind[13]. We shall follow he approach of Cauchy. The real numbers $\mathbf{R}$ is constructed from the rational numbers $\mathbf{Q}$ y the *completion* of $\mathbf{Q}$ with respect to the distance defined by the absolute value $|\ |$, i.e., or any two rational numbers a, b, let the distance $D_\infty(a, b) = |a - b|$, then $\mathbf{Q}$ becomes a netric space and $\mathbf{R}$ is the *completion* of $\mathbf{Q}$. However, there are other distance functions n $\mathbf{Q}$ which are compatible with the arithmetic operations, $+, -, \times$, in $\mathbf{Q}$. We will review he construction of the real numbers and investigate the completions with respect to other listance functions in $\mathbf{Q}$ in this section. We have the following definition for distance.

Definition 1.18. *Let* $\mathbf{S}$ *be a set. A non-negative numerical valued (see the discussion below) function* $D(a, b)$ *is said to be a distance if it satisfies the following conditions,*

(1) $D(a, b) = 0 \Longleftrightarrow a = b$
(2) $D(a, b) = D(b, a)$
(3) *(Triangle inequality)* $D(a, b) + D(b, c) \geq D(a, c)$ ∎

[12] French Mathematician 1789-1867.
[13] German Mathematian 1831-1916.

Discussion

(1) The term *numerical valued* usually means *real valued*. However, before we construc the real numbers, the term *real valued* has no meaning. Therefore, we shall use the term *numerical valued* to mean *rational valued* at the beginning.

It is easy to see that the function $D_\infty(a,b) = |a-b|$ is a distance function on the rationals $\mathbf{Q}$. In fact, there are many other distance functions on $\mathbf{Q}$ as seen in the following definition and theorem.

Definition 1.19. *Let $p \in \mathbf{Z}$ be a prime integer. Given any $a \neq 0 \in \mathbf{Q}$, let the integer ℓ be determined by the following condition*

$$a = p^\ell \frac{m}{n}, \qquad \text{where } m, n \text{ are integers such that } p \nmid m, \quad p \nmid n$$

Define the p-valuation V_p by the following formula

$$V_p(a) = \begin{cases} 0, & \text{if } a = 0 \\ p^{-\ell}, & \text{if } a \neq 0 \end{cases}$$

Define the p-distance D_p by $D_p(a,b) = V_p(a-b)$. ∎

Theorem 1.18. *The p-valuation satisfies the following conditions,*

(i) $V_p(a) \geq 0, V_p(a) = 0 \Longleftrightarrow a = 0$.
(ii) $V_p(a \cdot b) = V_p(a) \cdot V_p(b)$.
(iii) $V_p(a+b) \leq \max\,(V_p(a), V_p(b))$.

The function D_p satisfies the following conditions, and therefore D_p is a distance function,

(iv) $D_p(a,b) \geq 0, D_p(a,b) = 0 \Longleftrightarrow a = b$.
(v) $D_p(a,b) = D_p(b,a)$.
(vi) *(Strong triangle inequality)* $\max(D_p(a,b), D_p(b,c)) \leq D_p(a,c)$.

Proof: The conditions (i) & (ii) follows trivially from the definition 1.19.. For condition (iii), Let $a = p^{\ell_1}\frac{m_1}{n_1}$ and $b = p^{\ell_2}\frac{m_2}{n_2}$, say, $\ell_1 \leq \ell_2$. Then for some suitable ℓ_3, m_3, n_3, we have the following,

$$\begin{aligned} a + b &= p^{\ell_1}\left(\frac{m_1}{n_1} + p^{\ell_2-\ell_1}\frac{m_2}{n_2}\right) = p^{\ell_1}\frac{m_1 n_2 + p^{\ell_2-\ell_1} m_2 n_1}{n_1 n_2} \\ &= p^{\ell_1}\left(p^{\ell_3}\frac{m_3}{n_3}\right), \qquad \ell_3 \geq 0 \end{aligned}$$

namely

$$\begin{aligned} V_p(a+b) = p^{-\ell_1-\ell_3} \leq p^{-\ell_1} &= \max\,(p^{-\ell_1}, p^{-\ell_2}) \\ &= \max\,(V_p(a), V_p(b)) \end{aligned}$$

Therefore condition (iii) is verified. For the function D_p, conditions (iv) & (v) are obvious. In condition (iii), let us replace a, b in it by $a-b, b-c$. Then condition (vi) follows at once. ∎

Discussion

(1) Let us define an *absolute value* V as a map $V : \mathbf{Q} \to \{$ non-negative rationals $\}$ which satisfies the following,

(i) $V(a) \geq 0, V(a) = 0 \Longleftrightarrow a = 0$.
(ii) $V(a \cdot b) = V(a) \cdot V(b)$.
(iii)* $V(a+b) \leq (V(a) + V(b))$.

Then V is equivalence to either the usual absolute value $|\ |$ or a valuation V_p.

(2) The strong triangle inequality induces a non-intuitive geometry. For instance, under the strong triangle inequality, every triangle is isosceles; Let a, b, c be the vertices of a triangle. Suppose it is not isosceles. Then we must have $D(a,b) > D(b,c) > D(a,c)$ for some arrangement of a, b, c. But it contradicts to the strong triangle inequality max $(D(a,c), D(b,c)) \geq D(a,b)$.Similarly, any interior point of a circle is a center of the circle; Let a be the center of a circle of radius r, b any interior point and c a boundary point. Then we have the following

$$D(a,c) = r > D(a,b)$$
$$\max(D(a,b), D(b,c)) \geq D(a,c)$$

Therefore we must have

$$D(b,c) = r$$

In other words, b is a center. ∎

The following two theorems show the relations between the usual absolute value and all p-valuations.

Theorem 1.19. *Let V_∞ be the usual absolute value, i.e., $V_\infty(a) = |a|$. Then we have*

$$V_\infty(a) \prod_p V_p(a) = 1, \qquad \forall a \in \mathbf{Q}$$

Proof: Obvious. ∎

Theorem 1.20.(Independence of Absolute Values). *Let V_∞ be the usual absolute value and $V_{p_1}, V_{p_2}, \cdots, V_{p_n}$ be n distinct p-valuations. Given any $n+1$ rational numbers $a, a_1, a_2, \cdots, a_n \in \mathbf{Q}$ and $\epsilon > 0 \in \mathbf{Q}$, then there exist a rational $b \in \mathbf{Q}$ such that*

(i) $V_\infty(b-a) < \epsilon$,
(ii) $V_{p_i}(b - a_i) < \epsilon, \qquad \forall\, i$

Proof: We shall establish (ii) first. Let $m =$ the least common multiple of the denominators of $a_1, a_2, \cdots, a_n \in \mathbf{Q}$. Let a positive integer r be determined by the following condition,

$$p_i^{-s_i} = V_{p_i}(m)$$
$$p_i^{-r} < \epsilon p_i^{-s_i}, \qquad \forall\, i$$

It follows from the Chinese Remainder Theorem that there exists an integer c satisfying the following congruence equation,

$$c \equiv ma_i \pmod{p_i^r}, \qquad \forall\, i$$

namely,

$$V_{p_i}(c - ma_i) \leq p_i^{-r}, \qquad \forall\, i$$
$$V_{p_i}(\frac{c}{m} - a_i) \leq \epsilon, \qquad \forall\, i$$

Now we shall modify $\frac{c}{m}$ such that both conditions (i) & (ii) are satisfied. Let $q = \prod_i p_i^r$. Select integers u, v such that

$$|\frac{c}{m}\frac{1+uq}{1+vq} - a| < \epsilon$$
$$b = |\frac{c}{m}\frac{1+uq}{1+vq}|$$

The number b defined by the above equation will replace $\frac{c}{m}$. It is trivial to see

$$b - \frac{c}{m} = \frac{c}{m}\frac{(u-v)q}{1+vq}$$
$$V_{p_i}(b - \frac{c}{m}) \leq V_{p_i}(\frac{q}{m}) \leq \epsilon$$

Therefore we conclude

$$V_\infty(b-a) < \epsilon$$
$$V_{p_i}(b-a) = V_{p_i}((b - \frac{c}{m}) + (\frac{c}{m} - a))$$
$$\leq \max\left(V_{p_i}(b - \frac{c}{m}), V_{p_i}(\frac{c}{m} - a)\right)$$
$$\leq \epsilon$$

∎

The above theorems clarify the relations between all absolute values on $\mathbf{Q}$. For the remainder of this section, we shall concentrate on *one absolute value* V or rather its induced distance function D. For the general applications, we shall consider a set $\mathbf{S}$ and a distance function D on it. We have the following definition,

Definition 1.20. *Given a set* $\mathbf{S}$ *and a distance function* D *on it. A sequence* $\{a_1, a_2, \cdots, a_n, \cdots\} \in \prod_{i=1}^{\infty} \mathbf{S}$ *is said to be a fundamental sequence if the following condition is satisfied; for any* $\epsilon > 0 \in \mathbf{Q}$, *there is an integer* N $(= N(\epsilon))$ *such that for all* $n, m > N$. *we have*

$$D(a_n, a_m) < \epsilon$$

The set of all fundamental sequences is denoted by $\mathsf{F}(D)$. ∎

Definition 1.21. *Two fundamental sequences* $\{a_i\} = \{a_1, a_2, \cdots, a_n, \cdots\}$ *and* $\{b_i\} = \{b_1, b_2, \cdots, b_n, \cdots\}$ *are said to have a common limit, in symbol* $\{a_i\} \overset{D}{\sim} \{b_i\}$, *if the following condition is satisfied; for any* $\epsilon > 0 \in \mathbf{Q}$, *there is an integer* N $(= N(\epsilon))$ *such that for all* $m > N$, *we have*

$$D(a_m, b_m) < \epsilon$$

∎

Theorem 1.21. *The relation of having common limit,* $\overset{D}{\sim}$, *is a equivalence relation (cf Definition 1.4.) on the set of fundamental sequences* $\mathsf{F}(D)$.

Proof: We will check the three conditions for an equivalence relation.

(i) Obviously we have $\{a_i\} \overset{D}{\sim} \{a_i\}$.
(ii) Obviously $\{a_i\} \overset{D}{\sim} \{b_i\} \Longrightarrow \{b_i\} \overset{D}{\sim} \{a_i\}$.
(iii) Suppose that $\{a_i\} \overset{D}{\sim} \{b_i\}, \{b_i\} \overset{D}{\sim} \{c_i\}$. Then for any $\epsilon > 0 \in \mathbf{Q}$, there is an N such that for all $m > N$, we have

$$D(a_m, b_m) < \frac{\epsilon}{2}, \qquad D(b_m, c_m) < \frac{\epsilon}{2}$$

It then follows from the triangle inequality that

$$D(a_m, c_m) \leq D(a_m, b_m) + D(b_m, c_m) < \epsilon$$

∎

The following statement will define the real numbers $\mathbf{R}$ in a logical (while not intuitive) way.

Definition 1.22. *The quotient set of the set of all fundamental sequences* $\mathsf{F}(D)$ *with respect to the equivalence relation,* $\overset{D}{\sim}$, *is called the D-completion of* $\mathbf{S}$, *in symbol* $\mathbf{S}_D$.

Every element, i.e., every equivalence class, of $\mathbf{S}_D$ *will be called the limit of any sequence* $\{a_i\}$ *in this element. Especially, we have*

(i) *If* S= Q *and* $D = D_\infty$ *the distance induced by the usual absolute value* V_∞, *then* $\mathbf{S}_D$ *is the set of real numbers* **R**.

(ii) *If* S= Q *and* $D = D_p$ *the distance induced by the p-valuation* V_p, *then* $\mathbf{S}_D$ *is the set of p-adic numbers* $\mathbf{Q}_p$.

∎

In the above definition, every real number is in fact an equivalence class consisting of infinitely many fundamental sequences. To reconciliate the rational numbers, which are single numbers, with the real numbers, which are sets, we introduce the following theorem.

Theorem 1.22. *Let The map* $\varphi : \mathbf{S} \to \mathbf{S}_D$ *defined by* $\varphi(a) = [\{a, a, \cdots, a, \cdots\}] = [\{a\}]$. *Then we have*

(i) φ *is injective.*

(ii) *We may identify* S *and* φ (S) *by identifying* $a \in \mathbf{S}$ *and* $\varphi(a) \in \varphi$ (S).

Proof: It is obvious that $\{a, a, \cdots\}$ is a fundamental sequence. Let $a \neq b$, $|a - b| = \epsilon$. Then the two images $\varphi(a) = [\{a, a, \cdots\}] = [\{a\}], \varphi(b) = [\{b, b, \cdots\}] = [\{b\}]$ will satisfy

$$D(a, b) \not< \epsilon$$

Therefore $\{a\}$ and $\{b\}$ belong to two distinct equivalence classes. ∎

We may then identify a rational number a with the equivalence class of the fundamental sequence $\{a, a, \cdots\}$. Therefore, the set of rationals **Q** is a subset of the set of reals **R**.

Discussion

(1) Let us consider the **infinite decimal expansions** of the real numbers: Given any fundamental sequence $\{a_1, a_2, \cdots, a_n, \cdots\}$. Let $\epsilon_j = 10^{-j}$ and N_j be a positive integer given by Definition 1.20. as

$$D_\infty(a_n, a_m) = |a_n - a_m| < \epsilon_j = 10^{-j}, \qquad \forall\, n, m > N_j$$

It is clear that the above relation means that for large n, the integer part and the the first j decimal places of a_n are all identical. As j tends to ∞, a fixed infinite decimal expansion is presented which will be used to denote the equivalence class of the fundamental sequence. This way of using the infinite decimal expansions to represent real numbers started in Shang dynasty of China (about 1000 B.C.) and was completed in the book "*Nine Chapters on the Mathematical Arts*" (100 A.D.). As everybody knows, the infinite decimal expansions are not unique. For instance, the following two fundamental sequences belong to the same equivalence class,

$$\begin{array}{l}(1,\ 1,\ 1,\ \cdots, 1,\ \cdots)\\(0.9, 0.99, 0.999, \cdots, 0.99\cdots 9, \cdots)\end{array}$$

(2) Similar to the infinite decimal expansion, we will establish the **p-adic expansion** or the p-adic numbers $\mathbf{Q}_p$. Let us consider a rational number a first. Let

(1) $$a = p^{\ell}\frac{m}{n}, \qquad p \nmid m, \quad p \nmid n$$

We wish to approximate a. Let 0 be the first approximation of a. Then we have

$$D_p(a, 0) = p^{-\ell}$$

It follows from the above equation (1) that p, n are coprime and there exist r, s with

$$sn + rp = 1, \qquad sn \equiv 1 \pmod p, \qquad p \nmid s$$

Let t be the principal residue of the class $[sm]_p$. We then have

$$[s(m - nt)]_p = [sm - snt]_p = [sm - t]_p = [0]_p$$

namely

$$p \mid m - nt$$
$$a - tp^{\ell} = p^{\ell}(\frac{m}{n} - t) = p^{\ell}(\frac{m - nt}{n}) = p^{\ell+\ell'}\frac{m'}{n}$$
$$D_p(a, tp^{\ell}) \leq p^{-\ell-\ell'} < p^{-\ell}$$

Therefore tp^{ℓ} approximates a closer than 0. Now replacing a by $a - tp^{\ell}$, we will find a better approximation of a. Step by step, we find a power series expansion of $a = \sum_{i=-n}^{\infty} t_i p^i$ where all $t_i \in \{0, 1, 2, \cdots, p-1\}$ the set of the principal residues. For instance, let $p = 3$ and $a = -1/6$. Then we have

$$-\frac{1}{6} = 3^{-1} + 1 + 3 + 3^2 + \cdots + 3^n + \cdots$$

namely

$$D(-\frac{1}{6}, 3^{-1} + 1 + 3 + 3^2 + \cdots + 3^n + \cdots) = 3^{-n-1} \to 0$$

It is not hard to see $\mathbf{Q}_p = \{\sum_{i=-n}^{\infty} t_i p^i$ where all $t_i \in \{0, 1, 2, \cdots, p-1\}$ the set of the principal residues.$\}$. The p-adic expansion is unique and is different from the infinite decimal expansion. ∎

Instead of an abstract set **S**. let us consider the rational numbers **Q**. The other way to construct the real numbers **R** from the rational numbers **Q** is the method of *Dedekind cut* which is based on the inequality $\geq$. In this note, we will not consider the Dedekind cut method. However, we will show that the inequality $\geq$ can be generalized naturally to the real numbers **R**. We have the following definition,

Definition 1.23. *Let us consider the real numbers* $\mathbf{R}$ $(=\mathbf{Q}_{D_\infty})$*. Let* α, β *be two element of* $\mathbf{R}$ *with* $\alpha = [\{a_i\}]$ *and* $\beta = [\{b_i\}]$*. Suppose that for any* $\epsilon > 0 \in \mathbf{Q}$*, there is an* $N = (N(\epsilon))$*, such that*

$$a_i \geq b_i, \qquad \forall\, i > N$$

Then we say $\alpha \geq \beta$ *and* $\{a_i\} \geq \{b_i\}$*. If* $\alpha \geq \beta$ *and* $\alpha \neq \beta$*, then we say* $\alpha > \beta$*.* ∎

Theorem 1.23. *The above Definition 1.23. is well-defined, i.e., for the fundamental sequences* $\{a_i\}, \{a_i'\}$ *with the same limit (in other words,* $[\{a_i\}] = [\{a_i'\}]$*) and the fundamental sequences* $\{b_i\}, \{b_i'\}$ *with the same limit, then we always have,*

$$\{a_i\} \geq \{b_i\} \Longleftrightarrow \{a_i'\} \geq \{b_i'\}$$

Moreover, the inequality $\geq$ *satisfies the following three basic properties,*

(i) $\alpha \geq \beta, \beta \geq \alpha \Longleftrightarrow \alpha = \beta$.
(ii) $\alpha \geq \beta \geq \gamma \Longrightarrow \alpha \geq \gamma$.
(iii) *Given any* $\alpha, \beta \in \mathbf{Q}_D$*, we must have either* $\alpha \geq \beta$ or $\beta \geq \alpha$.

Proof: It is left as an exercise to the reader. ∎

Discussion

(1) We cannot generalize the inequality $\geq$ to $\mathbf{Q}_p$. For instance, in the preceding **Discussion**, we see

$$-\frac{1}{6} = 3^{-1} + 1 + 3 + 3^2 + \cdots + 3^n + \cdots$$
$$-\frac{1}{6} - (3^{-1} + 1) = 3 + 3^2 + \cdots + 3^n + \cdots$$

Can we believe that $-\frac{1}{6} \geq (3^{-1} + 1) = \frac{4}{3}$? ∎

Now we will extend the four Arithmetic operations, $+, -, \cdot, \div$, from the rationals $\mathbf{Q}$ to the completion $\mathbf{Q}_D$. Note that we may change a finitely many terms of a fundamental sequence $\{a_i\}$ without changing the equivalence class $[\{a_i\}]$. Therefore for any fundamental sequence $\{a_i\}$, if $[\{a_i\}] \neq [\{0\}]$, then only finitely many of the a_i's can be zeroes. After changing all of them to non-zeroes, we may assume that $a_i \neq 0, \forall\, ii$ as long as $[\{a_i\}] \neq 0$. The reader is asked to show that the following definition is well-defined.

Definition 1.24. *Let* α, β *be two elements of* $\mathbf{Q}_D$ *with* $\alpha = [\{a_i\}]$ *and* $\beta = [\{b_i\}]$*. Then* $\{a_i + b_i\}$*,* $\{a_i - b_i\}$ *and* $\{a_i b_i\}$ *are fundamental sequences. We define* $\alpha + \beta = [\{a_i + b_i\}]$*,* $\alpha - \beta = [\{a_i - b_i\}]$ *and* $\alpha\beta = [\{a_i b_i\}]$*. Moreover if* $[\{a_i\}] \neq [\{0\}]$ *with* $a_i \neq 0 \forall\, i$*, then* $\{\frac{b_i}{a_i}\}$ *is a fundamental sequence. We define* $\frac{\beta}{\alpha} = [\{\frac{b_i}{a_i}\}]$*.* ∎

We now want to extend the distance function D from the rationals $\mathbf{Q}$ to the completion $\mathbf{Q}_D$. There is a question about the value of the distance function D. In case that the distance function D is the usual absolute value, we know that the distance between two real numbers may not be rational any more. In this case we have to extend the value of the distance function D from the rationals $\mathbf{Q}$ to the reals $\mathbf{R}$. On the other hand, if the distance function D is the p-distance, then the values for $\mathbf{Q}_p$ stay rational (cf **Discussion** below). The following definition together with the next theorem verifies that the definition is well-defined.

Definition 1.25. *Let α, β be two elements of $\mathbf{Q}_D$ with $\alpha = [\{a_i\}]$ and $\beta = [\{b_i\}]$. Then $\{D(a_i, b_i)\}$, is a fundamental sequence and we define*

$$D(\alpha, \beta) = [\{D(a_i, b_i)\}]$$

∎

Discussion

(1) If the distance function D is the p-distance, let $\alpha = \{a_i\}$ be represented by $\sum_{i=-n}^{\infty} t_i p^i$ and $\beta = \{b_i\}$ be represented by $\sum_{i=-n}^{\infty} s_i p^i$, then $D(\alpha, \beta) = p^{-i}$ where i is the smallest integer with $t_i \neq s_i$ in the two preceding power series. Therefore the sequence $\{D(a_i, b_i)\}$ is stationary after some finitely many terms. ∎

Theorem 1.24. *Let us consider $\mathbf{Q}_D$. The preceding definition is well-defined, i.e.,*

(i) *$\{D(a_i, b_i)\}$ is a fundamental sequence.*
(ii) *For the fundamental sequences $\{a_i\}, \{a'_i\}$ with the same limit (in other words, $[\{a_i\}] = [\{a'_i\}]$) and the fundamental sequences $\{b_i\}, \{b'_i\}$ with the same limit, then we always have,*

$$[\{D(a_i, b_i)\}] = [\{D(a'_i, b'_i)\}]$$

Moreover, the extended function D is a distance (cf Definition 1.18.) on $\mathbf{Q}_D$.

Proof: (i) If $D = D_p$, then by the preceding **Discussion**, we know $\{D(a_i, b_i)\}$ is eventually stationary and hence fundamental. Let us assume that D is the usual absolute value. Let $d_i = D(a_i, b_i)$. We want to prove that $\{d_i\}$ is a fundamental sequence. Given any $\epsilon > 0 \in \mathbf{Q}$, there are positive integers N_a, N_b such that

$$D(a_m, a_n) < \frac{\epsilon}{2}, \qquad \text{if } m, n > N_a$$

$$D(b_m, b_n) < \frac{\epsilon}{2}, \qquad \text{if } m, n > N_b$$

Let $N = \max\ (N_a, N_b)$. Then for $m, n > N$, using the Triangle inequality, we have (assuming $D(a_m, b_m) \geq D(a_n, b_n)$),

$$\begin{aligned} D(d_m, d_n) &= |d_m - d_n| = |D(a_m, b_m) - D(a_n, b_n)| \\ &\leq |(D(a_m, a_n) + D(a_n, b_n) + D(b_n, b_m)) - D(a_n, b_n)| \\ &\leq |(D(a_m, a_n) + D(b_n, b_m))| \leq \epsilon \end{aligned}$$

(ii) The discussion for the case $D = D_p$ is left to the reader as an exercise. Let us assume that D is the usual absolute value. We shall use the notations of (i). Let $d'_i = D(a'_i, b'_i)$. Given any $\epsilon > 0 \in \mathbf{Q}$, there are positive integers M_a, M_b such that

$$D(a_n, a'_n) < \frac{\epsilon}{2}, \qquad \text{if } n > M$$
$$D(b_n, b'_n) < \frac{\epsilon}{2}, \qquad \text{if } n > M_b$$

Let $M = \max(M_a, M_b)$. Then for $n > M$, using the Triangle inequality, we have (assuming $d_n \geq d'_n$),

$$\begin{aligned} D(d_n, d'_n) &= |d_n - d'_n| = |D(a_n, b_n) - D(a'_n, b'_n)| \\ &\leq |(D(a_n, a'_n) + D(a'_n, b'_n) + D(b'_n, b_n)) - D(a'_n, b'_n)| \\ &\leq |(D(a_n, a'_n) + D(b'_n, b_n))| \leq \epsilon \end{aligned}$$

(iii) The discussion for the case $D = D_p$ is left to the reader as an exercise. Let us assume that D is the usual absolute value. The first two conditions of Definition 1.18. can be verified easily. Let us consider the third condition; the Triangle inequality. We shall use the notations of (i) and Definition 1.25.. Let α, β be given as before. Let $\gamma = \{c_i\}$ be a third fundamental sequence. Given any $\epsilon > 0 \in \mathbf{Q}$, there are positive integers L such that for all $m > L$ we have

$$D(a_m, b_m) \geq D(a_L, b_L) - \frac{\epsilon}{3}$$
$$D(b_m, c_m) \geq D(b_L, c_L) - \frac{\epsilon}{3}$$
$$D(a_L, c_L) \geq D(a_m, c_m) - \frac{\epsilon}{3}$$

namely

$$\begin{aligned} D(a_m, b_m) + D(b_m, c_m) &\geq D(a_L, b_L) + D(b_L, c_L) - \frac{2\epsilon}{3} \\ &\geq D(a_L, c_L) - \frac{2\epsilon}{3} \\ &\geq D(a_m, c_m) - \epsilon \end{aligned}$$

It then follows from Definition 1.23. and Definition 1.24. that

$$D(\alpha, \beta) + D(\beta, \gamma) \geq D(\alpha, \gamma)$$

∎

Discussion

(1) In Definition 1.22. we define *limit* in a non-intuitive way. Let $\alpha = [\{a_1, a_2, \cdots\}] = [\{a_i\}]$ be an equivalence class of fundamental sequences. Let us define $\alpha_i = [\{a_i, a_i, \cdots\}]$.

Then we shall identify $\alpha_i \in \mathbf{S}_D$ with $a_i \in \mathbf{S}$. It is trivial to see that for any $\epsilon > 0 \in \mathbf{Q}$, there is an integer N such that

$$D(\alpha, \alpha_i) < \epsilon, \qquad \forall\, i > N$$

namely, in the usual sense

$$\lim_{i\to\infty} a_i = \lim_{i\to\infty} \alpha_i = \alpha$$

∎

The final question studied in this section is if the completion $\mathbf{Q}_D$ is *completed*? i.e., if $(\mathbf{Q}_D)_D = \mathbf{Q}_D$? In other words, if we complete the completion $\mathbf{Q}_D$ with respect to the extended distance function D on $\mathbf{Q}_D$, do we get any extra elements? Let us introduce the following definition and theorem.

Definition 1.26. *Let* $\mathbf{S}$ *be a set with a distance function* D. *Let* $\{\alpha_i\}$ *be a fundamental sequence. If there is an element* α *with the property that for any* $\epsilon > 0 \in \mathbf{Q}$, *there is an integer* N *such that*

$$D(\alpha, \alpha_i) < \epsilon, \qquad \forall\, i > N$$

then we say the limit of the fundamental sequence $\{\alpha_i\}$ *exists and equals* α, *in symbol*

$$\lim_{i\to\infty} \alpha_i = \alpha$$

Moreover, if the limits of all fundamental sequences exist, then we say $\mathbf{S}$ *is completed with respect to* D. ∎

Discussion

(1) It is easy to see that if $\mathbf{S}$ is completed with respect to D, then $\mathbf{S}=\mathbf{S}_D$. ∎

Theorem 1.25. *The* D*-completion* $\mathbf{Q}_D$ *is completed.*

Proof: It amounts to proving that every fundamental sequence $\{\alpha_i\} \in \mathbf{Q}_D$ has a limit $\alpha \in \mathbf{Q}_D$. We shall construct the element α in the following way. Every element α_i can be represented as follows,

$$\alpha_i = [\{a_{i1}, a_{i2}, \cdots, a_{in}, \cdots\}]$$

where $\{a_{i1}, a_{i2}, \cdots, a_{in}, \cdots\}$ is a fundamental sequence in $\mathbf{S}$. Let $\epsilon_i = \dfrac{1}{i}$. Then there is an integer $N(i)$ such that

$$D(a_{im}, a_{in}) < \epsilon_i = \frac{1}{i}, \qquad \forall\, m, n > N(i)$$

We may replace $N(i)$ by a bigger integer if necessary. Therefore we may assume that $N(1) < N(2) < \cdots < N(i) < \cdots$. Let $a_i = a_{iN(i)}$. We claim

(i) $\{a_i\}$ is a fundamental sequence is $\mathbf{Q}_D$.

(ii) $\lim_{i\to\infty} \alpha_i = [\{a_i\}]$

Then we simply let $\alpha = [\{a_i\}]$ and our theorem will be proved. Let us prove the above two claims.

(i) Since $\{\alpha_i\}$ is a fundamental sequence in $\mathbf{Q}_D$, then given any $\epsilon > 0 \in \mathbf{Q}$, there is an integer N_1 such that for all $m, n > N_1$ the following

$$D(\alpha_m, \alpha_n) = [\{D(a_{mj}, a_{nj})\}] < \frac{\epsilon}{6} = [\{\frac{\epsilon}{6}, \frac{\epsilon}{6}, \cdots, \frac{\epsilon}{6}, \cdots\}].$$

In Definition 1.23., replacing ϵ by $\frac{\epsilon}{6}$, we know that there exists an integer $N_2(m,n)$ with

$$D(a_{m\ell}, a_{n\ell}) < \frac{\epsilon}{6} + \frac{\epsilon}{6} = \frac{\epsilon}{3}, \quad \forall\, \ell > N_2(m,n)$$

Let an integer N_3 satisfy

$$\frac{1}{N_3} < \frac{1}{\epsilon}$$

Let $N = \max(N_1, N_3)$. For any $m, n > N$, $\ell > \max(N_2(m,n), N(m), N(n))$, we have the following due to the Triangle inequality,

$$\begin{aligned} D(a_m, a_n) &= D(a_{mN(m)}, a_{nN(n)}) \\ &\leq D(a_{mN(m)}, a_{m\ell}) + D(a_{m\ell}, a_{n\ell}) + D(a_{n\ell}, a_{nN(n)}) \\ &< \frac{\epsilon}{3} + \frac{\epsilon}{3} + \frac{\epsilon}{3} = \epsilon. \end{aligned}$$

Therefore $\alpha = \{\alpha_i\}$ is a fundamental sequence in $\mathbf{Q}_D$.

(ii) We shall prove that $\lim_{i\to\infty} \alpha_i = \alpha$. Let us use the notations of (i). Given any $\epsilon_1 > 0 \in \mathbf{Q}$, let us replace ϵ in (i) by ϵ_1 and get an integer N. For any $n > N$, let

$$\begin{aligned} \ell &> \max(N, N(n)) \\ s &> \max(N_2(n,\ell), N(\ell), N(n)) \end{aligned}$$

Then we have the following due to the Triangle inequality,

$$\begin{aligned} D(a_{n\ell}, a_\ell) &= D(a_{n\ell}, a_{\ell N(\ell)}) \\ &\leq D(a_{n\ell}, a_{ns}) + D(a_{ns}, a_{\ell s}) + D(a_{\ell s}, a_{\ell N(\ell)}) \\ &< \frac{\epsilon_1}{3} + \frac{\epsilon_1}{3} + \frac{\epsilon_1}{3} = \epsilon_1. \\ \{D(a_{n\ell}, a_\ell)\} &< \epsilon_1 \\ D(\alpha_n, \alpha) &< \epsilon_1 \end{aligned}$$

Therefore $\lim_{i\to\infty} \alpha_i = \alpha$. ∎

Exercises

(1) Let $a_i = 1 + \frac{1}{2!} + \cdots + \frac{1}{i!}$. Show that $\{a_i\}$ is *not* a fundamental sequence in $\mathbf{Q}_p$ for any p (therefore the number e is not defined in $\mathbf{Q}_p$ for any p).

(2) Let x be an integer with $p \mid x$. Let $a_i(x) = 1 + x + \frac{x^2}{2!} + \cdots + \frac{x^i}{i!}$. Show that $\{a_i(x)\}$ is a fundamental sequence in $\mathbf{Q}_p$ (therefore the number e^x is defined in $\mathbf{Q}_p$).

(3) Let us use the *generalized binomial theorem*[14]: for any rational number ℓ, let

$$\binom{\ell}{i} = \frac{\ell(\ell-1)\cdots(\ell-i+1)}{i(i-1)\cdots 1}$$

Then we have:

$$(1+x)^\ell = 1 + \sum_{i=1}^{\infty} \binom{\ell}{i} x^i$$

Let $p = 3$. Given the p-adic expansion of $\frac{5}{2}$ as

$$\begin{aligned}\frac{5}{2} &= 3 + \frac{1}{1-3} = 3 + 1 + 3 + \cdots + 3^i + \cdots \\ &= 1 + 2\cdot 3 + 3^2 + \cdots + 3^i + \cdots\end{aligned}$$

prove:

$$\begin{aligned}(1+x)^{\frac{5}{2}} &\equiv (1+x)^{1+2\cdot 3+3^2+\cdots+3^i+\cdots} \pmod 3 \\ &\equiv (1+x)(1+x)^{2\cdot 3}(1+x)^{3^2}\cdots(1+x)^{3^i}\cdots \pmod 3 \\ &\equiv (1+x)(1+x^3)^2(1+x^{3^2})\cdots(1+x^{3^i})\cdots \pmod 3\end{aligned}$$

(4) Let the p-adic expansion of a rational number a be

$$a = a_{-m}p^{-m} + a_{-m+1}p^{-m+1} + \cdots + a_0 + a_1 p + \cdots$$

Find the p-adic expansion of $-a$.

(5) Find the first four coefficients of the following product in $\mathbf{Q}_7$,

$$(6 + 4\cdot 7 + 2\cdot 7^2 + 1\cdot 7 + \cdots)(3 + 0\cdot 7 + 0\cdot 7^2 + 6\cdot 7^3 + \cdots)$$

[14] The binomial triangle for integer powers was discovered by Chinese Mathematician Yang Hui in 'Hsiang Chieh Chiu Chang Ssu Fa' in 1261, later, it appeared in Arabic Mathematicain Al-Kashi's work in the 15-th century, the work of German Mathematicians Apian in 1527, Stifel in 1544, and the posthumous work of French Mathematician Pascal in 1665. The binomial coefficients of rational powers were due to English Mathematician Newton in 1665.

(6) Find the first four coefficients of the 5-adic expansion of $(3+2\cdot 5+3\cdot 5^2+1\cdot 5^3)^{-1}$ in $\mathbf{Q}_5$.

(7) Let $a \in \mathbf{Q}_p$. If the p-adic expansion of a has only finitely many non-zero coefficients then $a \in \mathbf{Q}$ and with denominator a power of p.

(8) Prove that the number 6 has a square root in $\mathbf{Q}_5$, i.e., the following equation is satisfied for some suitable element in $\mathbf{Q}_5$,

$$(a_0 + a_1\cdot 5 + a_2\cdot 5^2 + \cdots)^2 = 1 + 1\cdot 5, \qquad 0 \le a_i < 5.$$

(9) Prove that there is *no* square root for 7 in $\mathbf{Q}_5$.

(10) The following p-adic expansion

$$a = a_{-m}p^{-m} + a_{-m+1}p^{-m+1} + \cdots + a_0 + a_1 p + \cdots$$

is said to be *periodic* if there are positive integers r and N such that $a_{i+r} = a_i$ for all $i > N$. Prove that the expansion is periodic $\Longleftrightarrow a \in \mathbf{Q}$.

(11) Give a series as follows in $\mathbf{Q}_p$

$$\sum \alpha_n = \alpha_1 + \alpha_2 + \cdots$$

let its *partial sum* s_n be defined as

$$s_n = \alpha_1 + \alpha_2 + \cdots + \alpha_n$$

We say the series $\sum \alpha_n$ is a *convergent series* if $\lim_{n\to\infty} s_n$ exists. Prove that the series $\sum \alpha_n$ is a convergent series $\Longleftrightarrow \lim_{n\to\infty} \alpha_n = 0$.

(12) Prove that the equation $x^p - x = 0$ has p solutions in $\mathbf{Q}_p$.

CHAPTER II

Group Theory

§1 Definitions

Definition 2.1. *Let* G *be a* ***non-empty set*** *and* $*$ *a* ***binary operation*** *on* G, *i.e.,* $* : G \times G \to G$ *with* $* : (a, b) \mapsto a * b \in G$. *If for a subset* $H \subset G$, $a * b \in H$ *for all* $a, b \in H$, *then we say that* H *is* ***closed*** *under* $*$, *i.e.,* $*$ *induces a binary operation on* H. ∎

The operations $+, -, \cdot$, are binary operations on the set of integers **Z**, while $\div$ is not, because the result of a division $a \div b$ may not be an integer or may be ∞ if the divisor b is zero.

Definition 2.2. *Let* G *be a non-empty set with a binary operation* $*$. *If the following conditions are satisfied, then we say* (G,$*$) *is a* ***group***[1]. *Sometimes, we simply say* G *is a group without mention the operation* $*$ *if there is no confusion.*

(i) *Associative law:* $a * (b * c) = (a * b) * c$ *for all elements* $a, b, c \in G$.
(ii) *Existence of unit: there exists an element* e *with* $e * a = a * e = a$ *for any element* $a \in G$.
(iii) *Existence of inverse: given any element* $a \in G$, *there exists an element* $b \in G$ *with* $b * a = a * b = e$ *a unit. The element* b *is called the inverse of* a *with respect to* $*$.

∎

Discussion

(1) The meaning of the associative law is that the result of the operation $*$ is independent of the order of applying the operation $*$. Therefore we may define a trinary operation as follows

$$a * b * c = (a * b) * c = a * (b * c)$$

[1] The name 'group' was given by French Mathematician Galois 1811-1832.

In general, in any expression, we may omit the parenthesis ( ). Usually, we use a^n to express n terms multiplication $a * a * \cdots * a$. For our notations, we *define* $a^0 = e$ for any $a \in \mathbf{G}$.

(2) **Analysis of axioms:** It is easy to find a binary operation which is not associative. For instance, it is well-known that the multiplications of most *Lie algebras* are non-associative. Let us use the subtraction, $-$, on $\mathbf{Z}$ as an example: we define a binary operation $*$ on the set of integers $\mathbf{Z}$ with $a * b = a - b$. Then we see at once that 0 is the unit and a is the inverse of a. However, the associative law is not satisfied as indicated by

$$3 * (2 * 1) = 3 - (2 - 1) = 3 - 1 = 2 \neq (3 * 2) * 1 = (3 - 2) - 1 = 1 - 1 = 0$$

Let us consider another example. We define a binary operation $*$ on the set of integers $\mathbf{Z}$ with $a * b = a \cdot b$. Then $(\mathbf{Z}, *)$ satisfies conditions (i) & (ii) of Definition 2.1. only.

(3) **The uniqueness of unit**: Let e' be another unit. Then we have

$$e = e * e' = e'$$

Therefore the unit in a group is unique.

(4) **The uniqueness of inverse**: Let b, b' be inverses of a. Then we have

$$b = b * e = b * (a * b') = (b * a) * b' = e * b' = b'$$

Therefore the inverse of an element $a \in \mathbf{G}$ is unique. Usually we use a^{-1} to denote the inverse of a and a^{-n} to express the n terms multiplication $a^{-1} * a^{-1} * \cdots * a^{-1}$.

(5) The inverse of $a_1 * a_2$ is $a_2^{-1} * a_1^{-1}$. The inverse of $a_1 * a_2 * \cdots * a_n$ is $a_n^{-1} * \cdots * a_2^{-1} * a_1^{-1}$.

(6) Given an element $a \in \mathbf{G}$, if $a^n = e$ for some non-negative integer n, then $a^{-n} = e$ and there is a positive integer m with $a^m = e$. Let the *order* of a be defined by the smallest positive integer n with $a^n = e$, in symbol o(a)=n. If the only integer n which satisfies a^n is zero, then we say the order of a is infinite or ∞.

(7) The cardinality of a group $\mathbf{G}$, Card($\mathbf{G}$), is called the *order* o($\mathbf{G}$) of the group. If o($\mathbf{G}$) is finite, then $\mathbf{G}$ is said to be a *finite group*, otherwise $\mathbf{G}$ is an *infinite group*.

(8) Let $\mathbf{G}$ be a group and $\mathbf{S} \subset \mathbf{G}$ a subset of $\mathbf{G}$. If every element of $\mathbf{G}$ can be written as $s_1 * s_2 * \cdots * s_n$ where s_i or $s_i^{-1} \in \mathbf{S}$, then we say that $\mathbf{S}$ *generates* $\mathbf{G}$ or $\mathbf{S}$ is a *generating set* of $\mathbf{G}$. For example the set $\mathbf{G}$ always generates the group $\mathbf{G}$ and the groups $(\mathbf{Z}, +), (\mathbf{Z}_n, +)$ are generated by their units 1 and [1] respectively.

(9) If a group $\mathbf{G}$ is generated by a singleton $\mathbf{S} = \{s\}$, then we say that $\mathbf{G}$ is a *cyclic group*.

(10) If a group $\mathbf{G}$ is generated by a finite set, then $\mathbf{G}$ is said to be *finitely generated*. ∎

There are many examples of groups; the numerical groups, permutation groups, transformation groups, motion groups etc. Let us consider some examples.

Example 1.

(1) The set of all integers $\mathbf{Z}$ *is a group with respect to the usual summation* $+$*, while it is not a group with respect to the usual multiplication* $\cdot$ *, because not every element has an inverse with respect to multiplication. For instance, the number 2 has no multiplicative inverse.*

(2) As subsets of $\mathbf{Z}$*, we have two groups* $(\{0\},+),(\{1\},\cdot)$*. If the group* $\mathbf{G}$ *is a singleton, then the group is called a* ***unit group****. The preceding examples are unit groups.*

(3) The groups $(\mathbf{Z}_n,+),(\mathbf{Z},+)$ *are cyclic groups.* ∎

Example 2. (Permutation groups). *Consider the symmetric group* S_n *on* n *objects. Let the set of* n *objects be the set of integers* $\{1,2,\cdots,n\}$*,* S_n *be all bijective maps from this set to itself, and the binary operation* $*$ *be the* ***composition of maps****, i.e.,*

$$(\sigma * \delta)(i) = \sigma(\delta(i)), \qquad \forall\, \sigma,\delta \in S_n \quad 1 \le i \le n$$

We may write an element $\sigma \in S_n$ *as follows,*

$$\sigma = \begin{pmatrix} 1, & 2, & \ldots, & n \\ \sigma(1), & \sigma(2), & \ldots, & \sigma(n) \end{pmatrix}$$

Apparently, there are $(n!)$ *way of arranging the second row. Therefore, the order of* S_n*,* $o(S_n) = n!$*.* ∎

Example 3. (Rigid motion groups on the plane). *A* ***rigid motion*** *is a bijective map from the plane* $\mathbf{R}^2$ *to itself which preserves distances. It then follows that a rigid motion must map a triangle rigidly to another triangle. Take any line and any three points on it. If the images of the three points are not colinear, then it is easy to prove that one of the three distances between the three points will be altered. Therefore rigid motion will send a line to a line. Similarly, we can show that a rigid motion will not change the absolute value of the angle spanned by two lines. We have three special rigid motion groups: translation groups, reflection groups and rotation groups.*

(1) **Translation groups.** *The group* $\mathbf{G}$ *consists of* ***translations*** $\beta_{(a,b)}$ *with* $a,b \in \mathbf{R}$ *which are defined as follows,*

$$\beta_{(a,b)}(x,y) = (x+a, y+b), \qquad \forall\,(x,y) \in \mathbf{R}^2$$

It is easy to verify that

$$\beta_{(a,b)} * \beta_{(c,d)} = \beta_{a+c,b+d}$$
$$\beta_{(0,0)} * \beta_{(c,d)} = \beta_{(c,d)} * \beta_{(0,0)} = \beta_{(c,d)}$$
$$\beta_{(a,b)} * \beta_{(-a,-b)} = \beta_{(-a,-b)} * \beta_{(a,b)} = \beta_{(0,0)}$$

Therefore, $\beta_{(0,0)}$ *is the unit and* $\beta_{(-a,-b)}$ *is the inverse of* $\beta_{(a,b)}$*. Note that if* $(a,b) \neq (0,0)$*, then no point on the plane* $\mathbf{R}^2$ *is fixed by the translation* $\beta_{(a,b)}$*.*

(2) **Reflection groups.** *A rigid motion* γ *is said to be a non-trivial reflection if there is a line* ℓ *such that* $\gamma(x,y) =$ *the mirror image of* (x,y) *with respect to the line* ℓ*. In this case we call* ℓ *the the* ***reflection axis****. For simplicity, let the line* ℓ *be the* x*-axis. Then we have the following diagram for the mirror image,*

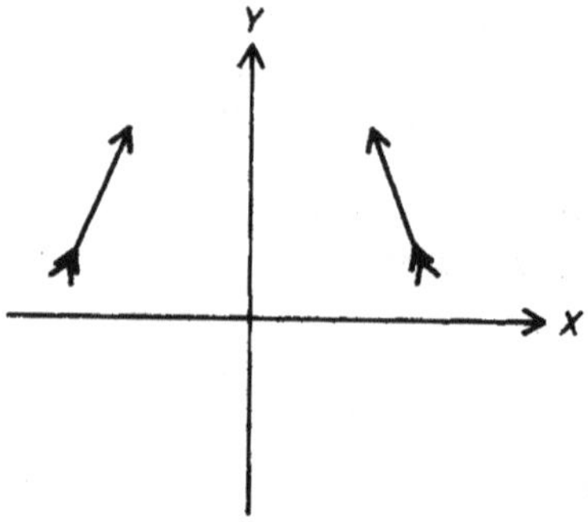

namely,

$$\gamma(x,y) = (-x,y)$$

A reflection group **G** *consists of a reflection γ and e=the identity map. Take a line passing through the origin. It is easy to see that a non-trivial reflection γ will change the sign of the angle. In fact, it will change the sign of every angle in the plane. Note that the reflection γ will fix every point on the line ℓ.*

(3) **Rotation groups.** *A rigid motion ρ on the plane* $\mathbf{R}^2$ *is called a non-trivial rotation if ρ fixes exactly one point and preserves the signs of angles, i.e., ρ preserves the orientation of the plane* $\mathbf{R}^2$. *We may take this point as the origin $(0,0)$ of our coordinate system. The x-axis will be moved to a line ℓ' passing through the origin. Let the angle spanned by the x-axis and ℓ' be θ with $0 \le \theta < 360$. Then it is not hard to see that every line passing through the origin will be rotated by the same angle θ. We shall write the rotation ρ as ρ_θ. Then we have the following equation,*

$$\rho_{\theta_1} * \rho_{\theta_2} = \rho_{[\theta_1+\theta_2]}$$

where $[\theta_1 + \theta_2]$ is the real number satisfying

$$0 \le [\theta_1 + \theta_2] < 360$$
$$360 \mid (\theta_1 + \theta_2) - [\theta_1 + \theta_2]$$

A rotation group **G** *consists of rotations with a common fixed point.*

(4) We claim: any rigid motion μ can be written as compositions of a translation, a rotation which fixes the origin, and a reflection which fixes the x-axis. Note that the identity e can be considered as a translation, rotation or reflection. Therefore our claim covers the case that the translation, rotation or reflection may be trivial. Let $\mu(0,0) = (a,b)$. Then we have

$$\beta_{(-a,-b)} * \mu(0,0) = \beta_{(-a,-b)}(a,b) = (0,0)$$

Let us write $\beta_{(-a,-b)} * \mu = \mu'$. Then μ' fixes the origin $(0,0)$. Let θ be the angle spanned by the x-axis and its image. Let $\mu'' = \rho_{(360-\theta)} * \mu'$. It is easy to see that μ'' will fix the x-axis. The y-axis will either be fixed by μ'' or turn a 180° angle. In the first case, it is easy to see that $\mu'' = e$ the identity. In the second case $\mu'' = \gamma$. Therefore, in the first case, we have

$$\mu'' = \rho_{(360-\theta)} * \mu' = \rho_{(360-\theta)} * \beta_{(-a,-b)} * \mu = e$$
$$\mu = \beta_{(a,b)} * \rho_\theta$$

In the second case, we have

$$\mu'' = \rho_{(360-\theta)} * \mu' = \rho_{(360-\theta)} * \beta_{(-a,-b)} * \mu = \gamma$$
$$\mu = \beta_{(a,b)} * \rho_\theta * \gamma$$

∎

Example 4. (Galilean transformation group and Lorentz transformation group). *Those are two important groups in Physics. For simplicity, let us assume that the space is 1-dimensional and represented by the x-axis, the time is represented by the t-axis. Let there be another coordinate system (x', t') moving against the first coordinate system uniformly with a velocity u. What is the coordinate (x', t') of an object with a fixed coordinate (x, y)? We shall assume that those two coordinate systems have the same origin $(0,0)$. Then in the classical Physics, we use the Galilean transformation group* $\mathsf{G} = \{g_u : u \in \mathbf{R}\}$ *to determine the coordinate (x', t') as follows,*

$$g_u \begin{pmatrix} x \\ t \end{pmatrix} = \begin{pmatrix} x - ut \\ t \end{pmatrix} = \begin{pmatrix} x' \\ t' \end{pmatrix}$$

It is easy to see that

$$g_u * g_v = g_{u+v}$$
$$t' = t$$

namely, time is not affected by the motion. However, it was discovered that the basic Maxwell equations of the electromagnetic theory are not invariant under the above Galilean transformation group. The transformation group which makes the Maxwell equations invariant is the Lorentz transformation group $\mathsf{L} = \{\ell_u : u \in \mathbf{R}\}$ *whose elements are defined as follows,*

$$\ell_u \begin{pmatrix} x \\ t \end{pmatrix} = \begin{pmatrix} \frac{x-ut}{\sqrt{1-u^2/c^2}} \\ \frac{t-ux/c^2}{\sqrt{1-u^2/c^2}} \end{pmatrix} = \begin{pmatrix} x' \\ t' \end{pmatrix}$$

where c is the speed of light. It is easy to see

$$\ell_u * \ell_v = \ell_w, \qquad w/c = \frac{u/c + v/c}{1 + uv/c^2}$$

It is easy to prove that if $|u|, |v| < c$, then $|w| < c$. An interpretation of the preceding fact is interesting in Physics; let us imagine that a spaceship departs from the earth with a speed u, and sends a probe straight ahead with a speed v relative to the spaceship. Measuring from the earth, the speed of the probe will be w. If both $u, v < c$, then $w < c$. For instance, if $u = v = (2/3)c$, then $w = (12/13)c$, instead of the common sense result of $(2/3)c + (2/3)c = (4/3)c$. This is the reason physicists tell us the speed of light can not be exceeded.

■

Example 5. (Matrices). *(1) Let the general linear group $GL(n, \mathbf{R})$ be the set $\{A : A\ n \times n$ real matrix with non-zero determinant $det(A).\}$ with the usual multiplication as the binary operation $*$. From Linear Algebra, we have*

$$A \cdot (B \cdot C) = (A \cdot B) \cdot C$$
$$I_n \cdot A = A \cdot I_n = A, \text{ where } I_n \text{ is the unit matrix.}$$
$$1 = det(I_n) = det(A \cdot A^{-1}) = det(A) \cdot det(A^{-1})$$

Therefore $A^{-1} \in GL(n, \mathbf{R})$, i.e., every element A has an inverse and $GL(n, \mathbf{R})$ is a group.

(2) Let the orthogonal group $O(n, \mathbf{R})$ be the set $\{A : A = (a_{ij})\ n \times n$ real matrix with $\sum_{j=1}^{n} a_{ij}a_{jk} = \delta_{ik}.\}$ with the usual multiplication as the binary operation $*$ where the Kronecker δ_{ik} is defined as

$$\delta_{ik} = \begin{cases} 1, & \text{if } i = k \\ 0, & \text{if } i \neq k \end{cases}$$

The orthogonal group is the group of all rigid motions which fixes the origin $(0, 0, \cdots, 0)$ of $\mathbf{R}^n$.

(3) A rotation ρ of $\mathbf{R}^n$ is a rigid motion which fixes the origin and preserve the orientation of the space $\mathbf{R}^n$. The group **Rot** *$(n, \mathbf{R}^n)$ of all rotations of $\mathbf{R}^n$ is the set $\{A : A = (a_{ij})\ n \times n$ real matrix with $\sum_{j=1}^{n} a_{ij}a_{jk} = \delta_{ik},\ det(A) = 1.\}$ with the usual multiplication as the binary operation $*$ where the Kronecker δ_{ik} is as defined in (2). We claim that if $n = 3$, then every rotation of $\mathbf{R}^3$ will have a fixed axis, namely a line passing through the origin with every point on it fixed by the rotation. Let us prove the above claim. Let*

$$A = \begin{pmatrix} a_{11}, & a_{12}, & a_{13} \\ a_{21}, & a_{22}, & a_{23} \\ a_{31}, & a_{32}, & a_{33} \end{pmatrix}$$

and the characteristic polynomial (cf §6, Chapter IV) of A be

$$det(\lambda I_3 - A) = det \begin{pmatrix} \lambda - a_{11}, & -a_{12}, & -a_{13} \\ -a_{21}, & \lambda - a_{22}, & -a_{23} \\ -a_{31}, & -a_{32}, & \lambda - a_{33} \end{pmatrix} = \lambda^3 + \alpha_1\lambda^2 + \alpha_2\lambda - det(A)$$
$$= \lambda^3 + \alpha_1\lambda^2 + \alpha_2\lambda - 1$$

We know from high school that every real polynomial of degree 3 must have a real root. Let one real root of the above equation be c_1 which is a characteristic value of the rotation A. Let v_1 be an associative characteristic vector. Then we have (cf Theorem 4.30),

$$A \cdot v_1 = c_1 \cdot v_1$$

From the length preserving property of a rigid motion, we conclude that length$(c_1 \cdot v_1) = |c_1|||v_1|| = ||v_1||$, $|c_1| = 1$ and $c_1 = \pm 1$. If $c_1 = 1$, then it is not hard to see that the line determined by the vector v_1 is a fixed axis. Let us assume that $c_1 = -1$. Then we have,

$$\begin{aligned} det(\lambda I_3 - A) &= \lambda^3 + \alpha_1\lambda^2 + \alpha_2\lambda - det(A) \\ &= \lambda^3 + \alpha_1\lambda^2 + \alpha_2\lambda - 1 \\ &= (\lambda + 1)(\lambda^2 + \beta_1\lambda - 1) \end{aligned}$$

The second part of the above equation is a quadratic polynomial with positive discriminant, $\beta_1^2 - 4(-1) = \beta_1^2 + 4$, and hence has two real roots c_2, c_3. Similar to the above arguments about c_1, we conclude that $c_2, c_3 = \pm 1$. On the other hand, we have $c_1 \cdot c_2 \cdot c_3 = 1$. Therefore one of c_1, c_2, c_3 must be 1, and its characteristic vector determines a fixed axis. Moreover, if a rotation ρ has two fixed axis, then two of the characteristic values c_1, c_2, c_3 must be 1. Since we always have $c_1 \cdot c_2 \cdot c_3 = 1$, then the third c_i must be 1 too. It is then easy to see that any rotation with two fixed axis must be the unit I_3. We thus establish that any non-trivial rotation of the space has one and only one fixed axis. ∎

Exercises

(1) Prove that every rigid motion σ on the plane $\mathbf{R}^2$ can be written as a product of reflections $\sigma = \gamma_1 * \gamma_2 * \cdots * \gamma_n$. Therefore the group of the rigid motions on the plane is generated by all reflections.

(2) Prove that the group of the rigid motions on the space $\mathbf{R}^3$ is generated by all reflections.

(3) Prove the *Special linear group* $SL(n, \mathbf{R}) = \{A : A \in GL(n, \mathbf{R}), det(A) = 1\}$ is a group with respect to the usual matrix multiplication.

(4) **Quaternions**[2] : Let the quaternions= $\{a_0 + a_1 i + a_2 j + a_3 k : a_i \in \mathbf{R}\}$ with the binary operation $*$ be defined as follows,

$$1 \cdot i = i, \quad 1 \cdot j = j, \quad 1 \cdot k = k, \quad i \cdot i = j \cdot j = k \cdot k = -1$$

$$i \cdot j = -j \cdot i = k, \qquad j \cdot k = -k \cdot j = i, \qquad k \cdot i = -i \cdot k = j$$

$$\begin{aligned} &(a_0 + a_1 i + a_2 j + a_3 k)(b_0 + b_1 i + b_2 j + b_3 k) \\ &\quad = a_0 b_0 + a_0 b_1 i + a_0 b_2 j + a_0 b_3 k + a_1 b_0 i + a_1 b_1 i^2 + a_1 b_2 ij + a_1 b_3 ik \\ &\quad + a_2 b_0 j + a_2 b_1 ji + a_2 b_2 j^2 + a_2 b_3 jk + a_3 b_0 k + a_3 b_1 ki + a_3 b_2 kj + a_3 b_3 k^2 \end{aligned}$$

---

[2] **Due to English Mathematician Hamilton 1843.**

Prove that the quaternions is a group.

(5) Multiplication table. Let $G = \{a_1, a_2, \cdots, a_n\}$ be a finite group. Then the matrix $(a_{ij}) = (a_i * a_j)$ is called the *multiplication table* of G. Show that every row and every column of it consists of distinct elements of G. One application of the multiplication table of G is for the use of some experiments. Suppose that we are experimenting on n plants with n fertilizers and n insecticides. Let us divide the garden plot into n rows and n columns. We will apply n fertilizers on n rows and n fertilizers on n columns of the plot. Use the multiplication table to arrange the plants on the n^2 squares so that every fertilizer and every insecticide have tested on every plant.

(6) Given arbitrary many groups G_i with unit e_i. Let us define the *direct sum* of G_i, $\oplus G_i$ as follows,

$$\oplus G_i = \{(a_i) : a_i \in G_i,\ a_i = e_i \text{ for all but finitely many } i\}$$

and the binary operation $*$ is defined as

$$(a_i) * (b_i) = (a_i * b_i)$$

Prove that the direct sum $\oplus G_i$ is a group.

(7) Let A, $B \in$ SL(2,R). If o(A), o$(B) < \infty$, then do we always have o$(AB) < \infty$?

(8) Given arbitrarily many groups G_i. Let us define the *direct product* of G_i, $\prod G_i$ as follows,

$$\prod G_i = \{(a_i) : a_i \in G_i\}$$

and the binary operation $*$ is defined as

$$(a_i) * (b_i) = (a_i * b_i)$$

Prove that the direct product $\prod G_i$ is a group.

(9) Let the *Full linear group* $FL(n, \mathbf{R}) = \{\text{all } n \times n \text{ real matrices}\}$. Prove that $FL(n, \mathbf{R})$ is a group with respect to the usual summation.

(10) Let I_n be the unit of $GL(n, \mathbf{R})$. Let the matrix $J \in GL(2n, \mathbf{R})$ be defined as

$$J = \begin{pmatrix} 0, & I_n \\ -I_n, & 0 \end{pmatrix}$$

Let A^t be the *transpose* of A. Let the *Symplectic group*, $S_p(2n, \mathbf{R})$, be defined as

$$S_p(2n, \mathbf{R}) = \{A : A \in GL(2n, \mathbf{R}), A^t J A = J\}$$

Prove that the Symplectic group is a group with respect to the usual multiplication.

§2 The Transformation Groups on Sets

The very first groups studied were finite permutation groups. Those are groups that act on sets. We have the following definition,

Definition 2.3. *Let* **S** *be a set and* **G** *be a group. If every element* $g \in \mathbf{G}$ *is a map* $g : \mathbf{S} \to \mathbf{S}$ *satisfying the following conditions*

(i) $(g_1 * g_2)(s) = g_1(g_2(s))$, $\forall\, g_i \in \mathbf{G}$ *and* $s \in \mathbf{S}$,
(ii) $e(s) = s$, $\forall\, s \in \mathbf{S}$ *and* e *is the unit of* **G**.

then we say **G** *is a transformation group (acting) on* **S**. ∎

All examples except the numerical groups discussed in the previous section are transformation groups on sets. Furthermore, any abstract group **G** can be understood as a transformation group (acting from the left) on the set **G**; let any element $g \in \mathbf{G}$ define a map $g : \mathbf{G} \to \mathbf{G}$ as follows,

$$g(a) = g * a, \qquad \forall\, a \in G$$

It is straight forward to show that the preceding conditions (i) & (ii) are satisfied.

Theorem 2.1. *Let a group* **G** *be a transformation group on a set* **S**. *Then every element* $g \in \mathbf{G}$ *is a bijective map of* **S**.

Proof: (i) The element g is injective: if $g(s_1) = g(s_2)$ for some $s_1, s_2 \in \mathbf{S}$, then $s_1 = e(s_1) = (g^{-1} * g)(s_1) = g^{-1}(g(s_1)) = g^{-1}(g(s_2)) = (g^{-1} * g)(s_2) = e(s_2) = s_2$.

(ii) The element g is surjective: let s_3 be any element in **S**. Let $s_4 = g^{-1}(s_3)$. Then we have $s_3 = e(s_3) = (g * g^{-1})(s_3) = g(s_4)$. ∎

The points of studying a transformation group **G** on a set **S** are the mutual influences. We shall introduce the following definition,

Definition 2.4. *Let* **G** *be a transformation group on a set* **S**. *The orbit, Orb(s), of an element* $s \in \mathbf{S}$ *is defined as*

$$Orb(s) \;=\{g(s) : g \in \mathbf{G}\}$$

*The orbit, Orb(***T***), of a subset* $\mathbf{T} \subset \mathbf{S}$ *is defined as*

$$Orb(\mathbf{T}) \;=\{g(\mathbf{T}) : g \in \mathbf{G}\}$$

∎

The term "orbit" suggests a motion. Indeed, in some applications of group theory, we may think that the group action $g : s \mapsto g(s)$ is induced by motion.

Example 6. (Orbits under rigid motions). *Let us consider Example 3. (i) Under the action of the group of all translations of* $\mathbf{R}^2$, *Orb(a,b)* $= \mathbf{R}^2$ *for any point* $(a, b) \in \mathbf{R}^2$. *All orbits coincide.*

(ii) Under the action of a reflection group $\Gamma = \{e, \gamma\}$, *we have*

$$Orb(a,b) = \begin{cases} \{(a,b)\}, & \text{if } (a,b) \in \text{the reflection axis.} \\ \{(a,b), \text{its mirror image}\}, & \text{if } (a,b) \notin \text{the reflection axis.} \end{cases}$$

All orbits either coincide or are disjoint.

(iii) Under the action of the group $\mathbf{Rot}(2,\mathbf{R})$, *of all rotations which fixed the origin, we have*

$$Orb(a,b) = \begin{cases} \{(a,b)\}, & \text{if } (a,b) = (0,0). \\ \{\text{ the circle with } (0,0) \text{ as center and passes } (a,b).\} \end{cases}$$

All orbits either coincide or are disjoint. ∎

Example 7. (Any conic curve is an orbit). *(i) Let us consider a parabola. For simplicity, we may change the coordinate of* $\mathbf{R}^2$ *such that the equation of the parabola is as follows,*

$$Y = X^2$$

Let a group $\mathbf{G}_1 = \{\tau_a : a \in \mathbf{R}\}$ *with action on* $\mathbf{R}^2$ *as follows,*

$$\tau_a(x,y) = (x+a, y+2ax+a^2), \qquad \forall\, (x,y) \in \mathbf{R}^2$$

Then it is easy to check

$$\begin{aligned}(\tau_a * \tau_b)(x,y) &= \tau_a(\tau_b(x,y)) = \tau_a((x+b, y+2bx+b^2)), \qquad \forall\, (x,y) \in \mathbf{R}^2 \\ &= (x+b+a, y+2bx+b^2+2a(x+b)+a^2) = \tau_{(a+b)}(x,y) \\ \tau_a * \tau_b &= \tau_{(a+b)}\end{aligned}$$

It is easy to see Orb((0,0))= $\{(a,a^2) : a \in \mathbf{R}^2\}$ *our original parabola. In general, Orb((c,d)) is a parabola defined by the following equation,*

$$Y = X^2 - (c^2 - d)$$

All orbits either coincide or are disjoint.

(ii) Let us consider a hyperbola. After changing the coordinate suitably, we may assume that its equation is as follows,

$$XY = 1$$

Let a group $\mathbf{G}_2 = \{\sigma_a : a \neq 0 \in \mathbf{R}\}$ *with action on* $\mathbf{R}^2$ *as follows,*

$$\sigma_a(x,y) = (ax, a^{-1}y), \qquad \forall\, (x,y) \in \mathbf{R}^2$$

Then it is easy to check

$$\begin{aligned}(\sigma_a * \sigma_b)(x,y) &= \sigma_a(\sigma_b(x,y)) = \sigma_a((bx, b^{-1}y)), \qquad \forall\, (x,y) \in \mathbf{R}^2 \\ &= (abx, (ab)^{-1}y) = \sigma_{(ab)}(x,y) \\ \sigma_a * \sigma_b &= \sigma_{(ab)}\end{aligned}$$

It is easy to see Orb((1,1))= $\{(a, a^{-1}) : a \neq 0 \in \mathbf{R}^2\}$ *our original hyperbola. In general, Orb((c,d)) is given as follows,*

$$Orb((c,d)) = \begin{cases} \{(0,0)\} & \text{if } (c,d) = (0,0) \\ \{X\text{-axis} \setminus (0,0)\} & \text{if } d = 0, c \neq 0 \\ \{Y\text{-axis} \setminus (0,0)\} & \text{if } c = 0, d \neq 0 \\ \{\text{ hyperbola defined by } XY = cd\} & \text{if } c \neq 0, d \neq 0 \end{cases}$$

All orbits either coincide or are disjoint.

(iii) Let us consider an ellipse. After changing the coordinate suitably, we may assume that its equation is as follows,

$$\frac{X^2}{a^2} + \frac{Y^2}{b^2} = 1, \qquad a \geq b > 0$$

Let a group $\mathbf{G}_3 = \{\delta_\theta : 0 \leq \theta < 2\pi\}$ with action on $\mathbf{R}^2$ as follows,

$$\delta_\theta(x,y) = (ar \cdot cos(\alpha + \theta), br \cdot sin(\alpha + \theta)), \qquad \forall\, (x,y) \in \mathbf{R}^2$$

where r and α are defined as follows

$$r = \sqrt{\frac{x^2}{a^2} + \frac{y^2}{b^2}}, \qquad \alpha = arctan\frac{y}{x}$$

Then it is easy to check

$$(\delta_{\theta_1} * \delta_{\theta_2})(x,y) = \delta_{[\theta_1+\theta_2]}(x,y), \qquad \forall\, (x,y) \in \mathbf{R}^2$$

where $[\theta_1 + \theta_2]$ is defined in Example 3. It is easy to see Orb($(a,0)$) = our original ellipse. In general, Orb((c,d)) is either the singleton $\{(0,0)\}$ or an ellipse defined by the following equation,

$$\frac{X^2}{(as)^2} + \frac{Y^2}{(bs)^2} = 1,$$

where s is given as

$$s = \sqrt{\frac{c^2}{a^2} + \frac{d^2}{b^2}}, \qquad \text{if } c \neq 0 \text{ or } d \neq 0$$

All orbits are either coincide or disjoint. ∎

Note that in all preceding examples we have the phenomena: all orbits either coincide or are disjoint. Indeed, this is a fundamental property of orbits. We have the following theorem.

Theorem 2.2. *Let* G *be a transformation group on a set* S. *Then* S *is a* ***disjoint union*** *of all orbits, i.e., all orbits either coincide or are disjoint and* S *is the union of them.*

Proof: (i) The set S is the union of all orbits: let $s \in \mathbf{S}$. Then we have $s = e(s) \Longrightarrow s \in$ Orb(s).

(ii) All orbits either are coincide or disjoint: suppose that we have Orb$(s_1) \cap$ Orb$(s_2) \neq \emptyset$ the empty set. Let s_3 be an element in the intersection. Then we must have $g_1, g_2 \in \mathbf{G}$ with

$$s_3 = g_1(s_1) = g_2(s_2)$$

Therefore, we have

$$s_1 = g_1^{-1} * g_2(s_2)$$

Now for any element $g(s_1) \in$ Orb(s_1), we have

$$g(s_1) = g * g_1^{-1} * g_2(s_2) = g'(s_2) \in \text{Orb}(s_2)$$

Thus we conclude Orb$(s_1) \subset$ Orb(s_2). Similarly, we can prove Orb$(s_2) \subset$ Orb(s_1). Therefore we conclude Orb$(s_1)=$ Orb(s_2).

∎

The above theorem shows that through the actions of the transformation group G, the set S is separated into many disjoint orbits. The following definition will generalize the concept.

Definition 2.5. *Let* G *be a transformation group on a set* S. *Let* T *be a subset of* S. *If the following condition is satisfied, then* T *is said to be an* ***invariant subset*** *of* S,

$$\forall\, t \in \mathbf{T}, \quad \forall\, g \in \mathbf{G} \Longrightarrow g(t) \in \mathbf{T}$$

∎

It is easy to see that any subset T is invariant $\Longleftrightarrow$ T is a union of orbits. For instance, in Example 3., for the rotation group **Rot(2, R)** acting on $\mathbf{R}^2$, every disc with the origin $(0, 0)$ as the center is an invariant subset of $\mathbf{R}^2$. If T is an invariant subset, then the group G can be viewed as a transformation group on T.

Since every element $g \in \mathbf{G}$ is a surjective map $g : \mathbf{S} \to \mathbf{S}$, then the preceding definition can be rewritten as,

Definition 2.5*. *Let* G *be a transformation group on a set* S. *Let* T *be a subset of* S. *If the following condition is satisfied,*

$$g(\mathbf{T}) = \mathbf{T}, \qquad \forall\, g \in \mathbf{G}$$

i.e., Orb(T)=T, *then* T *is said to be an* ***invariant subset*** *of* S.

∎

We have the following definition,

Definition 2.6. *Let* G *be a transformation group on a set* S. *If* $g(s) = s \,\forall\, s \in \mathbf{S} \Longrightarrow g = e$, *then* G *is said to be a faithful transformation group on* S. ∎

Any abstract group G is a faithful transformation group acting from the left on the set G.

Exercises

(1) Let a group G act from the left on the set G. Find Orb(g) for $g \in$ G.

(2) Let a group G act from *both sides* on the set G as follows,

$$g(g_1) = g * g_1 * g^{-1}$$

Prove that the group G is a transformation group on the set G.

(3) Let the symmetric group S_3 act on the set S_3 from both sides (Cf the preceding problem). Find Orb(g) for all $g \in S_3$.

(4) Let a group G= $\{\psi_a : a \in \mathbf{R}\}$ with action on $\mathbf{R}^2$ as follows,

$$\psi_a(x, y) = (x + a, e^{-a}y), \qquad \forall\, (x, y) \in \mathbf{R}^2$$

Find Orb((x, y)) for all $(x, y) \in \mathbf{R}^2$.

(5) Let a group G= $\{\beta_{(a,b)} : a, b \in \mathbf{Z}\}$ with action on $\mathbf{R}^2$ as follows,

$$\beta_{(a,b)}(x, y) = (x + a, y + b), \qquad \forall\, (x, y) \in \mathbf{R}^2$$

Find Orb((x, y)) for all $(x, y) \in \mathbf{R}^2$.

(6) The torus $T_2 = \{(a, b) : a, b \in \mathbf{R}, 0 \le a, b < 1\}$. Let a group G= $\{\mu_a : a \in \mathbf{R}\}$ with action on T_2 as follows,

$$\mu_a(x, y) = ((x + a)', (y + a)'), \qquad \forall\, (x, y) \in T_2$$

where $(x + a)'$ is the unique real number which satisfies $0 \le (x + a)' < 1, (x + a) - (x + a)' \in \mathbf{Z}$. Find Orb($(x, y)$) for all $(x, y) \in T_2$.

(7) The torus $T_2 = \{(a, b) : a, b \in \mathbf{R}, 0 \le a, b < 1\}$. Fix a number $c \in \mathbf{R}$. Let a group G= $\{\eta_a : a \in \mathbf{R}\}$ with action on T_2 as follows,

$$\eta_a(x, y) = ((x + ca)', (y + ca)'), \qquad \forall\, (x, y) \in T_2$$

where $(x + ca)'$ is the unique real number which satisfies $0 \le (x + ca)' < 1, (x + ca) - (x + ca)' \in \mathbf{Z}$. Find Orb($(x, y)$) for all $(x, y) \in T_2$. Discuss the properties of orbits if c is rational or irrational.

(8) Let the standard n-dimensional vector space V be represented vertically, i.e.,

$$V = \{\alpha = \begin{pmatrix} x_1 \\ x_2 \\ \cdot \\ \cdot \\ x_n \end{pmatrix} : x_i \in \mathbf{R}\}$$

Let the orthogonal group $O(n, \mathbf{R})$ act on V from the left, i.e.,

$$A(\alpha) = A\alpha$$

Find Orb(α) for all $\alpha \in V$.

(9) Let the orthogonal group $O(n-1, \mathbf{R})$ act on the set $O(n, \mathbf{R})$ as follows,

$$A(B) = \mathring{A}B$$

where

$$\mathring{A} = \begin{pmatrix} 1, & 0 \\ 0, & A \end{pmatrix}$$

Find Orb(B) for all $B \in GL(n.\mathbf{R})$.

(10) Let the direct product $G = GL(n, \mathbf{R}) \times GL(n, \mathbf{R})$ act on the set $FL(n, \mathbf{R})$ as follows,

$$(A, B)M = AMB^{-1}, \qquad \text{where } A, B \in GL(n, \mathbf{R}),\ M \in FL(n, \mathbf{R})$$

Find Orb(M) for all $M \in FL(n, \mathbf{R})$.

(11) Let the group $GL(n, \mathbf{R})$ act on the set $Symm(n, \mathbf{R})$ of all real symmetric matrices as follows,

$$A(S) = ASA^{-1}, \qquad \text{where } A \in GL(n, \mathbf{R}),\ T \in Symm(n, \mathbf{R})$$

Find Orb(T) for all $T \in Symm(n, \mathbf{R})$.

(12) Let **G** be a transformation group on a set **S**. Let us define a relation "$\sim$" as follows

$$a \sim b \Longleftrightarrow a, b \in \text{ same orbit.}$$

Show that "$\sim$" is an equivalence relation. The classes of the preceding "$\sim$" will be called *orbit classes*.

(13) Let us use the notations of exercises (5) & (6). Show that the map λ from the orbit classes of exercise (5) to T_2 defined as follows is bijective

$$\lambda(\text{Orb}((a, b))) = (a', b')$$

Therefore the orbit classes of exercise (4) can be naturally viewed as a torus.

(14) Let us use the notations of exercise (9). Show that the map χ from the orbit classes of exercise (9) to the $(n-1)$-dimensional unit sphere S^{n-1} defined as follows is bijective

$$\chi(\mathrm{Orb}((B))) = \begin{pmatrix} b_{11} \\ b_{12} \\ \cdot \\ \cdot \\ b_{1n} \end{pmatrix}, \qquad \text{where } B = \begin{pmatrix} b_{11} & b_{12} & \cdots & \beta_{1n} \\ b_{21} & b_{22} & \cdots & \beta_{2n} \\ \cdot & \cdot & \cdots & \cdot \\ \cdot & \cdot & \cdots & \cdot \\ b_{n1} & b_{n2} & \cdots & \beta_{nn} \end{pmatrix}$$

Therefore the orbit classes of exercise (9) can be naturally viewed as a $(n-1)$-dimensional unit sphere S^{n-1}.

(15) Let the *Special linear group with integer coefficients* $SL(2, \mathbf{Z})$ be defined as follows,

$$SL(2,\mathbf{Z}) = \{\begin{pmatrix} a, & b \\ c, & d \end{pmatrix} : a, b, c, d \in \mathbf{Z}, \det \begin{pmatrix} a, & b \\ c, & d \end{pmatrix} = 1\}$$

Show that $SL(2, \mathbf{Z})$ is a group with respect to the usual multiplication.

(16) Let us use the notations of exercise (15). Let $SL(2, \mathbf{Z})$ act on the upper complex plane $H = \{z : z \in \mathbf{C}, im(z) > 0\}$ as follows,

$$\begin{pmatrix} a, & b \\ c, & d \end{pmatrix}(z) = \frac{az+b}{cz+d}$$

Prove that there is a bijective map from the orbit classes to the shadow set (which contains the boundary point z with the real part non-negative $re(z) \geq 0$) in the following diagram,

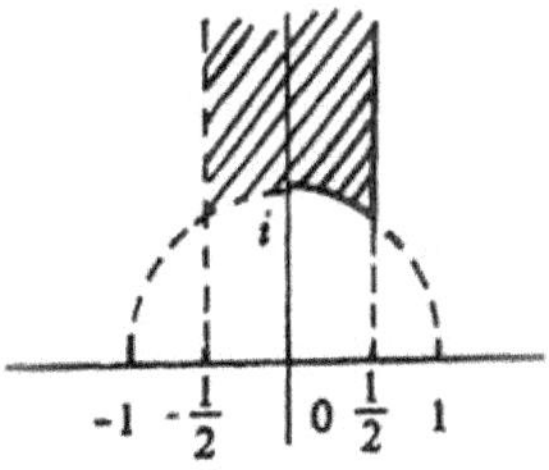

§3 Subgroups

In the preceding section, we study the influence of the transformation group **G** on the set **S**, i.e., the set **S** is separated into many disjoint orbits. In this section, we will study the influence of the set **S** on the group **G**. Let us introduce the concept of *subgroups*.

Definition 2.7. *Let* (**G**,$*$) *be a group and* **H**$\subset$**G** *be a subset of* **G**. *If* (**H**,$*$) *is a group, then it is said to be a subgroup of* (**G**,$*$). ∎

Discussion

(1) The group **G** is a subgroup of **G** and $\{e\}$ is a subgroup, the unit subgroup of **G**. Those two subgroups are called the *trivial* subgroups of **G**. All other subgroups are called the *proper subgroups* of **G**.

(2) Given a non-empty subset **H** of **G**, to check if (**H**,$*$) is a subgroup, it is unnecessary to check the associative law, because the law is *inherited* by **H** from **G**. It is necessary and sufficient to prove that for any two elements $a, b \in$**H**, the element $a * b^{-1} \in$**H**: the condition is certainly necessary. Let $b = a$. Then $e = a * a^{-1} \in$**H**. Let $a = e$. Then $a * b^{-1} = e * b^{-1} = b^{-1} \in$**H**. Therefore every element b has an inverse $b^{-1} \in$**H**. Replacing b by b^{-1}, we have $a * (b^{-1})^{-1} = a * b \in$**H**. Therefore **H** is closed under $*$. Therefore (**H**,$*$) is a subgroup. For simplicity, we say **H** is a subgroup of **G** if there is no confusion. ∎

Example 8. *All even integers* $2\mathbf{Z}$ *form a subgroup of* $(\mathbf{Z},+)$. *All odd integers do not form a subgroup, because the difference of two odd integers is not odd. Let* **S** *be any subset of* **G**. *Then the subset of all elements of the form* $\prod_{\text{finite}} a_i$ *where either* $a_i \in S$ *or* $a_i^{-1} \in S$ *is a subgroup of* **G**, *the subgroup* $\langle S\rangle$ *generated by* **S**. *For instance* $2\mathbf{Z} = \langle 2\rangle$ *and all multiples of* $3 = \langle 3\rangle = 3\mathbf{Z}$. *In general all multiple of* $n = \langle n\rangle = n\mathbf{Z}$. *Furthermore, it follows from the Euclidean algorithm that all subgroups of* $\mathbf{Z}$ *are of the form* $\langle n\rangle = n\mathbf{Z}$ *for some* $n \in \mathbf{Z}$. ∎

Example 9. (Subgroups of rigid motion groups). *We shall use the notations of Example 3.*

(1) Let us consider the translation group on $\mathbf{R}^2$. *Given a subgroup* **H**, *if there is an element* $\beta_{(a,b)}$ *which is not a unit, then* $(a, b) \neq (0,0)$ *and*

$$\beta^n_{(a,b)} = \beta_{na,nb} \neq \beta_{(ma,mb)} = \beta^m_{(a,b)}, \qquad \text{for } n \neq m$$

Therefore the only finite subgroup is the unit subgroup.

A subgroup **H** *is said to be a* **discrete subgroup** *if the set* P_H *of the parameters of the translations,* $P_H = \{(a,b) : \beta_{(a,b)} \in H\}$, *is a discrete set. Let us consider a discrete subgroup* **H** *which is not the unit subgroup* $\{e\}$.

Let (a_1, b_1) *be an element in* P_H *with the smallest positive length, i.e.,* $\sqrt{a_1^2 + b_1^2}$ *is positive and the smallest among the positive ones. Since* P_H *is discrete, then such element exists. Certainly,* $\beta^n_{(a_1,b_1)} = \beta_{(na_1,nb_1)} \in$**H**. *Furthermore, let us consider the line* ℓ

determined by (a_1, b_1) on $\mathbf{R}^2$, i.e., $\ell = \{(ra_1, rb_1) : r \in \mathbf{R}\}$. If there is any $(c,d) \in \ell$ with $(c,d) = (ra_1, rb_1) \neq (na_1, nb_1)\ \forall\ n \in \mathbf{Z}$ and $\beta_{(c,d)} \in \mathbf{H}$, then for some suitable n with $0 \le r - n < 1$, we have the element $\beta' = \beta_{(c,d)}\beta^{-n}_{(a_1,b_1)}$ with the parameter smaller, contradicting the selection of $\beta_{(a_1,b_1)}$. We conclude that all (c,d) must be multiples of (a_1, b_1). We have two possibilities:

(β) All elements in $\mathbf{H}$ are of the form $\beta_{(na_1,nb_1)}$.
(γ) There are some other elements.

Let us consider the second case.

Our treatment of the second case is similar to the material in §5 Complex Integers of Chapter I. Let (a_2, b_2) be an element in $P_H \setminus \ell$ with the smallest length, i.e., $\sqrt{a_2^2 + b_2^2}$ is the smallest. The set $\{n(a_1, b_1) + m(a_2, b_2) : n, m \in \mathbf{Z}\}$ is a lattice on the plane $\mathbf{R}^2$. Any element $(a_3, b_3) \in P_H \setminus \ell$ will be in one of the grids. It is an elementary geometric fact that the distance of any point in a parallelogram to one of the four vertices is less than the longer side (hint: use the two diagonals to divide the parallelogram into four parts.). Let the vertex for (a_3, b_3) be $n_0(a_1, b_1) + m_0(a_2, b_2)$. Then we have

$$\beta_{(a',b')} = \beta_{(a_3,b_3)} * \beta^{-n_0}_{(a_1,b_1)} * \beta^{-m_0}_{(a_2,b_2)}$$

$$\sqrt{a'^2 + b'^2} < \sqrt{a_2^2 + b_2^2}$$

It follows from the definition of (a_2, b_2) that $(a', b') = (0, 0)$, i.e.,

$$\beta_{(a_3,b_3)} = \beta^{n_0}_{(a_1,b_1)} * \beta^{m_0}_{(a_2,b_2)}$$

We therefore conclude that all discrete subgroups $\mathbf{H}$ of the translation group of $\mathbf{R}^2$ are of the following three types:

(α) $H = \{e\}$.
(β) $H = \{\beta_{(na_1,nb_1)} : n \in \mathbf{Z}\}$.
(γ) $H = \{\beta_{(na_1+ma_2,nb_1+mb_2)} : m, n \in \mathbf{Z}\}$.

(2) A reflection group consists of two elements e, γ. The only subgroups of it are the trivial ones, i.e., the whole group and the unit subgroup.

(3) Let us consider a finite non-trivial subgroup $\mathbf{H}$ of the rotation group $\mathbf{Rot}\ (2, \mathbf{R})$. Let the elements of $\mathbf{H}$ be presented as ρ_θ where θ is the angle of rotation and satisfies $0 \le \theta < 360$. Let θ_0 be the minimal positive angle of rotations of elements in $\mathbf{H}$. Then for any $\rho_\theta \in \mathbf{H}$, there are integer n and real number r such that

$$\theta = n\theta_0 + r, \qquad 0 \le < \theta$$

$$\rho_r = \rho_\theta * \rho_{\theta_0}^{-n} \in H$$

Therefore $r = 0$ due to the selection of θ_0. We conclude

$$\rho_r = e, \qquad \rho_\theta = \rho_{\theta_0}^n, \qquad H = \langle \rho_{\theta_0} \rangle$$

It is easy to see that the orbit of a point which is not the origin $(0,0)$ *is the set of a regular* n *sided polygon where* $n = o(H)$.

(4) *The symmetry group of a regular* n *sides polygon. The group* D_n *is also called the dihedral group of order* $2n$. *The group* D_n *is defined to be all bijective maps* μ *on the boundary of the regular polygon which preserve the distances of any two points. Then it is easy to see that* μ *will send one edge to another edge and one vertex to another vertex. Let the vertices of the regular* n*-polygon be labeled orderly as* $\{a_0, a_1, \cdots, a_{n-1}\}$ *with* a_0 *lying on the positive real axis and* $(0,0)$ *as the gravitational center of the polygon. Let* $\mu(a_i) = a_{\mu(i)}$. *Then we have* $\mu(2) =$ *the principle residue of* $\mu(1) \pm 1$ *mod(n)=*$[\mu(1) \pm 1]$. *We have*

(i) $\mu(2) = [\mu(1)+1]$. *Then it is easy to see that* $\mu(i+1) = [\mu(1)+i]$ *and* $\mu = \rho_\theta$ *where* $\theta = \mu(1)(360/n)$.

(ii) $\mu(2) = [\mu(1)-1]$. *Then it is easy to see that* $\mu(i+1) = [\mu(1)-i]$ *and* $\mu = \gamma * \rho_\theta$ *with* γ *the reflection which send* (x,y) *to* $(x,-y)$ *and* $\theta = [-\mu(1)](360/n)$.

Therefore, we conclude that D_n *is generated by the rotation* $\rho_{360/n}$ *and the reflection* γ. *It follows that* $o(D_n)$*=2n.* ∎

Let us investigate the influence of the set **S** on the group **G**. We have the following definition.

Definition 2.8. *Let* **G** *be a transformation group on a set* **S**. *Let* **T** *be a subset of* **S**. *The stabilizer of* **T**, *Stab(***T***), is defined as follows,*

$$Stab(\mathbf{T}) = \{g : g \in \mathbf{G}, g(\mathbf{T}) = \mathbf{T}\}$$

∎

Discussion

(1) Stab(**T**) is a subgroup of **G** for any subset **T**∈**S**: Let $g \in$ Stab(**T**). Then we have

$$\begin{aligned} &g(\mathbf{T}) = \mathbf{T} \\ &g^{-1} * g(\mathbf{T}) = g^{-1}(\mathbf{T}) \\ &e(\mathbf{T}) = g^{-1}(\mathbf{T}) \\ &g^{-1} \in \mathrm{Stab}(\mathbf{T}) \end{aligned}$$

If $h, g \in$ Stab(**T**), then $h * g^{-1} \in$ Stab(**T**). Therefore Stab(**T**) is a subgroup of **G**.

(2) Let **H** be a subgroup of **G**. Let the group **G** act on the set **G** from the left, i.e.,

$$g(g_1) = g * g_1$$

Then we have Stab(**H**)=**H**. Let us consider the group **H** acting on the set **G** from the left. Then we have,

$$\mathrm{Orb}(e) = \{h * e = h : h \in H\} = H$$

The orbits $\text{Orb}(g) = \{h * g : h \in H\} = H * g$ are called the *right-cosets* of **G** with respect to **H**. Similarly, we may consider the group act on the set **G** from the *right* as follows

$$h(g) = g * h^{-1}$$

The orbits thus produced $\{g * h^{-1} : h \in H\} = \{g * h : h \in H\} = g*\mathbf{H}$ are called the *left-cosets* of **G** with respect to **H**. Usually, we use *cosets* to mean left-cosets.

(3) Let a group **G** *conjugate* the set **G**, in other words, the group **G** act on the set **G** from both sides, i.e.

$$g(g_1) = g * g_1 * g^{-1}$$

Then $\text{Orb}(g_1)$ is called the conjugate class of g_1. ∎

The following is a trivial consequence of Theorem 2.2.,

Corollary. *A group* **G** *is the disjoint union of the cosets of* **G** *with respect to a subgroup* **H**. ∎

Theorem 2.3. *Any two cosets of a group* **G** *with respect to a subgroup* **H** *have the same cardinality.*

Proof: It suffices to prove that $\text{Orb}(e)$ and $\text{Orb}(g)$ have the same cardinality for any $g \in$ **G**. The element g induces a bijective map g: $\text{Orb}(e) \to \text{Orb}(g)$ as follows,

$$g(h) = g * h$$

∎

Definition 2.9. *The cardinality of the set of all (left-)cosets with respect to a subgroup* **H** *is called the index [***G***:* **H***] of* **H** *in* **G**. ∎

Theorem 2.4. (Lagrange Theorem). *Let* **G** *be a finite group. Then we have o(***G***)=[***G***:* **H***] o(***H***).* ∎

Corollary. *If o(***G***) is prime, then* **G** *is cyclic.* ∎

Let a group **G** act on the set **G** from left and **H** be a subgroup of **G**. Then we have Stab(**H**)=**H**. The distinct images of **H** under this action are precisely the (left-)cosets $g*$**H**. Therefore, we have: the cardinality of Orb(**H**)= [**G**: Stab(**H**)]. This is the formula we will generate in the following theorem.

Theorem 2.5. *Let a group* **G** *be a transformation group on a set* **S** *and* **T** *be a subset of* **S**. *Then we have : Card(Orb(***T***))=[***G***: Stab(***T***)] where Card(A)=the cardinality of the set A.*

Proof: It suffices to prove

$$g_1(\mathrm{T}) = g_2(\mathrm{T}) \Longleftrightarrow g_1 \in g_2 * \mathrm{Stab}(\mathrm{T}). \qquad \forall\, g_1, g_2 \in \mathrm{G}$$

($\Longrightarrow$) We have the following computations

$$g_1(\mathbf{T}) = g_2(\mathbf{T}), \qquad g_2^{-1}g_1(\mathbf{T}) = \mathbf{T}$$
$$g_2^{-1}g_1 \in \mathrm{Stab}(\mathbf{T}), \qquad g_1 \in g_2 * \mathrm{Stab}(\mathbf{T})$$

($\Longleftarrow$) Suppose that we have g_1, g_2 with

$$g_1 \in g_2 * \mathrm{Stab}(\mathbf{T}), \qquad g_1 = g_2 * h, \qquad h \in \mathrm{Stab}(\mathbf{T})$$
$$g_1(\mathbf{T}) = (g_2 * h)(\mathbf{T}), \qquad g_1(\mathbf{T}) = g_2(h(\mathbf{T})), \qquad g_1(\mathbf{T}) = g_2(\mathbf{T})$$

∎

We will use the preceding theorem to do some meaningful computations in the following example.

Example 10. (Finite subgroups of the space rotation group Rot $(3, R^3)$)[3].
The finite groups **G** *of the space rotation group* **Rot** $(3, \mathbf{R}^3)$ *are useful in crystallography. The reader is referred to Example 5. for some preliminary knowledge about space rotations. Let the finite group* **G** *act on the unit sphere* S^2*, i.e., the set of all points with distance 1 to the origin. As we pointed out in Example 5., every non-trivial rotation has exactly one axis of rotation, therefore every non-trivial rotation has precisely two poles, i.e., the fixed points, on the unit sphere* S^2. *Let* **S**= *the set of poles of all non-trivial rotations of* **G** *on the unit sphere* S^2.

(1) We claim that **G** *is a transformation group on the set* **S**. *Let* $s \in \mathbf{S}$, *i.e.,* s *is a pole of a non-trivial rotation* $g_1 \in \mathbf{G}$. *Let* $g_2 \in \mathbf{G}$. *Then we have*

$$g_1(s) = s$$
$$(g_2 * g_1 * g_2^{-1})(g_2(s)) = g_2(g_1(g_2^{\,*} * g_2(s))) = g_2(g_1(s)) = g_2(s)$$

Therefore $g_2(s)$ *is fixed by* $(g_2 * g_1 * g_2^{-1})$, *If it is the identity, then we have*

$$g_2 * g_1 * g_2^{-1} = e$$
$$g_2 * g_1 = g_2$$
$$g_1 = e$$

a contradiction. Therefore $g_2(s) \in \mathbf{S}$ *and* **G** *is a transformation group on* **S**. *From now on, we will treat* **G** *as a transformation group on the set* **S**.

(2) Let $s \in \mathbf{S}$ *and* v_s = *the cardinality of Orb*(s). *It follows from Theorem 2.5. that* $v_s = [\mathbf{G}: Stab(s)]$. *Let* $n_s = o(Stab(s))$ *and* $n = o(\mathbf{G})$. *Then we have*

$$n = o(\mathbf{G}) = n_s \cdot v_s$$

[3] See H. Weyl 'Symmetry'.

Let $s_1 = g_1(s) \in Orb(s)$. Then the map $\mathring{g}_1 : Stab(s_1) \to Stab(s)$ defined by $\mathring{g}_1(g) = g_1 * g * g_1^{-1}$ is obviously surjective. It is then easy to deduce that $n_{s_1} = n_s, v_{s_1} = v_s \; \forall s_1 \in Orb(s)$. For every $s \in \mathbf{S}$, there are $(n_s - 1)$ non-trivial rotations in $\mathbf{G}$ which take s as a pole. Every non-trivial rotation has precisely two poles. Therefore we have the following formula with the summation over one s from each orbit,

$$2(n-1) = \sum_{\text{orbits}} v_s(n_s - 1) = \sum_{\text{orbits}} (n - v_s)$$

Factoring out the above equation by n, we have

(1)
$$2 - \frac{2}{n} = \sum_{\text{each orbits}} (1 - \frac{v_s}{n}) = \sum_{\text{orbits}} (1 - \frac{1}{n_s})$$

Since $s \in \mathbf{S}$ is a pole of a non-trivial rotation, then we have

(2)
$$n_s = o(Stab(\mathbf{S})) \geq 2$$
$$1 - \frac{1}{n_s} \geq \frac{1}{2}$$

Comparing both sides of the equation (1), we conclude that there are at most three orbits, otherwise the right hand will be bigger then or equal to 2, and there are at least two orbits, otherwise the right hand side of the equation (1) will be less than 1.

We shall analyze the above numerical data to give us a clear idea about all possible finite rotation groups.

(3) Suppose that there are precisely two orbits. The equation (1) is explicitly as follows,

$$2 - \frac{2}{n} = (1 - \frac{1}{n_{s_1}}) + (1 - \frac{1}{n_{s_2}})$$
$$\frac{2}{n} = \frac{1}{n_{s_1}} + \frac{1}{n_{s_2}}, \quad n_{s_1}, n_{s_2} \leq n$$

Therefore we have

$$n = n_{s_1} = n_{s_2}$$
$$1 = v_{s_1} = v_{s_2}$$

It follows at once that $Stab(s_1)$=$\mathbf{G}$=$Stab(s_2)$ and s_1, s_2 are the common poles of all rotations. We may take s_1, s_2 as the north and south poles. The rotation group $\mathbf{G}$ is induced by the rotations on the horizontal plane (cf Example 9.).

(4) From now on, we assume that there are precisely three orbits. Let us fix the order of s_1, s_2, s_3 with

$$n_{s_1} \leq n_{s_2} \leq n_{s_3}$$

We claim $n_{s_1} = 2$: otherwise $n_{s_1} \geq 3$ and the right hand side of the equation (1) will be at least 2. Impossible. Furthermore, we claim $n_{s_2} \leq 3$: otherwise $n_{s_3} \geq n_{s_2} \geq 4$ and the right hand side of the equation (1) will be at least 2. Impossible.

(5) Let us consider the case $n_{s_1} = 2 = n_{s_2}$. Then from the equation (1), we deduce the following data,

$$n_{s_1} = n_{s_2} = 2, \qquad n_{s_3} = \frac{n}{2}$$
$$v_{s_1} = v_{s_2} = \frac{n}{2}, \qquad v_{s_3} = 2$$

It follows from the equation (1) that $n_{s_3} \geq 3$ or $n \geq 6$ in our present case. For simplicity, let us assume that s_3 is the north pole on the unit sphere. Then every non-trivial rotation which fixes the north pole will fix the south pole s_3'. Therefore $s_3' \in \mathbf{S}$ and $n_{s_3'} = n_{s_3} = \frac{n}{2} \geq 3$. Looking up the above data, we conclude that $\{s_3, s_3'\}$ must be one orbit, i.e., there is a rotation g which flips over the sphere to interchange the north and south poles. Let $\mathbf{H}$ be the rotation subgroup which fixes the north pole s_3 of order $o(H) = \frac{n}{2}$. Then $\mathbf{G} = H \cup g * \mathbf{H}$. We may view $\mathbf{G}$ as the groups of rotations and reflections of a regular planar $(\frac{n}{2})$-polygon, or the *dihedral group* $\mathbf{D}_{n/2}$. See the following diagram,

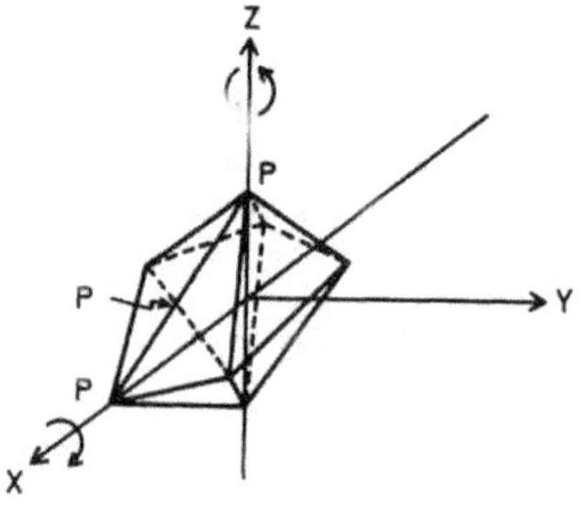

(6) Let us consider the other possibility $n_{s_1} = 2, n_{s_2} = 3$. Then from the equation (1), we deduce the following data,

$$n_{s_1} = 2, \quad n_{s_2} = 3, \quad \frac{1}{n_{s_3}} = \frac{1}{6} + \frac{2}{n}$$

The last equation shows that $n_{s_3} \leq 5$. Indeed, we have the following cases,

$$n_{s_1} = 2, \quad n_{s_2} = 3, \quad n_{s_3} = 3$$
$$n_{s_1} = 2, \quad n_{s_2} = 3, \quad n_{s_3} = 4$$
$$n_{s_1} = 2, \quad n_{s_2} = 3, \quad n_{s_3} = 5$$

(7) Let us consider the case $n_{s_1} = 2, n_{s_2} = 3, n_{s_3} = 3$. Then from the equation (1), we deduce the following data,

$$n_{s_1} = 2, \quad n_{s_2} = 3, \quad n_{s_3} = 3, \quad n = 12$$
$$v_{s_1} = 6, \quad v_{s_2} = 4, \quad v_{s_3} = 4$$

Similar to part (5), let us view s_3 as the north pole. Since $v_{s_3} = 4$, then there are 4 points in Orb(s_3). At least one of them, say s_{31}, is not the south pole. Since $n_{s_3} = 3$, there are three rotations (one of them is identity) fixing s_3. Let the three images of s_{31} be $\{s_{31}, s_{32}, s_{33}\}$. Then Orb$(s_3) = \{s_3, s_{31}, s_{32}, s_{33}\}$ with $\Delta s_{31}s_{32}s_{33}$ an equilateral triangle. Similarly, every three points must form an equilateral triangle. It is easy to see the four points form a regular tetrahedron and **G** is the group of all rotations which map the tetrahedron to itself.

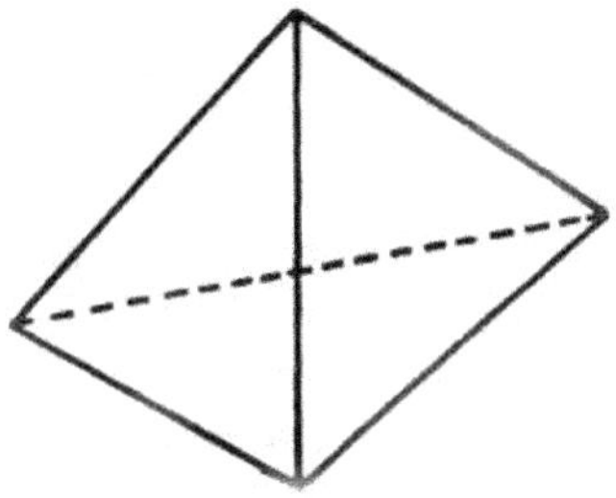

Orb(s_1) is the set of the projections of the six middle points of the six edges to the unit sphere (recall that $v(s_1) = 6$). Orb(s_2) is the set of the projections of the four centers of the four faces to the unit sphere (recall that $v(s_2) = 4$).

(8) Let us consider the case $n_{s_1} = 2, n_{s_2} = 3, n_{s_3} = 4$. Then from the equation (1), we deduce the following data,

$$n_{s_1} = 2, \quad n_{s_2} = 3, \quad n_{s_3} = 4, \quad n = 24$$
$$v_{s_1} = 12, \quad v_{s_2} = 8, \quad v_{s_3} = 6$$

Similar to part (5), let us view s_3 as the north pole. Since $v_{s_3} = 6$, then there are 6 points in Orb(s_3). At least one of them, say s_{31}, is not the south pole. Since $n_{s_3} = 4$, there are four rotations (one of them is the identity) fixing s_3. Let the four images of s_{31} be $\{s_{31}, s_{32}, s_{33}, s_{34}\}$. Now we have five out of six points of Orb(s_3) accounted for. Where is the sixth point? If it is not the south pole, then the four rotations will produce four more points in Orb(s_3). Therefore, the sixth point must be the south pole. We claim the four points $\{s_{31}, s_{32}, s_{33}, s_{34}\}$ must lie on the equator plane: otherwise, say they all lie on the upper space, the four rotations with one of them as fixed point will move s_3 outside Orb(s_3). The four points on the equator form a square. They, together with the two poles, form a regular octahedron. Finally **G** is the group of all rotations which map the octahedron to itself.

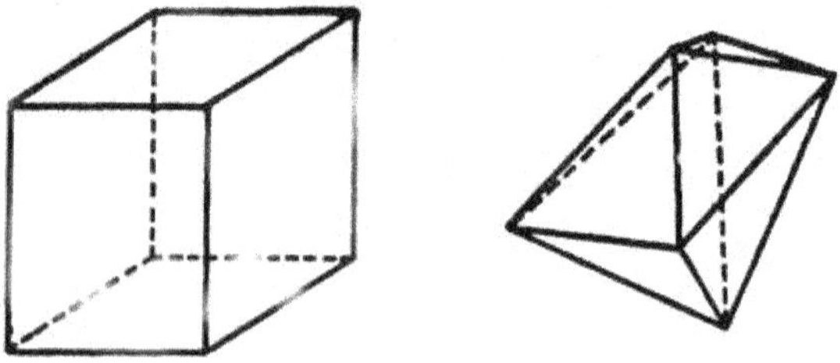

Orb(s_1) is the set of the projections of the twelve middle points of the twelve edges to the unit sphere (recall that $v(s_1) = 12$). Orb(s_2) is the set of the projections of the eight centers of the eight faces to the unit sphere (recall that $v(s_2) = 8$). We may take the dual of the octahedron, i.e., we take the set of the projections of the eight centers of the eight faces to the unit sphere as vertices. In this case we get a cube. Therefore, **G** is the group of all rotations which map the cube to itself.

(9) Let us consider the case $n_{s_1} = 2, n_{s_2} = 3, n_{s_3} = 5$. Then from the equation (1), we deduce the following data,

$$n_{s_1} = 2, \quad n_{s_2} = 3, \quad n_{s_3} = 5, \quad n = 60$$
$$v_{s_1} = 30, \quad v_{s_2} = 20, \quad v_{s_3} = 12$$

Similar to part (5), let us view s_3 as the north pole. Since $v_{s_3} = 10$, then there are 10 points in Orb(s_3). At least one of them s_{31} is not the south pole. Since $n_{s_3} = 5$, there are five rotations (one of them is the identity) fixing s_3. Let the five images of s_{31} be $\{s_{31}, s_{32}, s_{33}, s_{34}, s_{35}\}$ which lie on the same horizontal plane. Now we have six out of ten points of Orb(s_3) counted for. There must be another point r_{31} which produces five points in Orb(s_3). Counting s_3 in, we have eleven points. Where is the twelfth point? If it is not the south pole, then the five rotations will produce other five points in Orb(s_3). Therefore, the twelfth point must be the south pole. Since the antipode of s_3 is in Orb(s_3), then it must be true for every point in Orb(s_3). We may assume that r_{3i} is the antipode of s_{3i}. If all s_{3i}'s and r_{3i}'s are in the same plane, then replacing s_3 by s_{31}, it contradicts the preceding discussions. We therefore know that they are on two different horizontal planes.

Since s_{3i} and r_{3i} are antipodes, then they must lie on one plane in the upper space and one in the lower space. Let us say $\{s_{31}, s_{32}, s_{33}, s_{34}, s_{35}\}$ lies in the upper space. The edges connecting s_3 to s_{3i} are of the same length. Replacing s_3 by s_{31}, we see that $\Delta s_3 s_{31} s_{32}$ is equilateral. Now it is easy to see all points in Orb(s_3) are vertices of a regular icosahedron and G is the group of all rotations which map the icosahedron to itself.

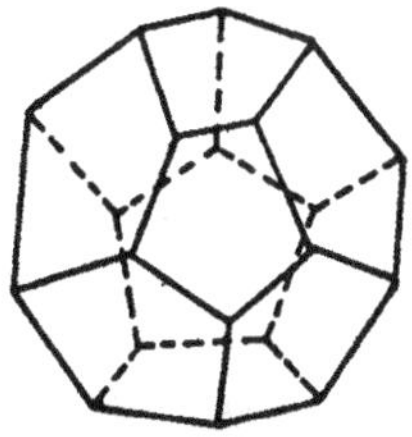

Orb(s_1) is the set of the projections of the thirty middle points of the thirty edges to the unit sphere (recall that $v(s_1) = 30$). Orb(s_2) is the set of the projections of the twenty centers of the twenty faces to the unit sphere (recall that $v(s_2) = 20$). We may take the *dual* of the icosahedron, i.e., we take the set of the projections of the twenty centers of the twenty faces to the unit sphere as vertices, then we get a dodecahedron. Therefore, G is the group of all rotations which map the dodecahedron to itself.

(10) In this way we get the rotation groups of the five regular polyhedrons which were known to the ancient Greeks. ∎

Exercises

(1) Let the *complex projective line* $\mathbf{P}^1_{\mathbf{C}} = \mathbf{C} \cup \infty$, here C is the set of complex numbers. Let G be the *fractional linear transformation group* defined as follows,

$$G = \left\{ x \mapsto \frac{ax+b}{cx+d} : det \begin{pmatrix} a, & b \\ c, & c \end{pmatrix} \neq 0 \right\}$$

Let $\mathbf{T} = 0, 1, \infty$. Find Stab(T).

(2) Let $\mathbf{G}_1$ be the translation group on the real plane $\mathbf{R}^2$ (cf Example 3. §1). Find Stab (T).

(3) Let G be a cyclic group (cf Discussion §1) of order n. Suppose that m is a positive integer with $m \mid n$. Prove that there is exactly one subgroup of order m and this subgroup is cyclic.

(4) Let the following elements be in $GL(2, \mathbf{R})$,

$$g_1 = \begin{pmatrix} 0, & -1 \\ 1, & 0 \end{pmatrix} \qquad g_2 = \begin{pmatrix} 0, & 1 \\ -1, & -1 \end{pmatrix}$$

Prove that $o(g_1)=4$, $o(g_2)=3$ and $o(g_1 * g_2)=\infty$.

(5) Let G be a group. The *commutator subgroup* $\mathbf{G}^{(1)}$ of G is defined to be the subgroup generated by $\{g_1^{-1} * g_2^{-1} * g_1 * g_2 : g_1, g_2 \in \mathbf{G}\}$. Show that if G is finite, then Card(class) $\geq$ Card($\mathbf{G}^{(1)}$) for any conjugate class of G.

(6) Let H_i be subgroups of G with finite indices for $i = 1, 2, \cdot, n$. Prove that $\cap H_i$ is a subgroup with finite index.

(7) Let H be a subgroup of G. Prove that the $g * H * g^{-1}$ is a subgroup of G.

(8) Let G be a group with all proper subgroups finite. Give examples to show that G may not be finite.

(9) Let G be a finitely generated group. Prove that every subgroup H with finite index is finitely generated.

(10) Let G be the multiplicative group generated by the real matrices A and B as follows,

$$A = \begin{pmatrix} 2, & 0 \\ 0, & 1 \end{pmatrix} \qquad B = \begin{pmatrix} 1, & 1 \\ 0, & 1 \end{pmatrix}$$

Let H be the set of all matrices in G with diagonal entries 1. Show that H is a subgroup which is not finitely generated.

(11) Let H be the subgroup of $\mathbf{Z} \times \mathbf{Z}$ generated by $(18, 12), (30, 24)$. Find $[\mathbf{Z} \times \mathbf{Z} : \mathbf{H}]$.

(12) Let H and K be subgroups of G of finite index. Prove

(i) $[\mathbf{G}:\mathbf{H} \cap \mathbf{K}] \leq [\mathbf{G}:\mathbf{H}]\ [\mathbf{G}:\mathbf{K}]$.

(ii) The above equality holds if and only if G= H$*$K.

(13) Let G be the group of rigid motions which transform a cube to itself. Let v be a vertex, and H=Stab(v). Find o(G), o(H) and [G:H].

(14) What are the possible orders of elements of the dihedral group $\mathbf{D}_5$? How many elements of the dihedral group $\mathbf{D}_5$ have order 2?

§4 Normal Subgroups and Inner Automorphisms

Let us consider a group **G** conjugate on itself, i.e., **G** act on **G** from both sides as follows,

$$g(g_1) = g * g_1 * g^{-1}$$

This action g satisfies an extra condition.

$$g(g_1 * g_2) = g * g_1 * g_2 * g^{-1} = g * g_1 * g^{-1} * g * g_2 * g^{-1} = g(g_1) * g(g_2)$$

Those maps are important and are named ***inner automorphisms***. We have the following related definitions,

Definition 2.10. *Let ρ: **G**→ **G'** be a map. If ρ preserves the group structure, i.e.,*

$$\rho(g_1 * g_2) = \rho(g_1) * \rho(g_2), \qquad \forall\, g_1, g_2 \in \mathbf{G}$$

then ρ is said to be a homomorphism. ∎

Note that the inner automorphisms g discussed above are homomorphisms.

Definition 2.11. *Let ρ be a homomorphism from a group **G** to a group **G'**. If ρ is injective, then we say ρ is an **into isomorphism**. If ρ is bijective, then we say ρ is an isomorphism and **G** is **isomorphic** to G'. If ρ is an isomorphism and G=**G'**, then we say ρ is an **automorphism**.* ∎

Discussion

(1) As we observed before that inner automorphisms are homomorphisms. It follows from Theorem 2.1. that all elements of a transformation group are bijective maps. Therefore the inner automorphisms are automorphisms. Those automorphisms are induced by elements ***in*** the group **G**, hence the name ***inner***.

(2) Let ρ be any homomorphism. Then we have

$$\rho(e) * \rho(g) = \rho(e * g) = \rho(g)$$

It follows from the uniqueness of unit in **G'** that $\rho(e) = e'$ the unit in **G'**. Furthermore, we have

$$\rho(g) * \rho(g^{-1}) = \rho(g * g^{-1}) = \rho(e) = e'$$

It follows from the uniqueness of the inverse of $\rho(g)$ in **G'** that $\rho(g^{-1}) = \rho(g)^{-1}$. ∎

The following definition combine the concepts of subgroups and invariant subset.

Definition 2.12. *Let the inner automorphism group* **Inn(G)** *be the transformation group of all inner automorphisms of a group* **G** *conjugating* **G** *into itself. A subgroup* **H** *which is an invariant subset is called an* ***invariant subgroup****. An invariant subgroup H is also called a* ***normal subgroup*** *of G, in symbol, $H \triangleleft$* **G***.* ∎

Definition 2.13. *(1) Let the center of* **G***, C(***G***), be the union of all singleton orbits under the conjugations of* **G***. Note that the center of* **G***, C(***G***), is an invariant set under the conjugations.*

(2) If **G***=C(***G***), then we say* **G** *is a* ***commutative group*** *or an* ***abelian*** *group. Otherwise we say* **G** *is a* ***non-commutative*** *group or a* ***non-abelian*** *group.* ∎

Discussion

(1) It follows from the definition that a subgroup **H** is a normal subgroup $\Longleftrightarrow g(H) = g * H * g^{-1} = \mathbf{H}\ \forall g \in \mathbf{G} \Longleftrightarrow g * H = H * g\ \forall g \in \mathbf{G}$. Sometimes we take the last condition as the definition for normal subgroup.

(2) It follows from the definition that an element g is in the center C(**G**) $\Longleftrightarrow g_1(g) = g_1 * g * g_1^{-1} = g\ \forall g_1 \in \mathbf{G} \Longleftrightarrow g_1 * g = g * g_1\ \forall g_1 \in \mathbf{G} \Longleftrightarrow g$ commutes with every element g_1 in **G**. Sometimes we take the last condition as the definition for the center.

(3) Suppose that g, g' are in the center C(**G**). Then we have $g_1(g^{-1} * g') = g_1(g^{-1}) * g_1(g') = g^{-1} * g'\ \forall g_1 \in \mathbf{G} \Longrightarrow g^{-1} * g' \in C(\mathbf{G})$. Therefore C(**G**) is a subgroup of **G**. We conclude that C(**G**) is an commutative normal subgroup of **G**.

(4) If a group **G** is commutative, then every subset is an invariant subset under conjugations and every subgroup is a normal subgroup. We know $n\mathbb{Z} \subset \mathbb{Z} \subset \mathbb{Q} \subset \mathbb{R} \subset \mathbb{C}$ are all commutative groups with respect to summation. If $\mathbf{G}_i$ are all commutative groups, then the direct sum $\oplus_i \mathbf{G}_i$ and the direct product $\prod_i \mathbf{G}_i$ are commutative groups. ∎

Example 11.

(1) In Example 1. §1 all numerical groups are commutative. In Example 3. §1, the translation group, the reflection groups and the rotation groups are commutative. We claim the group of all rigid motions on $\mathbf{R}^2$ *is non-commutative: it is easy to check that for* $(a, b) = (c, d) = (1, 0)$ *and* $\theta = 90°$*, we have*

$$\begin{aligned} \beta_{(1,0)} * \rho_{90}((1,0)) &= \beta_{(1,0)}((0,1)) = (1,1) \\ &\neq (-1,1) = \rho_{90}((1,1)) = \rho_{90} * \beta_{(1,0)}((1,0)) \end{aligned}$$

(3) The rotation group **Rot** $(3, \mathbf{R})$ *for the space* $\mathbf{R}^3$ *is non-commutative: recall Example 5., it is easy to see the following,*

$$\begin{pmatrix} 0, & 1, & 0 \\ -1, & 0, & 0 \\ 0, & 0, & 1 \end{pmatrix} \cdot \begin{pmatrix} 0, & 0, & -1 \\ 0, & 1, & 0 \\ 1, & 0, & 0 \end{pmatrix} = \begin{pmatrix} 0, & 1, & 0 \\ 0, & 0, & 1 \\ 1, & 0, & 0 \end{pmatrix}$$
$$\neq \begin{pmatrix} 0, & 0, & -1 \\ -1, & 0, & 0 \\ 0, & 1, & 0 \end{pmatrix} = \begin{pmatrix} 0, & 0, & -1 \\ 0, & 1, & 0 \\ 1, & 0, & 0 \end{pmatrix} \cdot \begin{pmatrix} 0, & 1, & 0 \\ -1, & 0, & 0 \\ 0, & 0, & 1 \end{pmatrix}$$ ∎

Example 12. (The orbit forms and conjugate classes of S_n).

(1) Orbit form. Let $\sigma \in S_n$. Let the subgroup $H=\langle\sigma\rangle$ be a transformation group acting on $\{1, 2, \cdots, n\}$. Clearly we have the following,

$$\text{Orb}(i) = (i, \sigma(i), \cdots, \sigma^{\ell-1}(i)), \quad \text{where } \sigma^{\ell}(i) = i, \text{ and } \sigma^{j}(i) \neq i, \ \forall 1 \leq j < \ell$$

We identify all cyclic permutations of an orbit. Recall that all orbits are disjoint. Any element σ can be written uniquely, up to a permutation of orbits and up to cyclic permutations of individual orbits, as follows,

$$(i, \sigma(i), \cdots)(j, \sigma(j), \cdots)\cdots(k, \sigma(k), \cdots)$$

Furthermore, any expression of disjoint union of orbits as above uniquely defines a permutation σ. Customarily, we sometimes omit the singleton orbits without any confusion. For instance, let $\sigma = (1, 2, 3)$. Then $\sigma(1) = 2$, $\sigma(2) = 3$, $\sigma(3) = 1$ and $\sigma(i) = i$, $\forall i > 3$.

(2) Conjugate classes. Let us consider some examples.

(i) *Let $\sigma = (1, 2), \rho = (1, 2, 3)$. Then the conjugation by ρ: $\rho(\sigma) = \rho * \sigma * \rho^{-1} = (2, 3)$.*
(ii) *Let $\sigma_1 = (2, 4)(1, 5), \rho = (1, 2, 3)$. Then the conjugation by ρ: $\rho(\sigma_1) = \rho * \sigma_1 * \rho^{-1} = (2, 4)(1, 5)$.*

For the above two examples, the conjugation by ρ is simply replacing the digits in σ (or σ_1) according to the permutation given by ρ. Indeed, this is the general phenomena which can be proven easily. Let us view the permutation ρ as a replacement of the digits $1, 2, \cdots, n$ by $1_\rho, 2_\rho, \cdots, n_\rho$ where $i_\rho = \rho(i)$. Then we have

$$((\rho(\sigma))(i_\rho) = \rho * \sigma * \rho^{-1}(\rho(i)) = \rho * \sigma(i) = \sigma(i)_\rho$$

namely, after replacing i by i_ρ, the conjugation $\rho(\sigma)$ will map it to $\sigma(i)_\rho$. Therefore, we conclude that two elements α, β of S_n belong to the same conjugate class if and only if α can be gotten from β by replacing the digits of the orbit form of β. Then every conjugate class is uniquely determined by the form of any element in it. For instance, in S_3, there are three conjugate classes with orbit forms ( )( )( ), ( )(,) and (, ,). In S_4, there are five conjugate classes ( )( )( )( ), ( )( )(,), ( )(, ,), (,)(,) and (, , ,). In S_n, the number of conjugate classes is the number of ways of writing n as summations of positive integers. ∎

Example 13. *Let us consider the following example due to Andre Weil. According to anthropologists, the marriage laws of many primitive societies share the following common rules;*

(i) *Every member of the society is assigned a **type**: marriage is only permissible among couples with the same **type**.*
(ii) *A person's type is decided by one's sex and the type of one's parents. Conversely, the type of one's parents can be determined by the sex and the type of this person.*

(iii) *The type of the brothers and the type of the sisters are always different.*

(iv) *The blood relations decide if two people have the same type: for instance, if a male and his grandfather have the same type, then all males and their grandfathers have the same types.*

(v) *It is always possible for some descendents of any two people to get married.*

We will use the group theory to study those marriage laws. Let the types of a primitive society be $\{1, 2, \cdots, n\}$. The passages of the types from the parents to the sons and daughters will be denoted by S and D in the following way,

$$S(i) = \text{ the type of son if the couple's type is } i$$
$$D(i) = \text{ the type of daughter if the couple's type is } i$$

The above (ii) means that both S and D are bijectives, hence S and D belongs to the symmetric group S_n on n numbers. We shall rewrite (iii) as follows

(iii)' $S(i) \neq D(i), \quad \forall\, i$, or $S^{-1}D(i) \neq i, \quad \forall\, i.$

Note that any blood relationship is an element in the subgroup $\mathbf{H}$ generated by $S, D = \langle S, D \rangle$; for instance, a male and a female whose fathers are brothers (they are the first cousins on the father side), the relation from the male to the female is "father's father's son' daughter", i.e., $DSS^{-1}S^{-1}$. Therefore, we have

$$DSS^{-1}S^{-1}(i) = DS^{-1}(i) \neq i, \qquad \forall\, i \text{ according to (iii)'}$$

therefore, they will never be allowed to get marriaged. Suppose that the mother of a male and the father of a female are siblings, the relation from the male to the female is "mother's father's son' daughter", i.e.,

$$DSD^{-1}S^{-1}$$

Therefore (iv) can be rewritten as

(iv)' *For any element $R \in \mathbf{H} = \langle S, D \rangle$ the subgroup generated by S, D, if $R \neq I$ the identity, then we always have,*

$$R(i) \neq i, \qquad \forall\, i.$$

Then it is trivial to see that a male is permissible to marriage his "mother's father's son' daughter" (his first cousin on mother side), if and only if

$$DSD^{-1}S^{-1} = I \iff \mathbf{H} \text{ is commutative.}$$

Note that the group $\mathbf{H}$ naturally act on the set $\{1, 2, \cdots, n\}$. It is easy to see that (v) can be rewritten as

(v)' *The orbit of any number i under the transformation group $\mathbf{H}$ is the whole set of numbers $\{1, 2, \cdots, n\}$.*

Therefore we may use group theory to do research in Anthropology. Let us consider the case of $n \leq 4$. If $n = 2$, then (iv)' implies $S = (1,2) = D$ which violates (iii)'. Impossible. If $n = 3$, then we must have $S = (1,2,3) = D^{-1}$ or $S = (3,2,1) = D^{-1}$. Therefore $\langle S, D\rangle$ is the cyclic group, and hence commutative. Let us consider the case $n = 4$. Let us disregard all rearrangements of the integers $1,2,3,4$ and the interchanging of $S \rightleftharpoons D$. Let P be the cyclic permutation $(1,2,3,4)$. A direct computation will establish that there are the following four possibilities:

(a) $S = P$, $D = I$ *the unit.*
(b) $S = P$, $D = P^2$.
(c) $S = P$, $D = P^3$.
(d) $S = (1,2)(3,4)$, $D = (1,3)(2,4)$.

In fact, Tarau uses (a) and Kariera uses (d). It is not hard to prove that if $n = 4$, then **H**$= \langle S, D\rangle$ *is always commutative. Hence if $n \leq 4$, a male is always permissible to marry his "mother's father's son' daughter" (his first cousin on mother side).* ∎

Let us come back to the study of centers of groups. An element $g \in \mathrm{C(G)} \iff g * g_1 * g^{-1} = g_1 \; \forall g_1 \in \mathbf{G} \iff$ the induced inner automorphism $g(g_1) = g_1$ is the identity. We have the following theorem.

Theorem 2.6. *Let* **G** *be a transformation group on a set* **S** *and* **G'** *be the permutation group on* **S**. *Then the natural map ρ :***G** $\to$ **G'** *with $\rho(g)(s) = g(s)$ $\forall s \in$* **S** *is a homomorphism and the set* **H**$= \{g : \rho(g) =$ *identity map*$\}$ *is a normal subgroup of* **G**.

Proof: It follows from the definition 2.3. that ρ is a homomorphism. The rest of the theorem is a special case of the following Theorem 2.7.. ∎

Before proving the next theorem, we introduce the following definition.

Definition 2.14. *Let ρ be a homomorphism from a group* **G** *to a group* **G'**. *Then the image of ρ, im(ρ), is defined as*

$$im(\rho) = \{g' : \rho(g) = g' \text{ for some } g \in \mathbf{G}\}$$

and the kernel of ρ, ker(ρ), is defined as

$$ker(\rho) = \{g : \rho(g) = e' \text{ where } e' \text{ is the unit for some } \mathbf{G'}\}$$

In other words, ker(ρ) is the pre-image of e' under the map ρ which can be written as $\rho^{-1}(e')$a. ∎

Theorem 2.7. *Let ρ be a homomorphism from a group* **G** *to a group* **G'**. *Then (1) the image of ρ, im (ρ) =***H'**, *is subgroup of* **G'**. *Furthermore, (2) let* **F'** *be a (resp. normal)* *subgroup of* **H'**, *then* **F**=*the pre-image of* **F'**$= \rho^{-1}$(**F'**)$= \{g : \rho(g) \in$ **F'**} *is a (resp. normal) subgroup of* **G**.

Proof: (1) The *image* of ρ=*im*(ρ)=**H'** is subgroup of **G'**: let $g_1', g_2' \in$ **H'**. Then there are $g_1, g_1 \in$ **G** with

$$\rho(g_1) = g_1', \qquad \rho(g_2) = g_2'$$

Therefore, we have

$$\begin{aligned}\rho(g_1 * g_2^{-1}) &= \rho(g_1) * \rho(g_2^{-1}) = \rho(g_1) * \rho(g_2)^{-1} \\ &= g_1' * g_2'^{-1} \Longrightarrow g_1' * g_2'^{-1} \in \mathbf{H'}.\end{aligned}$$

Moreover, $e' = \rho(e) \in$ **H'**, **F**$\neq \emptyset$, therefore, **H'** is a subgroup.

(2) Let the *pre-image* of **F'**$=\rho^{-1}$ (**F'**)=**F**. We have $\rho(e) = e' \in$ **F'**. Therefore, $e \in$ **F** and **F** *not* $= \emptyset$. Let $g_1, g_2 \in$ **F**. Then we have

$$\begin{aligned}\rho(g_1 * g_2^{-1}) = \rho(g_1) * \rho(g_2^{-1}) = \rho(g_1) * \rho(g_2)^{-1} \in \mathbf{H'} \\ \Longrightarrow g_1 * g_2^{-1} \in \mathbf{F}.\end{aligned}$$

Therefore, **F** is a subgroup.

Let **F'** be a normal subgroup of **H'**. Then for any $f \in$ **F**, we have,

$$\begin{aligned}&\rho(g) * \rho(f) * \rho(g)^{-1} \in \mathbf{F'}, && \forall g \in \mathbf{G} \\ \Longleftrightarrow\ &\rho(g * f * g^{-1}) \in \mathbf{F'}, && \forall g \in \mathbf{G} \\ \Longleftrightarrow\ &g * f * g^{-1} \in \mathbf{F}, && \forall g \in \mathbf{G}\end{aligned}$$

Therefore, **F** is a normal subgroup. ∎

Corollary 1. *The kernel ker*(ρ) *is a normal subgroup of* **G**. ∎

Corollary 2. *Theorem 2.6 is true.* ∎

Corollary 3. *The center of* **G** *is a normal subgroup.* ∎

Example 14. *If* $n \neq 2$, *then the center* $C(\mathbf{S}_n)$ *is the unit group. Certainly, we shall only consider the case* $n \geq 3$. *We have* $\rho \in C(\mathbf{S}_n) \Longleftrightarrow \rho(\sigma) = \rho * \sigma * \rho^{-1} = \sigma\ \forall\ \sigma \in \mathbf{S}_n$. *Let* $\rho \neq e$ *with the following orbit form*

$$\rho = (i, j, \cdots) \cdots (\cdots)$$

Let $k \neq i, j$ *and* $\sigma = (i, k)$. *Then it follows from Example 12. that*

$$\rho(\sigma) = (j, s) \neq (i, k) = \sigma$$

Therefore $\rho \notin C(\mathbf{S}_n)$. ∎

The above statements clarify the relations between the normal subgroups and the maps. The following theorem describe the relations between normal subgroups and the group binary operations,

Theorem 2.8. *Let* **H** *be a normal subgroup of* **G**. *Let us introduce a natural binary operation* $*$ *on the set of all (left) cosets of* **G** *with respect to* **H** *as follows,*

$$(g_1 * \mathbf{H}) * (g_2 * \mathbf{H}) = (g_1 * g_2) * \mathbf{H}$$

Then the set of all (left) cosets is a group which will be denoted as the quotient group **G/H**.

Proof: (1) Let us prove the operation $*$ is well-defined among the cosets. In other words, the operation is independent of the representative of every coset. Let g_3, g_4 be other representatives of the cosets $g_1 * \mathbf{H}, g_2 * \mathbf{H}$ respectively, i.e.,

$$g_3 * \mathbf{H} = g_1 * \mathbf{H}, \qquad g_4 * \mathbf{H} = g_2 * \mathbf{H}$$

Then we have $h_1, h_2 \in \mathbf{H}$ with

$$g_1 = g_3 * h_1, \qquad g_2 = g_4 * h_2$$

Moreover, since **H** is a normal subgroup, $g_4 * \mathbf{H} = \mathbf{H} * g_4$, then there must be h_3 with

$$h_1 * g_4 = g_4 * h_3$$

Therefore, we have

$$g_1 * g_2 = g_3 * h_1 * g_4 * h_2 = g_3 * g_4 * h_3 * h_2$$

namely, $g_1 * g_2$ and $g_3 * g_4$ belong to the same coset. We conclude that the operation $*$ is well-defined on the set of all cosets.

(2) It is easy to check the associative law for the operation $*$ on the set of all cosets. Moreover, $e*\mathbf{H}=\mathbf{H}$ is the unit and $g^{-1}*\mathbf{H}$ is the inverse of $g*\mathbf{H}$. Therefore $\mathbf{G/H}$ is a group. ∎

The following and two other *isomorphism theorems* (cf §7) are due to E.Noether[4].

Theorem 2.9. (First Isomorphism Theorem). *Let* ρ *be a surjective homomorphism from* **G** *to* **G'**. *Let* **H'** *be a normal subgroup of* **G'** *and* **H** *the pre-image of* **H**. *The induced map* $\overline{\rho} : \mathbf{G/H} \to \mathbf{G'/H'}$, *defined as follows, is an isomorphism,*

$$\overline{\rho}(g * \mathbf{H}) = \rho(g) * \mathbf{H'}$$

[4] German female Mathematician 1882-1935.

Proof: (1) Let us prove the map $\overline{\rho}$ is well-defined. In other words, the map $\overline{\rho}$ is independent of the representative of every coset. Let g_2 be other representatives of the coset $g_1 * \mathbf{H}$, namely

$$g_2 * \mathbf{H} = g_1 * \mathbf{H}$$

Then we have

$$\begin{aligned} &g_1^{-1} * g_2 \in \mathbf{H} \\ &\rho(g_1^{-1} * g_2) = \rho(g_1)^{-1} * \rho(g_2) \in \mathbf{H'} \\ &\rho(g_1) * \mathbf{H'} = \rho(g_2) * \mathbf{H'} \end{aligned}$$

(2) We claim that $\overline{\rho}$ is a homomorphism. Let g_1, g_2 be elements of $\mathbf{G}$. Then we have

$$\begin{aligned} \overline{\rho}((g_1 * \mathbf{H}) * (g_2 * \mathbf{H})) &= \overline{\rho}(g_1 * g_2 * \mathbf{H}) \\ &= \rho(g_1 * g_2) * \mathbf{H'} = \rho(g_1) * \rho(g_2) * \mathbf{H'} \\ &= (\rho(g_1) * \mathbf{H'}) * (\rho(g_2) * \mathbf{H'}) \\ &= \overline{\rho}(g_1 * \mathbf{H'}) * (\overline{\rho}(g_2 * \mathbf{H'}) \end{aligned}$$

(3) We claim that $\overline{\rho}$ is surjective. Let $g' * \mathbf{H'}$ be any coset in the quotient group $\mathbf{G'}/\mathbf{H'}$. Since ρ is surjective, then there is $g \in \mathbf{G}$ with $\rho(g) = g'$. Therefore, we have

$$\overline{\rho}(g * \mathbf{H}) = \rho(g) * \mathbf{H'} = g' * \mathbf{H'}$$

(4) We claim that $\overline{\rho}$ is injective. We have

$$\begin{aligned} \overline{\rho}(g_1 * \mathbf{H}) &= \overline{\rho}(g_2 * \mathbf{H}) \\ &\Longrightarrow \rho(g_1) * \mathbf{H'} = \rho(g_2) * \mathbf{H'} \\ &\Longrightarrow \rho(g_1) * \rho(g_2)^{-1} \in \mathbf{H'} \\ &\Longrightarrow \rho(g_1 * g_2^{-1}) \in \mathbf{H'} \\ &\Longrightarrow g_1 * g_2^{-1} \in \mathbf{H} \\ &\Longrightarrow g_1 * \mathbf{H} = g_2 * \mathbf{H} \end{aligned}$$

∎

Corollary. *Let ρ be a homomorphism from a group $\mathbf{G}$ to a group $\mathbf{G'}$. Then $\mathbf{G}/ker(\rho)$ and $im(\rho)$ are isomorphic groups.*

∎

Definition 2.15. *Let $\mathbf{H}$ be a normal subgroup of $\mathbf{G}$. The* **canonical map** *κ from $\mathbf{G}$ to $\mathbf{G}/\mathbf{H}$ is defined as*

$$\kappa(g) = g * \mathbf{H}, \qquad \forall\, g \in \mathbf{G}$$

∎

Discussion

(1) It is easy to see that the canonical map is surjective.

(2) Any homomorphism ρ from a group **G** to a group **G'** can be written as the composition of three homomorphisms as in the following diagram,

$$\begin{array}{ccc} G & \xrightarrow{\rho} & G' \\ \kappa\downarrow & & \uparrow i \\ G/\ker(\rho) & \xrightarrow{\overline{\rho}} & \mathrm{im}(\rho) \end{array}$$

where $\rho = i * \overline{\rho} * \kappa$ with i injective, $\overline{\rho}$ isomorphism and κ canonical. ∎

Exercises

(1) Find the center of the rigid motion group of the plane $\mathbf{R}^2$.

(2) Is the translation group a normal subgroup of the rigid motion group of the plane $\mathbf{R}^2$?

(3) Prove that the additive group $\mathbf{R}/\mathbf{Z}$ is isomorphic to the multiplicative group $\{e^{ix} : x \in \mathbf{R}\}$.

(4) The torus can be defined as $(\mathbf{R}\times\mathbf{R})/(\mathbf{Z}\times\mathbf{Z})$. Show that the torus is a commutative group.

(5) Discuss Example 12. for $n = 5$.

(6) Prove that if there is a subgroup with finite index, then there is a normal subgroup with finite index.

(7) Let a group **G** act on a set **S** and $b \in$ Orb(a). Prove that Stab(a) and Stab(b) are conjugated, i.e., there exists $g \in$ **G** with $g*$ Stab$(a)*g^{-1}$=Stab(b).

(8) Let a group **G** conjugate on itself. Let **H** be a subgroup of **G**. Then Stab(**H**) is called the *normalizer* of **H**. Prove that (1) the normalizer of **H** is a subgroup of G, (2) **H** is a normal subgroup of Stab(**H**), (3) Stab(**H**) is the maximal subgroup of G such that **H** is a normal subgroup of it.

(9) Count the number of homomorphisms $:\mathbf{Z}/m\mathbf{Z} \to \mathbf{Z}/n\mathbf{Z}$.

(10) Let **H** be a normal subgroup of **G**. Prove that **G**/**H** is commutative $\Longleftrightarrow$ **H**$\supset$the commutator subgroup (Cf §3 Problem 5).

(11) Let p be a prime integer. Count the number of all cyclic subgroups of $\mathbf{Z}/p^2 \oplus \mathbf{Z}/p^3$.

(12) Let **H** be a subgroup of **G**. If [**G**:**H**]=2, then **H** is a normal subgroup of **G**.

(13) Let **H** be a subgroup of **G**. If [**G**:**H**]=3, then there is a normal subgroup $\mathbf{H}^*$ with $[\mathbf{G}:\mathbf{H}^*]$=3 or 6.

(14) Let **G** be a group. A *maximal subgroup* of **G** is a subgroup **H** such that there is no subgroup of **G** properly between **H** and **G**. Let $\mathbf{Z}_{(p^n)} = \{m/p^i : m \in \mathbf{Z}\}$ for a prime number p. Let **G**= $\mathbf{Z}_{(p^n)}/\mathbf{Z}$. Show that there is no maximal subgroup of **G**.

(15) Suppose that there are maximal subgroups in a group **G**. Show that the intersection of all maximal subgroups is a normal subgroup of **G**.

---

§5 Automorphism Groups

Let us recall the following Definition 2.6. from §2,

Definition 2.6. *Let* **G** *be a transformation group on a set* **S**. *If* $g(s) = s\ \forall\ s \in \mathbf{S} \Longrightarrow g = e$, *then* **G** *is said a faithful transformation group on* **S**. ∎

If **G** is commutative, then all conjugations of **G** on **G** are the identity map. Furthermore, if **G** is not unit group, then **G** is not a faithful transformation group on **G**.

Recall from Theorem 2.6. that with a transformation group **G** acting on a set **S**, all elements of **G** whose actions on **S** are identity maps form a normal subgroup. We have the following theorem.

Theorem 2.10. *Let* **G** *be a transformation group on a set* **S**, $\mathbf{H}=\{g : g \in \mathbf{G}, g(s) = s\ \forall\ s \in \mathbf{S}\}$. *Then* $\mathbf{G}/\mathbf{H}$ *is a faithful transformation group on* **S** *with the action defined as follows,*

$$g * \mathbf{H}(s) = g(s), \qquad \forall s \in \mathbf{S}.$$

Proof: (1) We claim the definition is well-defined: let $g_1{*}\mathbf{H}=g_2{*}\mathbf{H}$. Then we have to prove that $g_1(s) = g_2(s)\ \forall\ s \in \mathbf{S}$. There exists a $h \in \mathbf{H}$ with $g_1^{-1} * g_2 = h$. Therefore, we have

$$g_1 * g_2(s) = h(s) = s$$
$$g_1(s) = g_1 * g_1^{-1} * g_2(s) = g_2(s)$$

(2) We claim that $\mathbf{G}/\mathbf{H}$ is a transformation group on **S**. We have

$$\begin{aligned} &e * \mathbf{H}(s) = e(s) = s \\ &(g_1 * \mathbf{H}) * (g_2 * \mathbf{H})(s) = (g_1 * g_2 * \mathbf{H})(s) \\ &\quad = (g_1 * g_2)(s) = g_1(g_2((s))) \\ &\quad = g_1 * \mathbf{H}(g_2 * \mathbf{H}(s)) \end{aligned}$$

(3) We claim that $\mathbf{G}/\mathbf{H}$ is a faithful transformation group on **S**. Let $g{*}\mathbf{H}$ act on **S** as the identity map. Then we have

$$g * \mathbf{H}(s) = g(s) = s, \qquad \forall\ s \in \mathbf{S}$$
$$g \in \mathbf{H}, \qquad g * \mathbf{H} = \mathbf{H}$$

∎

If the set $\mathbf{S}$ is a group, then it is more natural to consider the transformation group of homomorphisms $\mathbf{G}$ on $\mathbf{S}$. Note that all homomorphisms in $\mathbf{G}$ must be automorphisms of $\mathbf{S}$ (cf Theorem 2.1.). We have the following definition.

Definition 2.17. *Let the set* $\mathbf{S}$ *be a group. If* $\mathbf{G}$ *is a transformation group on the set* $\mathbf{S}$ *and all elements of* $\mathbf{G}$ *are homomorphisms of the group* $\mathbf{S}$*. Then* $\mathbf{G}$ *is said a* ***transformation group on the group*** $\mathbf{S}$. ∎

Theorem 2.11. *Let* $\mathbf{S}$ *be a group. Then all automorphisms of* $\mathbf{S}$ *form a group, namely the automorphism group* $\mathbf{Aut(S)}$ *of* $\mathbf{S}$.

Proof: It is obvious that the identity map e is an automorphism, therefore, $e \in \mathbf{Aut(S)}$ and $\mathbf{Aut(S)}$ is not the empty set. Let $\sigma \in \mathbf{Aut(S)}$. Then σ is a bijective map, and its inverse map σ^{-1} exists and is bijective. We claim that σ^{-1} is a group homomorphism: let s_1, s_2 be two elements of $\mathbf{S}$ with s_3, s_4 satisfying

$$\sigma^{-1}(s_1) = s_3, \qquad i.e., \ \sigma(s_3) = s_1,$$
$$\sigma^{-1}(s_2) = s_4, \qquad i.e., \ \sigma(s_4) = s_2.$$

Then we have the following implications,

$$\sigma(s_3 * s_4) = \sigma(s_3) * \sigma(s_4) = s_1 * s_2$$
$$\sigma^{-1}(s_1 * s_2) = s_3 * s_4 = \sigma^{-1}(s_1) * \sigma^{-1}(s_2)$$

We conclude that σ^{-1} is a homomorphism and in $\mathbf{Aut(S)}$. It is easy to show that $\mathbf{Aut(S)}$ is a group. ∎

Example 15. *(1)* $\mathbf{Aut(Z)}=\{e, -e\}$. *Let* $\sigma \in \mathbf{Aut(Z)}$. *Note that only* $\{1\}$ *and* $\{-1\}$ *are the singleton generators of* $\mathbf{Z}$. *Therefore we have*

$$\sigma(1) = \begin{cases} 1 \\ -1 \end{cases}$$

If $\sigma(1) = 1$, *then we have* $\sigma(n) = \sigma(1 + \cdots + 1) = \sigma(1) + \cdots + \sigma(1) = n$. *Therefore* $\sigma = e$. *If* $\sigma(1) = -1$, *then we have* $\sigma(n) = \sigma(1 + \cdots + 1) = \sigma(1) + \cdots + \sigma(1) = -n$. *Therefore* $\sigma = -e$.

(2) $\mathbf{Aut(Z_n)}$ *is isomorphic to* $\mathbf{Z}_n^\times$. *Given* $\sigma \in \mathbf{Aut(Z_n)}$, *let us define a map* π *from* $\mathbf{Aut(Z_n)}$ *to* $\mathbf{Z}_n^\times$ *as* $\pi(\sigma) = \sigma(1)$. *Recall that* $\mathbf{Z}_n^\times = \{[m]_n : m, n \text{ coprime}\}$. *It is easy to see that* $\sigma(1) = [m]_n$ *must generate* $\mathbf{Aut(Z_n)}$. *Therefore we must have the following with suitable integers* a, b,

$$a[m]_n = [1]_n, \qquad am = 1 - bn$$
$$m, n \text{ are coprime.}$$

Hence $\pi(\sigma) \in \mathbf{Z}_n^\times$. *Given any* $[m]_n \in \mathbf{Z}_n^\times$, *define* $\sigma(\ell) = \ell[m]_n$. *Then it is easy to see* $\sigma \in \mathbf{Aut(Z)}$ *and* $\pi(\sigma) = [m]_n$. ∎

Example 16. *Let us consider the direct sum $\mathbb{Z} \times \mathbb{Z} = \mathbb{Z}^2$. For the convenience of our notations, let us write the elements of $\mathbb{Z}^2$ as vertical vectors, i.e.,*

$$\mathbb{Z}^2 = \{\begin{pmatrix} a_1 \\ a_2 \end{pmatrix} : a_1, a_2 \in \mathbb{Z}\}$$

Let $\sigma \in \mathbf{Aut}(\mathbb{Z}^2)$ with

$$\sigma(\begin{pmatrix} 1 \\ 0 \end{pmatrix}) = \begin{pmatrix} b_{11} \\ b_{21} \end{pmatrix}, \quad \sigma(\begin{pmatrix} 0 \\ 1 \end{pmatrix}) = \begin{pmatrix} b_{12} \\ b_{22} \end{pmatrix}$$

Let us define a map π from $\mathbf{Aut}(\mathbb{Z}^2)$ to $\mathbf{GL(2,R)}$ with

$$\pi(\sigma) = \begin{pmatrix} b_{11}, & b_{12} \\ b_{21}, & b_{22} \end{pmatrix}$$

Then it is easy to see that the images of π are all 2×2 integer coefficients matrices which are invertible. Therefore, the images have determinants ± 1. Under this identification, we may take $\mathbf{Aut}(\mathbb{Z}^2)$ as $(\mathbf{SL(2,\mathbb{Z})}) \cup (-\mathbf{SL(2,\mathbb{Z})})$. Similarly, we may identify $\mathbf{Aut}(\mathbb{Z}^n)$ as $(\mathbf{SL(n,\mathbb{Z})}) \cup (-\mathbf{SL(n,\mathbb{Z})})$. ∎

Theorem 2.12. *Given a group **G**, the inner automorphism group $\mathbf{Inn(G)}$ is a normal subgroup of the automorphism group $\mathbf{Aut(G)}$.*

Proof. Let $g, g_1 \in \mathbf{G}$ and $\sigma \in \mathbf{Aut(G)}$. Then we have

$$\begin{aligned}(\sigma * g_1 * \sigma^{-1})(g) &= \sigma * g_1(\sigma^{-1}(g)) = \sigma(g_1 * \sigma^{-1}(g) * g_1^{-1}) \\ &= \sigma(g_1) * \sigma(\sigma^{-1}(g)) * \sigma(g_1)^{-1} = \sigma(g_1) * g * \sigma(g_1)^{-1}\end{aligned}$$

Therefore, $\sigma * g_1 * \sigma^{-1} \in \mathbf{Inn(G)}$. ∎

Exercises

(1) Let $\mathbf{G} = \mathbb{Z}/6\mathbb{Z}$. Show that $\mathbf{Aut(G)} \neq \mathbf{Inn(G)}$.
(2) Prove that $\mathbf{Aut(Q)} = \{e, -e\}$.
(3) Prove that $\mathbf{Aut(R)} \supsetneq \mathbf{GL(1,R)} \simeq \mathbf{R}$.
(4) Let $\mathbf{G} = \mathbb{Z} + \mathbb{Z}i + \mathbb{Z}j + \mathbb{Z}k$ be the quaternions with integer coefficients (Cf Exercise §1). Find $\mathbf{Aut(G)}$.
(5) Let $\mathbf{G} = \prod_i \mathbb{Z}_{p_i^{r_i}}$, where p_i's are distinct prime numbers and r_i's are positive integers. Find $\mathbf{Aut(G)}$.
(6) Let $\mathbf{G} = \prod_{i=1}^r \mathbb{Z}_p$, where p is a prime number and r is a positive integer. Show that $o(\mathbf{Aut(G)}) = (p^r - 1)(p^r - p) \cdots (p^r - p^{r-1})$.
(7) Prove that $\mathbf{Aut(S_n)} = \mathbf{Inn(S_n)} = \mathbf{S_n}$ for all $n \neq 6$.
(8) Let τ be an automorphism of a group **G** such that $\tau^2 =$ identity and $\tau(g) = g$ if and only if $g = e$. Show that (i) **G** is commutative, (ii) $\tau(g) = g^{-1}\ \forall g \in \mathbf{G}$.

§6 p-Groups and Sylow Theorems

In this section we will study the necessary *or* sufficient numerical conditions on the subgroups of a finite group. Let a group **G** acting on a set **S**. Since all orbits either coincide or are disjoint, then with the usual notation, Card(A)= the cardinality of the set A, we have the following

Orbit Equation: Card(**S**)=$\sum$ Card(orbit).

Let a group **G** conjugate itself. Then the above equation becomes

Class Equation: o(**G**)=$\sum$ Card(class).

Definition 2.16. *Let p be a prime. A group* **G** *is said to be a p-group if* o(**G**)=p^n *for some positive integer n.* ∎

Theorem 2.13. *Let* **G** *be a p-group. Then the center of* **G** *is not the unit group.*

Proof. Let **C** be a conjugate class and $g \in \mathbf{C}$. It follows from Theorem 2.5. that

$$\text{Card}(\mathbf{C})=[\mathbf{G}:\text{Stab}(g)]$$

Since we have

$$o(\mathbf{G}) = p^n$$

Therefore we conclude

$$\text{Card}(\mathbf{C}) = p^m, \qquad m \geq 0$$

The class equation becomes

$$p^n = \sum p^m$$

Since the unit e is with conjugate class $\{e\}$, then one of m's in the above equation is 0. The left hand side of the above equation is divisible by p, therefore there must be another $m = 0$. The corresponding element g will have a singleton conjugate class and hence be in the center. ∎

Theorem 2.14. *Let* **G** *be a p-group of order* p^n. *Then there is a sequence* $\{\mathbf{G}_i\}$, *where* $\mathbf{G}_i$ *is a proper normal subgroup of* **G**, *which satisfies the following conditions,*

$$\mathbf{G} = \mathbf{G}_n \triangleright \mathbf{G}_{n-1} \triangleright \cdots \triangleright \mathbf{G}_{i+1} \triangleright \mathbf{G}_i \triangleright \cdots \triangleright \mathbf{G}_0 = \{e\}$$

Furthermore, $\mathbf{G}_{i+1}/\mathbf{G}_i$ *is a cyclic group of order p.*

Proof. (1) We claim the existence of $\mathbf{G}_1$. Let **C** be the center of **G**. It follows from the preceding theorem that **C** is not the unit subgroup. Let $\mathbf{C}_1$ be the cyclic group generated by any non-unit in **C**. Then we have

$$o(\mathbf{C}_1) \mid o(\mathbf{G})$$

Therefore $o(g)=o(C_1)=p^m$. Let G_1 be the subgroup generated by $g^{p^{m-1}}$. Then G_1 is a normal subgroup of G of order p.

(2) Let us use mathematical induction on the exponent n. If $n = 1$, then we are trivially done. Suppose that our theorem is true for all groups of order p^{n-1}. From (1), we have a normal subgroup G_1. Let G^* be G/G_1. Then $o(G_1) = p^{n-1}$ and there is a sequence as follows

$$\begin{array}{c} G \\ \downarrow \kappa \\ G/G_1 = G^* = G^*_{n-1} \triangleright G^*_{n-2} \triangleright \cdots \triangleright G^*_{i+1} \triangleright G^*_i \triangleright \cdots \triangleright G^*_0 = \{e\} \end{array}$$

where G^*_{i+1}/G^*_i is a cyclic group of order p and κ is the canonical map. Let G_{i+1} be the pre-image of G^*_i under κ. It follows from Theorem 2.7. that G_i is a normal subgroup of G and the following sequence

$$G = G_n \triangleright G_{n-1} \triangleright \cdots \triangleright G_{i+1} \triangleright G_i \triangleright \cdots \triangleright G_0 = \{e\}$$

Furthermore, $G_{i+1}/G_i \simeq G^*_i/G^*_{i-1}$ is a cyclic group of order p. ∎

Discussion

(1) Theorem 2.14. can be rephrased to say that all p-groups are *nilpotent groups*. For the definition of *nilpotent group*, the reader is referred to any book on *group theory*.

(2) The sequence of subgroups is said to be a *composition series*. For a definition and discussion, the reader is referred to the next section.

(3) The theorem will be applied in *field theory*. The reader is referred to Chapter V.

(4) It is easy to see $o(G_i)=p^i$. ∎

The above theorem shows that the numeral $o(G)$ affects the structure of the group G if $o(G)=p^n$. The following three *Sylow theorems*[5] shows that the numeral $o(G)$ affect the structure of the group G in general.

Theorem 2.15. (First Sylow Theorem). *Let* G *be a group with* $o(G)=mp^n$ *where* $p \nmid m$. *Then there is a subgroup* H *of* G *with* $o(H)= p^n$.

Proof: (1) Let S be the set of all subsets of G with cardinality p^n. We claim $p \nmid$ Card(S)=the cardinality of S. Clearly we have

$$\mathrm{Card}(S) = \binom{mp^n}{p^n} = \frac{mp^n}{p^n} \cdots \frac{(mp^n - i)}{(p^n - i)} \cdots \frac{(mp^n - p^n + 1)}{1}$$

It is easy to see that $p^\ell \mid mp^n - i \iff p^\ell \mid i \iff p^\ell \mid p^n - i$. Therefore our claim is proved.

(2) Let the group G act on the set S from the left, namely $g(A) = g * A, \quad \forall A \in S$, then G is a transformation group on S. We have the following **orbit equation**.

$$\mathrm{Card}(S) = \sum \mathrm{Card}(\text{orbit}).$$

[5] Norwegian Mathematician 1832-1918.

Since the left hand side is not divisible by p, then at least one Card(orbit) is not divisible by p. Let the orbit be C and $A \in$ C.

(3) Let **H**=Stab(A). Then **H** is a subgroup of **G**. We claim o(**H**)$\geq p^n$. It follows from Theorem 2.5. that

$$\text{Card(C)=[G:Stab}(A)]=[\mathbf{G}:\mathbf{H}]$$
$$\text{o}(\mathbf{G})=[\mathbf{G}:\mathbf{H}]\ \text{o}(\mathbf{H})$$

Comparing the above equations, we conclude that $p^n \mid$ o(**H**) and o(**H**)$\geq p^n$.

(4) We claim o(**H**)$= p^n$. It suffices to show that o(**H**)$\leq p^n$. Since **H**=Stab(A), then **H** is a transformation group on the set A. For any element $a \in A$, we have

$$\mathbf{H}(a) = \{h * a : h \in \mathbf{H}\} \subset A$$

It is clear that o(**G**)=Card(**H**(a))$\leq$ Card(A)$= p^n$. ∎

Discussion

(1) The above theorem proves the existence of the subgroups of order p^n. Any subgroup of the maximal order p^n will be called a *Sylow p-subgroup* of **G**. ∎

Theorem 2.16. (Second Sylow Theorem). *Let* **G** *be a group with o(***G***)=mp^n where $p \nmid m$. Then all Sylow p-subgroups of* **G** *form an orbit under the inner automorphisms of* **G** *acting on* **G**.

Proof: (1) Let **H** be any Sylow p-subgroup. Let **S**$= \{g_1 * \mathbf{H}, g_2 * \mathbf{H}, \cdots, g_m * \mathbf{H}\}$ be the set of cosets of **G** with one element being **H**, say $g_1 *$**H**=**H**. Let **G** act from the left on **S**. Then there is only one orbit. We claim

$$g * \mathbf{H} * g^{-1} = \text{Stab}(g^*\mathbf{H}), \qquad \forall g \in \mathbf{G}$$

We may assume that $g_i = g$. Note the following

$$\text{Stab}(\mathbf{H})=\mathbf{H}, \qquad [\mathbf{G}:\text{Stab}(g_i*\mathbf{H})] = m$$
$$\text{o}(\text{Stab}(g_i*\mathbf{H})) = p^n, \quad g_i * \mathbf{H} * g_i^{-1} * (g_i * \mathbf{H}) = g_i * \mathbf{H}$$
$$g_i * \mathbf{H} * g_i^{-1} \subset \text{Stab}(g_i*\mathbf{H}), \quad \text{o}(g_i*\mathbf{H} * g_i^{-1}) = \text{o}(\text{Stab}(g_i*\mathbf{H}))$$

Therefore our claim is proved.

(2) Any Sylow p-subgroup **H'** of **G** acts on **S** from the left. Let the stabilizer of g_i***H** under the action of **H'** be denoted by $\overline{\text{Stab}}(g_i * \mathbf{H})$. We have the following orbit equation,

$$\text{Card}(\mathbf{S}) = m = \sum \text{Card(orbit)}$$

Since $p \nmid m$, then there must be an orbit, say C, with cardinality coprime to p. Let g_i***H**$\in$ C. Therefore we have

$$\text{Card(C)}=[\mathbf{H}':\overline{\text{Stab}}(g_i * \mathbf{H})] = p^t, \ 0 \leq t \leq n$$

Therefore $t = 0$ and C is a singleton which shows that H'⊂ Stab(g_i*H). Since the cardinalities of both sides of the preceding inclusion are equal, we conclude,

$$\text{H'}=\text{Stab}(g_i * \text{H}) = g_i * \text{H} * g_i^{-1}$$

∎

Theorem 2.17. (Third Sylow Theorem). *Let G be a group with o(G)=mp^n where $p \nmid m$. Let the number of all Sylow p-subgroups of G be r. Then we have (1) $r \mid m$ (2) $r \equiv 1 (mod\ p)$.*

Proof: (1) Let H be a Sylow p-subgroup of G and T=$\{g * \text{H} * g^{-1} : g \in \text{G}\}$. Let G act on T by conjugating. According to the preceding theorem, T is the set of all Sylow p-subgroups and the only orbit under the conjugating of G. Let the stabilizer of H under this action be Stab*(H). Then it follows from Theorem 2.5. that

$$r = [\text{G}:\text{Stab}^*(\text{H})]$$

It is obvious that Stab*(H)⊃ H. Therefore $p^n \mid$ o(Stab*(H)) and $r \mid m$.

(2) (i) Let us use the notations of the proof of the preceding theorem. Recall that H is a Sylow p-subgroup and S=$\{g_1 * \text{H}, g_2 * \text{H}, \cdots, g_m * \text{H}\}$ is the set of cosets of G, with one element being H, say g_1*H=H. Let G act from the left on S. Then there is only one orbit. We have

$$g_i * \text{H} * g_i^{-1} = \text{Stab}(g_j * \text{H}) \Longleftrightarrow$$
$$g_i * \text{H} * g_i^{-1} * g_j * \text{H} = g_j * \text{H} \Longleftrightarrow$$
$$\text{H} * g_i^{-1} * g_j * \text{H} = g_i^{-1} * g_j * \text{H} \Longleftrightarrow$$
$$\text{H} = \text{Stab}(g_i^{-1} * g_j * \text{H})$$

Therefore there is a bijection between the cosets fixed by H and the cosets fixed by $g_i * \text{H} * g_i^{-1}$. Note that it follows from the preceding theorem that all Sylow p-subgroups are of the form $g_i * \text{H} * g_i^{-1}$. Thus each Sylow p-subgroup fixes the same number k cosets. We have

$$m = r \cdot k$$

(2) (ii) Let us only consider the action of H on S. As pointed in (2) (i), there are k cosets fixed by H. Namely, there are k elements of S with singleton orbits. Let the stabilizer of g_i*H be Stab'(g_i*H). Then we have

$$\text{Card}(\text{Orb}(g_i * \text{H}))=[\text{H}:\text{Stab'}(g_i * \text{H})] = p^q, \quad 0 \leq q \leq n$$

The orbit equation will become

$$r \cdot k = m = k + \alpha \cdot p \equiv k \ (\text{mod}\ p)$$

Since $p \nmid m$, then $p \nmid k$. Let us factor out k from the above equation and get

$$r \equiv 1 \ (\text{mod}\ p)$$

∎

Exercises

(1) Let G be a group with o(G)= 100. Show that there is a normal subgroup with order 25.
(2) Let p, q be prime integers with $p < q, q \not\equiv 1 \pmod{p}$. Let G be a group of order pq. Prove that G is a cyclic group.
(3) Prove that groups of order 15, 35 or 77 must be cyclic.
(4) Let p be a prime integer. Prove that a group of order p^2 must be commutative.
(5) Let p be a prime integer. Prove that a group of order $2p$ must either be a cyclic group or isomorphic to the dihedral group D_p, the symmetric group of a regular p sided polygon.
(6) Suppose that G is a simple group with o(G)< 60. Prove that o(G) is a prime integer.
(7) Find the Sylow 2-subgroups and Sylow 3-subgroups of the symmetric groups S_3, S_4 and S_5.
(8) Let G be a group of order 168. How many elements are there with order 7?
(9) Let n be an odd integer. Prove that any group of order $2n$ must contain a subgroup of order n.
(10) Let G be a group of order mp^n with p prime, $p \nmid m$ and $n > 0$. Let H be a subgroup of G with $p \mid$ o(H). Prove that any Sylow p-subgroup of H is of the form L∩H where L is a suitable Sylow p-subgroup of G.
(11) Prove that every group of order p^2 is abelian.
(12) Let G be a group of order p^2. Find **Aut**(G).
(13) Let p be a prime number. Show that the center of every non-commutative group of order p^3 is isomorphic to $\mathbb{Z}/p\mathbb{Z}$.
(14) Show that every group of order 567 has a normal subgroup of order 27.
(15) Show that every group of order 352 has a normal subgroup of order 16.
(16) Is every group of order 21 cyclic?
(17) Suppose that G is a group of order mn with m, n coprime. Let H and K be normal subgroups of G with o(H)=m and o(K)=n. Prove that $h * k = k * h$ for all $h \in$H and $k \in$K.
(18) Let G be a p-group of order p^n. Suppose that for any $m < n$, there is a unique subgroup of order p^m. Prove G is cyclic.

§7 Jordan-Hölder Theorem

In the first part of this section, we will study the relation between the homomorphisms and the group structure of a group G. Recall Theorem 2.9. (First Isomorphism Theorem). Let N be a normal subgroup of G. Then we have the canonical map κ: G $\to$ G/N. Let H be a subgroup of G. It is easy to see that κ(H)= $\{\kappa(h) : h \in$ H$\}$ is a subgroup of G/N.

The following theorem establish a canonical map between the pre-image $\kappa^{-1}(\kappa(\mathbf{H}))$ and $\kappa(\mathbf{H})$.

Theorem 2.18. (Second Isomorphism Theorem). *Let* N *be a normal subgroup of* G *and let* H *be a subgroup of* G. *Let* κ: $\mathbf{G} \to \mathbf{G}/\mathbf{N}$ *be the canonical map. Then we have*

(1) $\mathbf{H}\cap\mathbf{N}=\mathbf{H}\cap\kappa^{-1}(0) \lhd \mathbf{H}$.

(2) $\mathbf{H}*\mathbf{N}=\{h*n : h \in \mathbf{H}, n \in \mathbf{N}\} = \kappa^{-1}(\kappa(\mathbf{H}))$ *is a subgroup of* G.

(3) $\mathbf{N} \lhd \mathbf{H}*\mathbf{N}$.

Moreover, $\mathbf{H}*\mathbf{N}/\mathbf{N}$ *and* $\mathbf{H}/\mathbf{H}\cap\mathbf{N}$ *are canonically isomorphic to* $\kappa(\mathbf{H})$. *Therefore there is a natural isomorphism* ρ: $\mathbf{H}*\mathbf{N}/\mathbf{N}\to\mathbf{H}/\mathbf{H}\cap\mathbf{N}$ *defined by*

$$\rho(h*n*\mathbf{N}) = h*(\mathbf{H}\cap\mathbf{N})$$

Proof: (1) The proof is routine and left to the reader as an exercise.

(2) We have

$$g \in \kappa^{-1}(\kappa(\mathbf{H})) \Longleftrightarrow \kappa(g) = \kappa(h) \quad \text{for some } h \in \mathbf{H}.$$
$$\Longleftrightarrow h^{-1}*g \in \mathbf{N} \quad \text{for some } h \in \mathbf{H}. \Longleftrightarrow g = h*n \quad \text{for some } h \in \mathbf{H},\ n \in \mathbf{N}.$$

(3) The proof is routine and left to the reader as an exercise.

(4) For the last part, we observe that under the canonical map κ, both $h*n*\mathbf{N}$ and $h*(\mathbf{H}\cap\mathbf{N})$ will be mapped to $\kappa(h)$. ∎

We have the following theorem.

Theorem 2.19. (Third Isomorphism Theorem). *(1) Let* κ *be the canonical map* $\mathbf{G}\to\mathbf{G}/\mathbf{N}=\mathbf{G}'$. *Let* S= *the set of the subgroups of* G *which contains* N, *and* S'= *the set of the subgroups of* G'. *Then we have*

(i) $\mathbf{S}=\{\kappa^{-1}(\mathbf{H}'): H' \in \mathbf{S}'\}$.

(ii) κ *induces a bijective map from* S *to* S' *by* $\kappa(\kappa^{-1}(\mathbf{H}'))=\mathbf{H}'$.

Moreover, $\kappa^{-1}(\mathbf{H}')$ *is a normal subgroup of* G $\Longleftrightarrow$ H' *is a normal subgroup of* G'.

(2) Let κ, κ' *be surjective homomorphisms as in the following diagram,*

$$\begin{array}{ccccc} \mathbf{G} & \xrightarrow{\kappa} & \mathbf{G}' & \xrightarrow{\kappa'} & \mathbf{G}'' \\ {\scriptstyle\bar{\kappa}}\downarrow & & & & \\ \mathbf{G}/\mathbf{K} & & & & \end{array}$$

Let $\ker(\kappa')=\mathbf{K}'$ *and* $\kappa^{-1}(\mathbf{K}')=\mathbf{K}$. *Then the the image,* $\bar{\kappa}(\mathbf{G})$, *of the canonical map* $\bar{\kappa}$: $\mathbf{G} \to \mathbf{G}/\mathbf{K}$ *is isomorphic to* G'. *In other words, we have* $\mathbf{G}/\mathbf{K}\simeq(\mathbf{G}/\mathbf{N})/(\mathbf{K}/\mathbf{N})$.

Proof: (1) Let us assume that H is a subgroup and use Theorem 2.18. Then we have

$$\mathbf{H}\supset\mathbf{N} \Longleftrightarrow \mathbf{H}=\mathbf{H}*\mathbf{N} \Longleftrightarrow \mathbf{H} = \kappa^{-1}\kappa(\mathbf{H})$$

Therefore (i) is proved and (ii) follows trivially. The last part of (1) is a routine checking.
(2) The proof is straight forward and left to the reader as an exercise. ∎

In the remaining part of this section, we will study the topics of the composition series and ***solvable groups***, which play a prominent role in the ***Galois theory*** of solving equations (Cf Chapter V). Historically, the study of solvable groups established the significance of group theory. For this purpose, we will define

Definition 2.17. *Let G be a group. A sequence of **distinct** subgroups* $G=G_n \triangleright G_{n-1} \triangleright \cdots \triangleright G_i \triangleright \cdots \triangleright G_0 = \{e\}$ *is called a **normal series** of G. The set of the quotient groups* $\{G_{i+1}/G_i$*: i=0,*$\cdots$*,n-1*$\}$ *is called the set of the quotient groups **belongs** to the normal series* $G=G_n \triangleright G_{n-1} \triangleright \cdots \triangleright G_i \triangleright \cdots \triangleright G_0 = \{e\}$. *Let* $G= G^*_m \triangleright G^*_{m-1} \triangleright \cdots \triangleright G^*_i \triangleright \cdots \triangleright G^*_0 = \{e\}$ *be a second normal series of G, if every* G_i *appears in the second normal series, then we say the second normal series is a **refinement** of the first normal series. If a normal series has no refinement other than itself, then it is called a **composition series**.* ∎

Example 17. *(1) Let* $G=\mathbb{Z}\oplus\mathbb{Z}\oplus\mathbb{Z}$, $N=\mathbb{Z}\oplus 0\oplus 0$ *and* $H=\{(x,x,0) : x \in \mathbb{Z}\}$. *Let* κ *be the canonical map:* $G\rightarrow G/N$. *Then it is easy to see that* $H+N=\mathbb{Z}\oplus\mathbb{Z}\oplus 0$ *and* $H\cap N=(0,0,0)$. *The map* ρ *in Theorem 2.18. is simply* $\rho((x,y,0)) = (x,x,0)$.

(2) Let $G=\mathbb{Z}\oplus\mathbb{Z}\oplus\mathbb{Z}$, $N=\mathbb{Z}\oplus 0\oplus 0$, $G'=\mathbb{Z}\oplus\mathbb{Z}$, $G''=\mathbb{Z}$, *and* κ, κ' *be defined as* $\kappa((x,y,z)) = (y,z), \kappa'((y,z)) = (z)$. *Then* $K'=\mathbb{Z}\oplus 0$ *and* $K= \mathbb{Z}\oplus\mathbb{Z}\oplus 0$. *All statements in Theorem 2.19. can be verified easily.*

(3) Let $G=G_n \triangleright G_{n-1} \triangleright \cdots \triangleright G_i \triangleright \cdots \triangleright G_0 = \{e\}$ *be a normal series of G. If every quotient group* G_{i+1}/G_i *is a cyclic group of prime order, then the normal series is a composition series.*

(4) In Theorem 2.14., we showed that for any p-group of order p^n, *there is a sequence of distinct subgroups* $G=G_n \triangleright G_{n-1} \triangleright \cdots \triangleright G_i \triangleright \cdots \triangleright G_0 = \{e\}$ *with* $o(G_{i+1}/G_i)=p$. *Therefore all members of the set of quotient groups are isomorphic. This is a composition series for G.*

(5) Let us consider the additive group $G=\mathbb{Z}/6\mathbb{Z}$. *Let* $G_2 = 2\mathbb{Z}/6\mathbb{Z}$ *and* $G_3=3\mathbb{Z}/6\mathbb{Z}$ *be two subgroups of G. Then there are the following three normal series in G,*

$$G \triangleright \{e\}, \qquad G \triangleright G_2 \triangleright \{e\}, \qquad G \triangleright G_3 \triangleright \{e\}$$

where the last two are composition series with the sets of quotient groups $\{G/G_2, G_2/\{e\}\}$ *and* $\{G/G_3, G_3/\{e\}\}$ *respectively. Note that those two sets of quotient groups are isomorphic to* $\{\mathbb{Z}/2\mathbb{Z}, \mathbb{Z}/3\mathbb{Z}\}$ *and* $\{\mathbb{Z}/3\mathbb{Z}, \mathbb{Z}/2\mathbb{Z}\}$ *respectively. In a way, the existence of those two composition series is due to the two ways of factorization of* $6 = 3\cdot 2 = 2\cdot 3$. ∎

In fact, the above Example 17. (5) indicates a general phenomenon. We will first state the main theorem as follows, and then prove it after we established the lemma and Schreier's theorem.

Theorem 2.20. (Jordan[6]-Hölder[7] Theorem). *Suppose that a group G has a composition series. Then there is a bijective map between the sets of quotient groups of any two composition series such that the corresponding quotient groups are isomorphic.*

We shall establish the following lemma.

Lemma. *Let G be a group with two subgroups* H_1 *and* H_2. *Let* $H_1 \triangleright N_1$ *and* $H_2 \triangleright N_2$. *Then we have the following,*

(1) $N_1 * (H_1 \cap H_2) \triangleright N_1 * (H_1 \cap N_2)$ *and* $N_2 * (H_1 \cap H_2) \triangleright N_2 * (H_1 \cap H_2)$.
(2) $(H_1 \cap H_2) \triangleright (H_1 \cap N_2) * (H_2 \cap N_1)$.
(3) *The following three quotient groups are isomorphic,*

$$N_1 * (H_1 \cap H_2) / N_1 * (H_1 \cap N_2)$$
$$N_2 * (H_1 \cap H_2) / N_2 * (H_1 \cap H_2)$$
$$H_1 \cap H_2 / (H_1 \cap N_2) * (H_2 \cap N_1)$$

Proof: (1) It is easy to see that $H_1 \cap H_2$ and $H_1 \cap N_2$ are subgroups of H_1 and $H_1 \cap H_2 \triangleright H_1 \cap N_2$. It follows from Theorem 2.18. that both $N_1 * (H_1 \cap H_2) and N_1 * (H_1 \cap N_2)$ are subgroups of H_1. It is a direct checking and left to the reader to show

$$N_1 * (H_1 \cap H_2) \triangleright N_1 * (H_1 \cap N_2)$$
$$N_2 * (H_1 \cap H_2) \triangleright N_2 * (H_1 \cap H_2).$$

(2) It follows from (1) that $H_1 \cap H_2 \triangleright H_1 \cap N_2$ and $H_1 \cap H_2 \triangleright H_2 \cap N_1$. It is routine to show $(H_1 \cap H_2) \triangleright (H_1 \cap N_2) * (H_2 \cap N_1)$.

(3) We will show that the two quotient groups are isomorphic,

$$N_1 * (H_1 \cap H_2) / N_1 * (H_1 \cap N_2), \quad H_1 \cap H_2 / (H_1 \cap N_2) * (H_2 \cap N_1).$$

By symmetry, our claim (3) will then be proved.

Let us define a map $\rho : N_1 * (H_1 \cap H_2) \rightarrow H_1 \cap H_2 / (H_1 \cap N_2) * (H_2 \cap N_1)$ as follows,

$$\rho(n_1 * h) = h * (H_1 \cap N_2) * (H_2 \cap N_1), \quad \text{for } n_1 \in N_1, h \in H_1 \cap H_2$$

It is routine to check ρ is a *well-defined surjective homomorphism.* We claim that $\ker(\rho) = \rho^{-1}(e) = N_1 * (H_1 \cap N_2)$.

[6] French Mathematician 1838-1922.
[7] German Mathematician 1859-1937.

It is easy to see $\ker(\rho) \supset N_1 * (H_1 \cap N_2)$. Let us prove the other inclusion. Let $n_1 * h \in \ker(\rho)$, namely,

$$\begin{aligned}
&\rho(n_1 * h) = h * (H_1 \cap N_2) * (H_2 \cap N_1) = (H_1 \cap N_2) * (H_2 \cap N_1) \\
\Longrightarrow & h \in (H_1 \cap N_2) * (H_2 \cap N_1) \\
\Longrightarrow & h = h_1 * h_2, \text{ where } h_1 \in (H_1 \cap N_2), h_2 \in (H_2 \cap N_1) \subset N_1 \\
\Longrightarrow & h \in N_1 * (H_1 \cap N_2) \\
\Longrightarrow & n_1 * h \in N_1 * (H_1 \cap N_2)
\end{aligned}$$

Therefore we have $\ker(\rho) = N_1 * (H_1 \cap N_2)$. It follows from Corollary of Theorem 2.9. that the two quotient groups $N_1 * (H_1 \cap H_2) / N_1 * (H_1 \cap N_2)$, and $H_1 \cap H_2 / (H_1 \cap N_2) * (H_2 \cap N_1)$ are isomorphic. ∎

We have the following theorem,

Theorem 2.21. (Schreier's Theorem). *For any two normal series of a group G, there are refinements such that there is a bijective map of the two sets of the quotient groups of the two refinements with the corresponding quotient groups isomorphic.*

Proof: Let the two normal series of G be given as follows,

$$G = G_n \triangleright G_{n-1} \triangleright \cdots \triangleright G_i \triangleright \cdots \triangleright G_0 = \{e\}$$
$$G = G'_m \triangleright G'_{m-1} \triangleright \cdots \triangleright G'_i \triangleright \cdots \triangleright G'_0 = \{e\}$$

Let

$$G_{ij} = G_{i-1} * (G_i \cap G'_j)$$
$$G'_{ji} = G'_{j-1} * (G'_j \cap G_i)$$

Then we have the following two matrices,

$$\begin{pmatrix} G_{n,m}, & G_{n,m-1}, & \cdots, & G_{n,0} \\ G_{n-1,m}, & G_{n-1,m-1}, & \cdots, & G_{n-1,0} \\ \cdots & \cdots & \cdots & \cdots \\ \cdots & \cdots & \cdots & \cdots \\ G_{0,m}, & G_{0,m-1}, & \cdots, & G_{0,0} \end{pmatrix} \quad \begin{pmatrix} G'_{m,n}, & G'_{m,n-1}, & \cdots, & G'_{m,0} \\ G'_{m-1,n}, & G'_{m-1,n-1}, & \cdots, & G'_{m-1,0} \\ \cdots & \cdots & \cdots & \cdots \\ \cdots & \cdots & \cdots & \cdots \\ G'_{0,n}, & G'_{0,n-1}, & \cdots, & G'_{0,0} \end{pmatrix}$$

Note the following relations between the first column and the last column of either matrix,

$$\begin{aligned}
G_{i,m} &= G_{i-1} * (G_i \cap G'_m) = G_{i-1} * (G_i \cap G) \\
&= G_{i-1} * G_i = G_i = G_i * (G_{i+1} \cap G'_0) \\
&= G_{i+1,0} \\
G'_{i,n} &= G'_{i+1,0}
\end{aligned}$$

It follows from the preceding Lemma that every group in either matrix is a normal subgroup of the preceding group and

$$G_{i,j}/G_{i,j-1} \simeq G'_{i,j}/G'_{i,j-1}$$

Now we simply connect successive rows in one long line in either matrix to form a sequence and then delete any group which is identical with its precedent. We get two refinements of the given normal series which have isomorphic sets of quotient groups. ∎

Proof of **Theorem 2.20. (Jordan-Hölder Theorem).:**

Let two composition series be given for a group G. Since both series have no refinements other than themselves, Theorem 2.20. follows from Schreier's theorem. ∎

Discussion

(1) In general there may not be any composition series in a group G. For instance, let G be $\mathbb{Z}$ and $\mathbb{Z} \triangleright n_1\mathbb{Z} \triangleright \cdots \triangleright n_m\mathbb{Z} \triangleright \{0\}$ be a normal series where $n_i \mid n_{i+1}$. Then more normal subgroups can be inserted between $n_m\mathbb{Z}$ and $\{0\}$.

(2) The unique factorization theorem of integers follows from the Jordan-Hölder Theorem. Let $n = \prod_{i=1}^{m} p_i$ be a prime factorization of a positive integer n. Then it gives a composition series $\mathbb{Z}/n\mathbb{Z} \triangleright p_1\mathbb{Z}/n\mathbb{Z} \triangleright \cdots \triangleright \prod_{i=1}^{q} p_i\mathbb{Z}/n\mathbb{Z} \triangleright \{0\}$ with the set of quotient groups isomorphic to $\{\mathbb{Z}/p_1\mathbb{Z}, \cdots, \mathbb{Z}/p_q\mathbb{Z}, \cdots, \mathbb{Z}/p_m\mathbb{Z}\}$. Therefore, it follows that any two prime factorizations must be unique up to a reordering of the prime factors. ∎

Definition 2.18. *A group G having no proper normal subgroup, namely the only normal subgroups are G and $\{e\}$, is called a **simple group**.* ∎

Lemma. *A normal series of a group G is a composition series if and only if all quotient groups of this series are simple groups.*

Proof: Let the normal series be $G = G_n \triangleright G_{n-1} \triangleright \cdots \triangleright G_i \triangleright \cdots \triangleright G_0 = \{e\}$. Then it follows from the first isomorphism theorem that the necessary and sufficient condition for the impossibility to insert a normal subgroup between G_{i+1} and G_i is the quotient group G_{i+1}/G_i having no normal subgroup except itself and the unit group. In other words, the quotient group is simple. ∎

Lemma. *If there is a composition series in a group G, then any normal series can be refined to a composition series.*

Proof: It follows from the preceding Lemma that the composition series is with quotient groups simple. It follows from Schreier theorem that the normal series can be refined to a new normal series with quotient groups corresponding to the quotient groups of the composition series which are simple. Therefore the refinement is a composition series. ∎

Definition 2.19. *Suppose that there is a composition series for a group G. Then the length, $\ell(G)$, is defined to be the number of quotient groups of the composition series. Note that any two composition series must have the same number of quotient groups.* ∎

Definition 2.20. *If there is a normal series in a group G with all quotient groups commutative, then the group G is said to be solvable.* ∎

Theorem 2.22. *Let H be a normal subgroup of G. Then G is solvable if and only if both H and G/H are solvable.*

Proof: ($\Longrightarrow$) It follows from Theorem 2.21. that the normal series $G \triangleright H \triangleright \{e\}$ can be refined to $G{=}G_n \triangleright G_{n-1} \triangleright \cdots \triangleright G_i\ ({=}H) \triangleright \cdots \triangleright G_0 = \{e\}$ with all quotient groups commutative. It follows from the first isomorphism theorem that

(1) $$(\text{bf } G_j/H)/(G_{j-1}/H) \simeq \text{bf } G_j/G_{j-1}, \quad j = n, \cdots, i+1$$

Therefore, we have the following two normal series with quotient groups commutative,

(2) $$H{=}G_i \triangleright G_{i-1} \triangleright \cdots \triangleright \{e\}$$
(3) $$G/H{=}G_n/H \triangleright G_{n-1}/H \triangleright \cdots \triangleright G_i/H\ ({=}H/H)$$

($\Longleftarrow$) On the other hand if we are given normal series (2) & (3) with all quotient groups commutative, then we have the following normal series,

$$G{=}G_n \triangleright G_{n-1} \triangleright \cdots \triangleright G_i\ ({=}H) \triangleright \cdots \triangleright G_0 = \{e\}$$

It follows from equation (1) above that all quotient groups are commutative. ∎

Exercises

(1) Find all composition series for D_4.
(2) Find all composition series for S_3 and S_4.
(3) Find all composition series for any finite rotation group of $\mathbf{R}^3$.
(4) Prove that if order of G is 30, then G is not simple.
(5) Let p, q be primes. Prove that any group of order p^2q is solvable.
(6) Let H be a cyclic normal subgroup of a group G with G/H cyclic. Is G cyclic?
(7) Let G be a group of order 825. Prove that G is solvable.
(8) Let G be a finite group. Show that G is solvable if and only of there is a composition series $G{=}G_n \triangleright G_{n-1} \triangleright \cdots \triangleright G_i \triangleright \cdots \triangleright G_0 = \{e\}$ with G_i/G_{i-1} cyclic group of prime order for all i.
(9) Let $G^{(1)}$ be the commutator subgroup (Cf Exercise (11) §3) or the first derived group of G. Inductively, we define the n-th derived group of G, $G^{(n)}$, as $G^{(n)} = (G^{(n-1)})^{(1)}$. Prove that G is solvable if and only if for some k we have $G^{(k)} = \{e\}$.
(10) Prove that no group of order 72 is simple.

(11) Prove that no group of order 80 is simple.
(12) Let G be a solvable group of order 30. Find the set of quotient groups of a composition series of G.
(13) Let G be a simple group of order 168. Find the number of elements of order 7.
(14) Describe , up to isomorphism, all commutative group of order 72.
(15) Prove that a group of order 12 is solvable.
(16) Prove that a group of order 105 is solvable.

---

§8 Symmetric Group S_n

We come back to study one of the concrete groups $\mathbf{S}_n$ (Cf Example 2.). We have the following lemma.

Lemma. *Let G be a faithful transformation group on n digits $\{1, 2, \cdots, n\}$. Then there is an injective homomorphism ρ: $\mathbf{G} \to \mathbf{S}_n$ with $\rho(g)(i) = g(i)$ for all $g \in \mathbf{G}$.*

Proof: The group G is a set of bijective maps of the n digits while $\mathbf{S}_n$ is the set of *all* bijective maps. ∎

Let G be any finite group of order n. Let G act from the left on the set S. Then G is a faithful transformation group on the set G. Therefore we have the following,

Corollary. *Let G be a finite group of order n. Then there is a injective homomorphism ρ: $\mathbf{G} \to \mathbf{S}_n$. In other words, G may be consider as a subgroup of $\mathbf{S}_n$.* ∎

Theorem 2.23. *The group $\mathbf{S}_n$ is generated by a set of 2-cycles $\{(1,2), (1,3), \cdots, (1,n)\}$.*

Proof: It follows from the definition that we may consider $\mathbf{S}_{n-1} \subset \mathbf{S}_n$. In fact, $\mathbf{S}_{n-1} = \{\rho : \rho \in \mathbf{S}_n, \rho(n) = n\}$. If $n = 1$, then $\mathbf{S}_1$ is the unit group. We usually use the convention that the empty set generates the unit group. Therefore our theorem is true in this case. It is obvious that $\mathbf{S}_2$ is generated by $\{(1,2)\}$. Let us consider the case $n \geq 3$. Let us use mathematical induction and assume that the theorem is true for $n-1$. Let $\rho \in \mathbf{S}_n$. There are the following two possibilities,

$$\rho(n) = \begin{cases} n \\ i \neq n \end{cases}$$

In the first case, $\rho \in \mathbf{S}_{n-1}$ and can be written as a product of 2-cycles in $\{(1,2), (1,3), \cdots, (1, n-1)\}$. Therefore we are done. In the second case, let

$$\sigma = (1,n) * (1,i) * \rho, \qquad \text{or } \rho = (1,i) * (1,n) * \sigma$$

Then we have

$$\sigma(n) = (1,n) * (1,i) * \rho(n) = (1,n) * (1,i)(i) = (1,n)(1) = n$$

Therefore, $\sigma \in \mathbf{S}_{n-1}$ and can be written as a product of 2-cycles in $\{(1,2),(1,3),\cdots,(1,n-1)\}$ and ρ can be written as a product of 2-cycles in $\{(1,2),(1,3),\cdots,(1,n)\}$. ∎

The following discussion will useful in the theory of equations.

Definition 2.21. *(1) Let $X_1, X_2, \cdots, X_n$ be n variables and*

$$f = \prod_{n \geq i > j \geq 1} (X_i - X_j)$$

For any $\sigma \in \mathbf{S}_n$, let us define $\sigma(f)$ as follows,

$$\sigma(f) = \prod_{n \geq i > j \geq 1} (X_{\sigma(i)} - X_{\sigma(j)})$$

*It is elementary to see $\sigma(f) = \pm f$. If $\sigma(f) = f$, then we say σ is an **even permutation**. If $\sigma(f) = -f$, then we say σ is an **odd permutation**.*

*(2) The set of all even permutations of $\mathbf{S}_n$ is called the **alternating group** $\mathbf{A}_n$.* ∎

It is easy to see the following rules for multiplication,

$$\begin{aligned}
(\text{odd permutation}) * (\text{odd permutation}) &= (\text{even permutation}) \\
(\text{odd permutation}) * (\text{even permutation}) &= (\text{odd permutation}) \\
(\text{even permutation}) * (\text{odd permutation}) &= (\text{odd permutation}) \\
(\text{even permutation}) * (\text{even permutation}) &= (\text{even permutation})
\end{aligned}$$

We have the following theorem.

Theorem 2.24. *The alternating group $\mathbf{A}_n$ is a normal subgroup of $\mathbf{S}_n$. $[\mathbf{S}_n : \mathbf{A}_n] = 2$.*

Proof: It is trivial to check that the identity permutation e is an even permutation and it follows from the above rules that the inverse of an even permutation is even. Therefore $\mathbf{A}_n$ is a subgroup of $\mathbf{S}_n$.

It is evident that $(1,2)$ is an odd permutation. Let ρ be any odd permutation. Then $(1,2) * \rho$ is an even permutation. Therefore we have

(1) $$\mathbf{S}_n = \mathbf{A}_n \cup \rho * \mathbf{A}_n$$

Thus we conclude $[\mathbf{S}_n : \mathbf{A}_n] = 2$.

Let us consider the right cosets. Then we have

(2) $$\mathbf{S}_n = \mathbf{A}_n \cup \mathbf{A}_n * \rho$$

Since both the equations (1) & (2) are disjoint unions, then we must have

$$\rho * \mathbf{A}_n = \mathbf{A}_n * \rho$$
$$\mathbf{A}_n = \rho * \mathbf{A}_n * \rho^{-1}$$

Therefore $\mathbf{A}_n$ is a normal subgroup of $\mathbf{S}_n$. ∎

Theorem 2.25. *Let $n \geq 3$. The alternating group $\mathbf{A}_n$ is generated by $\{(1,2,3), \cdots, (1,2,i), \cdots, (1,2,n)\}$.*

Proof: The proof is similar to the proof of Theorem 2.23.. Note that $\mathbf{A}_3=\{e, (1,2,3), (1,3,2)\}$ with $(1,3,2) = (1,2,3)^2$. Therefore our theorem is true for $n = 3$. Let us assume that $n \geq 4$. Let us use mathematical induction and assume that the theorem is true for $n-1$. Let $\rho \in \mathbf{A}_n$. There are the following two possibilities,

$$\rho(n) = \begin{cases} n \\ i \neq n \end{cases}$$

In the first case, $\rho \in \mathbf{A}_{n-1}$ and can be written as a product of 3-cycles in $\{(1,2,3), (1,2,4), \cdots, (1,2,n-1)\}$. Therefore we are done. In the second case, let

$$\sigma = (1,2,n) * (1,2,i)^{-1} * \rho, \qquad \text{or } \rho = (1,2,i) * (1,2,n)^{-1} * \sigma$$

Then we have

$$\sigma(n) = (1,2,n) * (1,2,i)^{-1} * \rho(n) = (1,2,n) * (1,2,i)^{-1}(i)$$
$$= (1,2,n)(2) = n$$

Therefore, $\sigma \in \mathbf{A}_{n-1}$ and can be written as a product of 3-cycles in $\{(1,2,3), \cdots, (1,2,n-1)\}$ and ρ can be written as a product of 3-cycles in $\{(1,2,3), \cdots, (1,2,n)\}$. ∎

The following theorem is useful in group theory and equation theory.

Theorem 2.26. *The groups $\mathbf{A}_n$ are simple groups except $n = 4$.*

Proof: (1) For $n \leq 3$, we have $\mathbf{A}_1$, $\mathbf{A}_2$ unit groups and $\mathbf{A}_3$ a group of order 3. It is easy to see that they are simple groups.

(2) From now on we will assume that $n \geq 5$. Let $\mathbf{N}$ be a normal subgroup which is not the unit group $\{e\}$. By Theorem 2.25., it suffices to show that $\mathbf{N} \ni (i,j,k)$ for all $i \neq j \neq k$.

(3) We claim there is at least one $(i,j,k) \in \mathbf{N}$. In other words, let $\rho \in \mathbf{N}$ such that ρ has the shortest total length of all non-singleton orbits, and $\rho \neq e$, then we claim $\rho = (i,j,k)$. Let the following be the orbit form of ρ,

$$\rho = (a_1, \cdots, a_m)(b_1, \cdots, b_\ell) \cdots$$

(3) (i) If the lengths of the orbits, $m, \ell, \cdots$ are not equal, say $\ell > m$, then $\rho^m \in \mathbf{N}$ will fix all $a_1, \cdots, a_m$. It is easy to check $\rho^m \neq e$ and with a shorter total length of all non-singleton orbits, a contradiction. Therefore we conclude that if ρ has more than one non-singleton orbits, then all non-singleton orbits must be of the same length m. Namely,

$$\rho = (a_1, \cdots, a_m)(b_1, \cdots, b_m) \cdots (d_1, \cdots, d_m)$$

(3) (ii) Suppose that the common length is $m > 3$. Let

$$\sigma = (a_3, a_4) * (a_1, a_2) * \rho * (a_1, a_2)^{-1} * (a_3, a_4)^{-1} \in \mathbf{N}$$

Then we have

$$\sigma * \rho(a_1) = a_1, \qquad \sigma * \rho \neq e, \qquad \sigma * \rho \in \mathbf{N}$$

It is easy to check that $\sigma * \rho$ has a shorter total length of all non-singleton orbits, a contradiction.

(3) (iii) Suppose that the common length $m = 3$ and with more than one non-singleton orbits. Let

$$\sigma = (a_3, b_1) * (a_1, a_2) * \rho * (a_1, a_2)^{-1} * (a_3, b_1)^{-1} \in \mathbf{N}$$

Then we have

$$\sigma * \rho(a_1) = a_1, \qquad \sigma * \rho \neq e, \qquad \sigma * \rho \in \mathbf{N}$$

It is easy to check that $\sigma * \rho$ is with a shorter total length of all non-singleton orbits, a contradiction.

(3) (iv) Suppose the common length is $m = 2$. Let

$$\rho = (c_1, c_2)(c_3, c_4) \cdots (c_{2r-1}, c_{2r})$$

We have two cases (a) $r \geq 3$, (b) $r = 2$. Let us show the impossibility of case (a) first. Let

$$\begin{aligned} \sigma &= (c_1, c_2, c_3) * \rho * (c_1, c_2, c_3)^{-1} \\ &= (c_2, c_3)(c_1, c_4)(c_5, c_6) \cdots (c_{2r-1}, c_{2r}) \\ \sigma * \rho &= (c_1, c_3)(c_2, c_4) \in \mathbf{N} \end{aligned}$$

It is obvious that $\sigma * \rho$ has a shorter total length of all non-singleton orbits, a contradiction. Now let us show the impossibility of case (b). Note that we assume $n \geq 5$. Therefore there is a distinct c_5. Let

$$\begin{aligned} \sigma &= (c_1, c_2, c_5) * \rho * (c_1, c_2, c_5)^{-1} = (c_2, c_5)(c_3, c_4) \\ \sigma * \rho &= (c_1, c_5, c_2) \in \mathbf{N} \end{aligned}$$

It is obvious that $\sigma * \rho$ is with a shorter total length of all non-singleton orbits, a contradiction.

Therefore there is one $(i,j,k) \in \mathbf{N}$.

(4) Without losing generality, we may assume $(1,2,3) \in \mathbf{N}$. Then the following computation shows that $(i,2,3) \in \mathbf{N}$,

$$(1,i,j) * (1,2,3) * (1,i,j)^{-1} = (i,2,3)$$

Similarly we may replace 2 by j and 3 by k. Therefore $(i,j,k) \in \mathbf{N}$ for all distinct i,j,k. ∎

Corollary. *If $n = 3$ or $n \geq 5$, then $\mathbf{S}_n \triangleright \mathbf{A}_n \triangleright \{e\}$ is a composition series of $\mathbf{S}_n$.* ∎

The following theorem is a foundation for Galois theory of the solvability of equations.

Theorem 2.27. *The permutation group $\mathbf{S}_n$ is solvable if $n \leq 4$ and non-solvable if $n \geq 5$.*

Proof: (1) The theorem is trivial for $n = 1, 2$. For $n = 3$, $\mathbf{S}_3 \triangleright \mathbf{A}_3 \triangleright \{e\}$ is a composition series of $\mathbf{S}_3$ with quotient groups of orders 2, 3 respectively. Therefore $\mathbf{S}_3$ is solvable. Let us consider $\mathbf{S}_4$. Let $\mathbf{K}$ be the following *Klein 4-group*,

$$\mathbf{K} = \{e, (1,2)(3,4), (1,3)(2,4), (1,4)(2,3)\}$$

It is easy to check that $\mathbf{K}$ is a subgroup and consists of two conjugate classes. Therefore $\mathbf{K}$ is a normal subgroup of $\mathbf{S}_4$ and $\mathbf{A}_4$. We have a normal series $\mathbf{S}_4 \triangleright \mathbf{A}_4 \triangleright \mathbf{K} \triangleright \{e\}$ with commutative quotient groups of orders 2, 3, 4 respectively.

(2) Let us consider tha case $n \geq 5$. It follows from the preceding Lemma that $\mathbf{S}_n \triangleright \mathbf{A}_n \triangleright \{e\}$ is a composition series of $\mathbf{S}_n$. It is easy to see that any normal series of $\mathbf{S}_n$ must be either of length 1 or 2 and with the set of quotient groups either $\{\mathbf{S}_n\}$ or $\{\mathbf{A}_n, \{e\}\}$. It suffices to show $\mathbf{S}_n$ and $\mathbf{A}_n$ are non-commutative. We have,

$$(1,2,3)(2,3,4) \neq (2,3,4)(1,2,3)$$

∎

Exercises

(1) Prove that $\mathbf{A}_n$ is the only subgroup of $\mathbf{S}_n$ of index 2.
(2) Find all subgroups of $\mathbf{S}_4$ which are isomorphic to $\mathbf{S}_3$.
(3) Find an element of $\mathbf{S}_7$ with maximal order.
(4) Find the cardinality of the conjugate class of $(1,2,3,4,5)$ in $\mathbf{S}_5$.
(5) Prove that $\mathbf{S}_n$ is generated by $(1,2)$ and $(1,2,\cdots,n)$.
(6) Let n be an even integer. Prove that $\mathbf{S}_n$ is generated by all n-cycles $(i_1, i_2, \cdots, i_n)$.
(7) Let p be a prime factor of a positive integer n with $p < n$. Show that there does not exist a subgroup $\mathbf{H}$ with $[\mathbf{S}_n{:}\mathbf{H}] = p$.

(8) Let $\mathbf{G}$ be a finite group. Prove that $\mathbf{G}$ has a normal subgroup of index 2 if and only if there is an injective homomorphism τ: $\mathbf{G} \to \mathbf{S}_n$ for some n such that $\tau(\mathbf{G}) \not\subseteq \mathbf{A}_n$.
(9) Is every finite group $\mathbf{G}$ isomorphic to a subgroup of $\mathbf{A}_n$ for some n?
(10) Find the center of $\mathbf{S}_4$.
(11) Show that $\mathbf{S}_5$ does not contain a commutative subgroup of order 20.
(12) Does $\mathbf{S}_6$ contain a subgroup of order 10?
(13) Find all conjugate classes of $\mathbf{S}_5$.
(14) Find the number of conjugates of $(1,2,3)(4,5,6)$ in $\mathbf{S}_6$.
(15) Find Sylow 2-subgroups and Sylow 3-subgroups of $\mathbf{A}_4$.
(16) Show that $\mathbf{A}_4$ has no subgroup of order 6.
(17) Find an element of maximal order in $\mathbf{S}_{10}$.
(18) Consider the scheme as illustrated by following diagram, a person traces from the left to the right on a horizontal line, if the person meets a vertical line, then the person must follow the vertical line until encountering a horizontal line and following the horizontal line to the right. We require that no two vertical lines are at the same horizontal distance. In the following diagram,

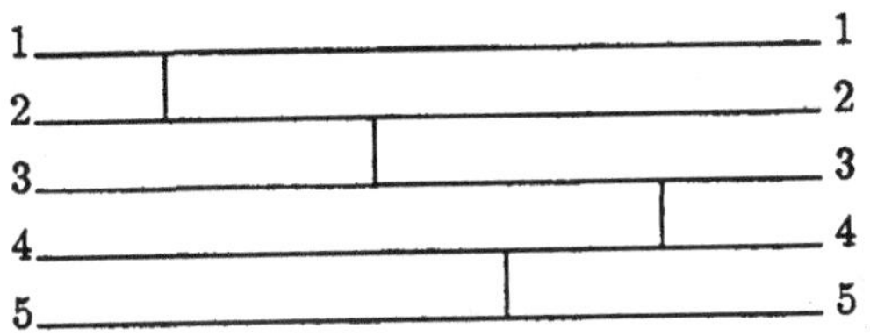

we have $1 \to 4$, $2 \to 1$, $3 \to 2$, $4 \to 5$, $5 \to 3$. Prove that

(i) Any one of the above schemes always produces a bijective map.
(ii) Any bijective map can be represented by one of the above schemes.

CHAPTER III

Polynomials

§1 Fields and Rings

In Arithmetic we use the rational numbers **Q**, in Analytic Geometry, Analysis, Calculus, Topology we use the real numbers **R**, in Complex Analysis and solving real equations we use the complex numbers **C** extensively. In fact many of our mathematical reasonings simply involve the *common* algebraic structure of the sets of numbers **Q**, **R**, **C**. To treat all sets of numbers uniformly, and to crystallize the mathematical reasonings, we abstract the important ingredients of the sets of numbers and define a new concept *field*[1] as follows.

Definition 3.1. *Let* $\mathbf{K}$ *be a set with two operations* $+, \cdot$. *If the following conditions are satisfied,*

(1) *$(\mathbf{K}, +)$ is a commutative group. Customarily we use 0 to denote the additive identity.*
(2) *$(\mathbf{K}^\times, \cdot)$ is a commutative group where $\mathbf{K}^\times = \mathbf{K}\backslash\{0\} = \{k : k \in \mathbf{K}, k \neq 0\}$. Customarily we use 1 to denote the multiplicative identity.*
(3) *Associative law: for any $a, b, c \in \mathbf{K}$, we have*

$$a \cdot (b + c) = a \cdot b + a \cdot c$$
$$(b + c) \cdot a = b \cdot a + c \cdot a$$

then $\mathbf{K}$ *is called a field.* ∎

Discussion

(1) With any field $\mathbf{K}$ as the set of numbers, we may solve a non-trivial linear equation. Namely, let $a \neq 0, b \in \mathbf{K}$, a linear equation as follows with variable x,

$$a \cdot x = b$$

[1] Galois initiated the study of abstract fields.

can be solved uniquely with $x = a^{-1} \cdot b$.

(2) Let **K** be a field. Then (**K**,+) and ($\mathbf{K}^\times$,·) are groups. Therefore, **K** and $\mathbf{K}^\times$ are non-empty sets, $0 \neq 1$, and Card(**K**)≥ 2.

(3) Let **K** be a field. Note that $0 \cdot a, a \cdot 0$ are not obviously defined. However, by the associative law, we have

$$0 \cdot a = (0+0) \cdot a = 0 \cdot a + 0 \cdot a$$
$$0 \cdot a - 0 \cdot a = 0 \cdot a$$
$$0 = 0 \cdot a$$

Similarly, we have $a \cdot 0 = 0$. Thus it is trivial to check that the multiplication, · , is commutative in the whole set **K**.

(4) Let **K** be a field and **F**⊂ **K**. If **F** is a field with respect to the same addition + and multiplication · of **K**, then **F** is said to be a *subfield* of **K**. ∎

Example 1. *(1) The rational numbers* **Q**, *the real numbers* **R**, *the complex numbers* **C** *are fields. The preceding examples of fields have infinitely many elements and are called infinite fields. The integers* **Z** *is not a field.*

(2) We claim that the residue classes $\mathbf{Z}_n$ *is a field if and only if n is a prime number.*

($\Longrightarrow$) If $n = 1$, then $\mathbf{Z}_n$ is a singleton and hence not a field. If $n \geq 2$ and n is not a prime, $n = a \cdot b$ with $a, b > 1$, then $[a]_n, [b]_n \in \mathbf{Z}_n^\times$ while $[a]_n \cdot [b]_n = [0]_n \notin \mathbf{Z}_n^\times$. Therefore $(\mathbf{Z}_n^\times, \cdot)$ is not a group and $\mathbf{Z}_n$ is not a field.

($\Longleftarrow$) Let n be a prime number. Clearly, $[1]_n$ is the unit for $\mathbf{Z}_n^\times$. For any element $[m]_n \in \mathbf{Z}_n^\times$, we must have $n \nmid m$ and there exist b_1, b_2 such that,

$$b_1 \cdot m + b_2 \cdot n = 1$$
$$[b_1]_n \cdot [m]_n = [1]_n$$

Therefore, $[b_1]_n$ is the multiplicative inverse of $[m]_n$. It is then easy to verify that $\mathbf{Z}_n^\times$ is a field. For a prime number p, $\mathbf{Z}_p$ is a field of finitely many elements and is called a finite field. A finite field is also called a Galois field. ∎

Example 2. (Hamming Code. Self-Correcting Message). *A finite field is abstract in nature while very useful in applications. Let us consider an example of self-correcting message. Usually, a message consists of a sequence of 0 and 1 corresponding to the off and on of switches. We shall consider $\{0, 1\} = \mathbf{Z}_2$ the finite field of two elements. Due to the "noise" and human errs, there might be errs in the messages reaching the receivers. To compensate for the mistakes, which happen infrequently, one way is to send every digit twice. For instance, for a message $a_1a_2a_3a_4$, one should send $a_1a_1a_2a_2a_3a_3a_4a_4$. If the pair of digits are identical, then it is safe to assume that there is no err. If a pair takes two values 0, 1, then one knows that there is an err at that place. Even then one can not be sure what is the true value at that place. To correct the possible err, one should send triples. For instance, for a message $a_1a_2a_3a_4$, one should send $a_1a_1a_1a_2a_2a_2a_3a_3a_3a_4a_4a_4$. If a triple of digits are identical, then it is safe to assume that there is no err. If a triple* takes two values 0, 1, then one knows that there is an err at that place. The digit which happen twice is likely to be the true value. This is a self-correcting message with triple workloads. Using the finite field $\mathbb{Z}_2$, we will deduce a easier way to send self-correcting message.

Let us assume that the messages will be sent in a group of four digits. Note that in $\mathbb{Z}_2$ the arithmetic laws are as follows,

$$0+0=0, \quad 0+1=0+1=1, \quad 1+1=0$$
$$0\cdot 0=0, \quad 0\cdot 1=0\cdot 1=1, \quad 1\cdot 1=1$$

Further note that $2^4-1=7$. Let us consider

$$\mathbb{Z}_2^7 = \oplus_{i=1}^7 \mathbb{Z}_2$$

Let us consider the following matrix A,

$$A = \begin{pmatrix} 1, & 1, & 1 \\ 1, & 1, & 0 \\ 1, & 0, & 1 \\ 0, & 1, & 1 \\ 1, & 0, & 0 \\ 0, & 1, & 0 \\ 0, & 0, & 1 \end{pmatrix}$$

Let $v=(c_1, c_2, c_3, c_4, c_5, c_6, c_7)$. We define the multiplication $v \cdot A$ as the usual matrix multiplication, namely,

$$v \cdot A = (u_1, u_2, u_3)$$
$$u_1 = c_1 + c_2 + c_3 + c_5$$
$$u_2 = c_1 + c_2 + c_4 + c_6$$
$$u_1 = c_1 + c_3 + c_4 + c_7$$

Given any message (a_1, a_2, a_3, a_4), we determine auxiliary digits (b_1, b_2, b_3) by the condition $(a_1, a_2, a_3, a_4, b_1, b_2, b_3) \cdot A = (0,0,0)$, namely,

$$-b_1 = a_1 + a_2 + a_3$$
$$-b_2 = a_1 + a_2 + a_4$$
$$-b_3 = a_1 + a_3 + a_4$$

Therefore, one sends out the message $a_1a_2a_3a_4b_1b_2b_3$. The receiver will get the message $v' = (a_1'a_2'a_3'a_4'b_1'b_2'b_3')$ and form the multiplication $v' \cdot A = (u_1', u_2', u_3')$. If $u_1' = u_2' = u_3' = 0$, then there is no err. Otherwise, (u_1', u_2', u_3') must be i-th row of A, and the err appears at the i-th place of the received message. An interchange $0 \rightleftarrows 1$ will correct the received message. For instance, if $(u_1', u_2', u_3') = (1,0,1)=$ 3rd row of A, then a_3 is erred.

The mathematical reason for the above is very simple. Suppose that the err is infrequent, then there is at most one err for the 7 digits message which happens at the i-th place. Therefore, we have

$$\begin{aligned} v' &= v + (0, \cdots, 0, 1, 0, \cdots, 0) \\ v' \cdot A &= v \cdot A + (0, \cdots, 0, 1, 0, \cdots, 0) \cdot A \\ &= (0,0,0) + i\text{-}th \text{ row of } A \\ &= i\text{-}th \text{ row of } A \end{aligned}$$

It is easy to generalize the above method to messages with more digits. For any n, let A be a suitable $n \times (2^n - 1)$ matrix. Let $m = 2^n - 1 - n$. For any message $a_1 a_2 \cdots a_m$, we shall find auxiliary digits $b_1 b_2 \cdots b_n$ such that $(a_1, a_2, \cdots, a_m, b_1, \cdots, b_n) \cdot A = (0, 0, \cdots, 0)$. We may construct a way of sending self-correcting messages as above. For instance, for $n = 4$ then $m = 11$. For $n = 5$, then $m = 26$. Indeed, this method is better than sending triples. ∎

Although many interesting number systems are fields, there are many others which are *not* fields. For instance, $\mathbf{Z}$, $\mathbf{Z}_4$ and $\mathbf{GL(n, R)}$ for $n \geq 2$ are not fields. We have the following definition for a wide category of number systems,

Definition 3.2. *Let* $\mathbf{R}$ *be a set with two operations* $+, \cdot$. *If the following conditions are satisfied,*

1. $(\mathbf{R}, +)$ *is a commutative group. Customarily we use* 0 *to denote the additive identity.*
2. $(\mathbf{R}, \cdot)$ *is associative, namely for any* $a, b, c \in \mathbf{R}$, *we have*

$$a \cdot (b \cdot c) = (a \cdot b) \cdot c$$

   and there is the multiplicative unit 1, *namely* $1 \cdot a = a \cdot 1 = a$ *for all element* $a \in \mathbf{R}$.

3. *Associative law: for any* $a, b, c \in \mathbf{R}$, *we have*

$$\begin{aligned} a \cdot (b + c) &= a \cdot b + a \cdot c \\ (b + c) \cdot a &= b \cdot a + c \cdot a \end{aligned}$$

then $\mathbf{R}$ *is called a* ***ring***[2]. *Furthermore if the multiplication,* $\cdot$ *,is commutative, then it is called a* ***commutative ring***. ∎

Discussion

(1) In some books, the definition of ring in some book may differ from the above in that it is not required the existence of the multiplicative unit 1. In this case the ring in our definition will be called ***a ring with unit***. We shall use the term ***ring without unit*** to

[2] The concepts of 'ring' and 'ideal' (cf §5) were due to German Mathematician Kummer 1810-1893.

indicate an algebraic object satisfying all conditions of our definition of ring except the existence of multiplicative unit. For instance, $2\mathbf{Z}$ is a ring without unit.

(2) With any ring $\mathbf{R}$ as the set of numbers, we may not solve a non-trivial linear equation. For instance, let $\mathbf{R} = \mathbf{Z}_4$ and $a = [2]_4, b = [3]_4$. Then we can not solve the following linear equation with variable x,

$$a \cdot x = b$$

(3) Similar to the discussion about fields, we have no trouble to prove $0 \cdot a = a \cdot 0 = 0$ for any ring.

(4) In a ring we may have $0 = 1$. Then it is easy to see $a = 1 \cdot a = 0 \cdot a = 0$ for all $a \in \mathbf{R}$. In other words, the ring $\mathbf{R}$ is a singleton and will be called ***null ring***.

(5) Let $\mathbf{R}$ be a commutative ring. If we have

$$a \cdot b = 0, \qquad a \neq 0, \quad b \neq 0$$

then we say that a, b are ***zero-divisors*** of $\mathbf{R}$. If $\mathbf{R}$ is not a null ring and without zero-divisors, then $\mathbf{R}$ is called an ***integral domain***.

(6) Let $\mathbf{R}$ be a non-null ring. If $a \cdot b = b \cdot a = 1$, then a is said to be a multiplicative inverse of b and b is said to be a multiplicative inverse of a. Note that the multiplicative inverse of a, if it exists, is unique and will be denoted by a^{-1}.

(4) Let $\mathbf{R}$ be a ring and $\mathbf{S} \subset \mathbf{R}$. If $\mathbf{S}$ is a ring with respect to the same addition $+$ and multiplication $\cdot$ of $\mathbf{R}$, then $\mathbf{S}$ is said to be a ***subring*** of $\mathbf{R}$. ∎

Example 3. *(1) Every field is a ring and an integral domain. The fields $\mathbb{Q}, \mathbb{R}, \mathbb{C}$ are all rings and integral domains.*

(2) The integers $\mathbf{Z}$ is an integral domain. The ring $\mathbf{FL(n, R)}$ *is not an integral domain $\iff n \geq 2$.*

(3) The ring $\mathbf{Z}_n$ is an integral domain $\iff$ n is prime $\iff$ $\mathbf{Z}_n$ is a field. If n is not a prime, then there exist $0 < a, b < n$ with $a \cdot b = n$, we have

$$[a]_n \cdot [b]_n = [a \cdot b]_n = [n]_n = [0]_n$$

Therefore, $\mathbf{Z}_n$ is not an integral domain. ∎

Exercises

(1) Find all units in the ring $\mathbf{Z}[\sqrt{2}]$.
(2) Find all units in the ring $\mathbf{Z}[i]$.
(3) Let us use the notations of Example 2.. Find the matrix A for $n = 4, m = 11$. Suppose that the probability of the err at a place is 10^{-5}. Compute the probability of the inapplicability of the method provided by Example 2..

(4) Prove that **FL(n, R)** is a commutative ring.
(5) Prove that **FL(n, R)** is not an integral domain for $n \geq 2$.
(6) The *characteristic of a field* **K**, χ (**K**), is defined as follows

$$\chi(\mathbf{K}) = \begin{cases} n, & \text{if there is a smallest positive integer n with } \sum_{i=1}^{n} 1 = 0 \\ 0, & \text{otherwise} \end{cases}$$

Prove that if χ (**K**)> 0, then it must be a prime.
(7) Let the characteristic of **K**=χ (**K**)= $p > 0$. Then for any $a, b \in \mathbf{K}$, we must have,

$$(a+b)^p = a^p + b^p$$

(8) Let $\mathbf{R}_i$ be rings. We define the *direct sum* $\oplus_i \mathbf{R}_i = \{\{r_i\} : r_i \in \mathbf{R}_i, r_i = 0$ except finitely many i$\}$. We define the summation + and the multiplication $\cdot$ as follows,

$$\{r_i\} + \{s_i\} = \{r_i + s_i\}$$
$$\{r_i\} \cdot \{s_i\} = \{r_i \cdot s_i\}$$

Prove that $\oplus$ $\mathbf{R}_i$ is a ring without unit if there are infinitely many i. Furthermore, if there is more then one $\mathbf{R}_i$, then $\oplus$ $\mathbf{R}_i$ is not an integral domain.
(9) Let $\mathbf{R}_i$ be rings. We define the *direct product* $\prod_i \mathbf{R}_i = \{\{r_i\} : r_i \in \mathbf{R}_i,\}$. We define the summation + and the multiplication $\cdot$ as follows,

$$\{r_i\} + \{s_i\} = \{r_i + s_i\}$$
$$\{r_i\} \cdot \{s_i\} = \{r_i \cdot s_i\}$$

Prove that $\oplus$ $\mathbf{R}_i$ is a ring. Furthermore, if there is more then one $\mathbf{R}_i$, then $\oplus$ $\mathbf{R}_i$ is not an integral domain.
(10) **Semi-group ring.** A set with a binary operation (**G**, $\cdot$) is said to be a *semi-group* if the binary operation is associative and if there is a unit. Let **R** be a ring and **G** be a semi-group. Let the *semi-group ring* **R**[**G**]= $\{\sum r_i g_i : r_i \in \mathbf{R}, g_i \in \mathbf{G}, r_i =$ 0 except finitely many i$\}$. We define the summation + and the multiplication $\cdot$ as follows,

$$\sum r_i g_i + \sum s_i g_i = \sum (r_i + s_i) g_i$$
$$\sum r_i g_i \cdot \sum s_i g_i = \sum \Big(\sum_{g_j \cdot g_k = g_i} (r_j \cdot s_k) \Big) g_i$$

Prove that the above definitions are well-defined, and **R**[**G**] is a ring.
(11) Prove that the commutative law for summation in the definition of ring can be deduced from other conditions in the definition.
(12) Show that if $1 - a \cdot b$ is an invertible element in a ring **R**, then $1 - b \cdot a$ is invertible.
(13) Show that if $a, b, 1 - a \cdot b$ are invertible in a ring **R**, then $a - b^{-1}, a^{-1} - (a - b^{-1})^{-1}$ are invertible.

(14) Prove that the set $\mathbf{Q}[\sqrt[3]{2}] = \{a + b\sqrt[3]{2} + c\sqrt[3]{4} : a, b, c \in \mathbf{Q}\}$ forms a ring with respect to the usual summation and multiplication in the real numbers $\mathbf{R}$. Find the inverse of $-2 + \sqrt[3]{2} + 3\sqrt[3]{4}$.

(15) Show that all continuous functions defined on the closed interval $[0, 1]$ form a ring with respect to the usual summation and multiplication. Fing a zero-divisor in this ring.

(16) Prove that the set of matrices $\mathbf{C}$ forms a field where

$$\mathbf{C} = \left\{ \begin{pmatrix} a, & b \\ -b, & a \end{pmatrix} : a, b \in \mathbf{R} \right\}$$

---

§2 Polynomial Rings and Quotient Fields

The discussions of linear and quadratic polynomials with integer coefficients formed a cornerstone of mathematics in the ancient civilizations of Babylon, Egypt, Greece, China and India. We shall introduce the following definition.

Definition 3.3. *Let* $\mathbf{R}$ *be a ring and* x *be a symbol. We define the polynomial ring of one variable over* $\mathbf{R}= \mathbf{R}[x] = \{a_0 + a_1 x + \cdots + a_m x^m : a_i \in \mathbf{R}, m$ *non-negative integer*$\}$ *where* $x^0 = 1$ *and*

$$\sum r_i x^i + \sum s_i x^i = \sum (r_i + s_i) x^i$$

$$\sum r_i x^i \cdot \sum s_i x^i = \sum (\sum_{j+k=n} (r_j \cdot s_k)) x^n$$

∎

Discussion

(1) It is easy to see that $\mathbf{R}[x]$ is a ring. The unit for summation 0 for $\mathbf{R}$ is the unit for summation for $\mathbf{R}[x]$. The unit for multiplication 1 is the unit for multiplication for $\mathbf{R}[x]$.

(2) Let $\mathbf{Z}_+$ be the semi-group (Cf Exercise 8, §1) of all non-negative integers. Then $\mathbf{R}[x]$ is nothing but the semi-group ring $\mathbf{R}[\mathbf{Z}_+]$.

(3) If $\mathbf{R}$ is a commutative ring, then $\mathbf{R}[x]$ is a commutative ring.

(4) The polynomial ring of n variables[3] $x_1, x_2, \cdots, x_n$ over a ring $\mathbf{R}$. Let $\mathbf{R}^{(1)}= \mathbf{R}[x_1]$, $\mathbf{R}^{(2)}= \mathbf{R}^{(1)}[x_2]$ and inductively $\mathbf{R}^{(n-1)}= \mathbf{R}^{(n-2)}[x_{n-1}]$. We define *The polynomial ring of* n *variables* $x_1, x_2, \cdots, x_n$ over a ring $\mathbf{R}= \mathbf{R}[x_1, x_2, \ldots, x_n] = \mathbf{R}^{(n-1)}[x_n]$. It is easy to see that $\mathbf{R}[x_1, x_2, \ldots, x_n]$ is a ring. In fact, we have

$$\mathbf{R}[x_1, x_2, \ldots, x_n] = \{ \sum_{finite} r_{m_1 m_2 \cdots m_n} x_1^{m_1} x_2^{m_2} \cdots x_n^{m_n} : r_{m_1 m_2 \cdots m_n} \in \mathbf{R},$$

$$m_1, m_2, \cdots, m_n \text{ non-negative integers}\}$$ ∎

[3]It was introduced systematically by Chinese Mathematician Chu Shih-Chieh in 1303, and French Mathematician Bezout 1730-1783.

Example 4. (Solving a system of simultaneous linear equations). *Let n simultaneous linear equations in n variables be given as follows,*

$$r_{11}x_1 + r_{12}x_2 + \cdots + r_{1n}x_n = c_1$$
$$r_{21}x_1 + r_{22}x_2 + \cdots + r_{2n}x_n = c_2$$
$$\cdots\cdots\cdots$$
$$\cdots\cdots\cdots$$
$$r_{n1}x_1 + r_{n2}x_2 + \cdots + r_{nn}x_n = c_n$$

where $r_{ij} \in \mathbf{R}$ a ring. It is easy to see that the methods we learned in high school about solving the above system depended only on the arithmetic laws obeyed by the summation $+$ and multiplication $\cdot$. Therefore, for any ring $\mathbf{R}$, we have the same rules. For instance, if the following coefficient determinant is invertible,

$$\det \begin{pmatrix} r_{11}, & r_{12}, & \cdots & r_{1n} \\ r_{21}, & r_{22}, & \cdots & r_{2n} \\ \cdots, & \cdots, & \cdots & \cdots \\ \cdots, & \cdots, & \cdots & \cdots \\ r_{n1}, & r_{n2}, & \cdots & r_{nn} \end{pmatrix}$$

then the values of the variables are uniquely determined by the quotients of two determinants as in Cramer's rule. ∎

Example 5. (Affine Geometry). *(1) Let us consider the case $\mathbf{R} = \mathsf{R}$ the set of real numbers. It is easy to see that $\mathsf{R}[x] = \mathsf{R}[x-a]$, namely any polynomial in x is a polynomial in $(x-a)$ and vice versa. If we write the ring with the variable x, then it means we select the origin of the coordinate system as represented by $x = 0$. If we use another variable $x - a$, then we make a translation on the real line to move the origin from 0 to a.*

(2) Let us consider the case $\mathbf{R} = \mathsf{R}$ the set of real numbers. It is easy to see that $\mathsf{R}[x_1, x_2] = \mathsf{R}[x_1 - a, x_2 - b]$. If we write the ring with the variables x_1, x_2, then it means we select the origin of the coordinate system as represented by $x_1 = 0, x_2 = 0$. If we use another variable $x_1 - a, x_2 - b$, then we make a translation on the real line to move the origin from $(0,0)$ to (a,b).

(3) Let us consider the case $\mathbf{R} = \mathsf{R}$ the set of real numbers. Let

$$\begin{cases} y_1 = x_1 \\ y_2 = x_2 + x_1^n \end{cases} \qquad \begin{cases} x_1 = y_1 \\ x_2 = y_2 - y_1^n \end{cases}$$

Then it is easy to see $\mathsf{R}[x_1, x_2] = \mathsf{R}[y_1, y_2]$. Let us use the coordinates represented by $x_1 = c, x_2 = d$ as the usual flat coordinate system. Then the new coordinate system represented by $y_1 = c, y_2 = d$ consists of all vertical lines and some curves.

(4) Given the affine space $\mathbf{R}^n$, *we may consider the ring* $\mathbf{A}$ *of all polynomial functions on* $\mathbf{R}^n$. *We may write* $\mathbf{A} = \mathbf{R}[x_1, x_2, \cdots, x_n]$ *which means we select the origin and a coordinate system. If we write* $\mathbf{R}[x_1, x_2, \cdots, x_n] = \mathbf{R}[y_1, y_2, \cdots, y_n]$, *then we are given two coordinate systems of* $\mathbf{R}^n$ *and their relations.*

(5) The problem of finding all $y_1, y_2, \cdots, y_n$ *such that* $\mathbf{R}[x_1, x_2, \cdots, x_n] = \mathbf{R}[y_1, y_2, \cdots, y_n]$ *for* $n \geq 3$ *is a hard and interesting mathematical problem.*

(6) In fact, we may replace the real field $\mathbf{R}$ *by any field* $\mathbf{F}$ *and consider the* n*-dimensional affine space* $\mathbf{F}^n$ *over* $\mathbf{F}$. *In the same way, we have the polynomial ring* $\mathbf{F}[x_1, x_2, \cdots, x_n]$ *of* n *variables over* $\mathbf{F}$. *Especially, we are interested in the case* $\mathbf{F} = \mathbf{C}$. *Then we get complex affine geometry.* ∎

Now we shall study an arithmetic theory of rings. By this we mean to associate non-negative integers to polynomials as follows,

Definition 3.4. *Let* $\mathbf{R}[x]$ *be a polynomial ring of one variable over a ring* $\mathbf{R}$. *Let us define degree of a polynomial* $f(x)$, $deg(f)$, *as*

$$deg(f) = \begin{cases} -\infty, & \text{if } f=0 \\ \max\{i : f(x) = \sum_j a_j x^j, a_i \neq 0\}, & \text{if } f \neq 0 \end{cases}$$

Let us define order of a polynomial $f(x)$ *at* $x = 0$, $ord_x(f)$, *as*

$$ord_x(f) = \begin{cases} \infty, & \text{if } f=0 \\ \min\{i : f(x) = \sum_j a_j x^j, a_i \neq 0\}, & \text{if } f \neq 0 \end{cases}$$

∎

Discussion

(1) In the polynomial ring of *one* variable over a ring $\mathbf{R}$, the degree def(f) is independent of the variable x. In the case of more than one variable, we may define the degree deg(f) too. For instance, in Example 5. we have

$$\begin{cases} y_1 = x_1 \\ y_2 = x_2 + x_1^n \end{cases} \qquad \begin{cases} x_1 = y_1 \\ x_2 = y_2 - y_1^n \end{cases}$$

Then it is easy to see $\mathbf{R}[x_1, x_2] = \mathbf{R}[y_1, y_2]$. However, we have,

$$\deg_{x_1, x_2}(y_2) = n \neq 1 = \deg_{y_1, y_2}(y_2), \quad \text{if } n \neq 1$$

In other words, in an affine space of dimension ≥ 2, there is no intrinsic concept of linearity. ∎

Theorem 3.1. *Let* $\mathbf{D}$ *be an integral domain. The function deg(*f*) has the following properties,*

(1) *If* $f \neq 0$, *then* deg(f) ≥ 0.
(2) deg($f(x)g(x)$) = deg($f(x)$)+deg($g(x)$).
(3) deg($f(x) + g(x)$) $\leq$ max(deg($f(x)$),deg($g(x)$)).

Similarly, the function $ord_x(f)$ has the following properties,

(1)' *If $f \neq 0$, then $ord_x(f) \geq 0$.*
(2)' $ord_x(f(x)g(x)) = ord_x(f(x)) + ord_x(g(x))$.
(3)' $ord_x(f(x) + g(x)) \geq \min(ord_x(f(x)), ord_x(g(x)))$.

Proof: The theorem can be proved by direct checking. Let us only prove (2). Let f(x), g(x) be expressed as follows,

$$f(x) = a_0 + a_1x + a_2x^2 + \cdots + a_nx^n, \quad a_n \neq 0, \quad deg(f) = n$$
$$g(x) = b_0 + b_1x + b_2x^2 + \cdots + b_nx^n, \quad b_m \neq 0, \quad deg(g) = m$$

Then we have,

$$f(x)g(x) = a_0b_0 + (a_1b_0 + a_0b_1)x + \cdots + a_nb_mx^{n+m}$$

Since **D** is an integral domain, then $a_nb_m \neq 0$ and

$$deg(f(x)g(x)) = n + m = deg(f(x)) + deg(g(x)).$$

∎

Discussion

(1) Let **D** be any integral domain. Then **D**[x]= **D**[y] if and only if $y = ax + b$ with a invertible: since $x, y \in$ **D**[x]=**D**[y], then $y = f(x), x = g(y)$ and $x = g(f(x))$. Computing the degrees on the both sides, we conclude that deg($f(x)$)= deg($g(x)$)= 1. Let $y = ax + b, x = cy + d$. Then we have $x = cax + cb + d$. Thus a is invertible. On the other hand, if $y = ax + b$ with a invertible, then we have $x = a^{-1}(y - b)$ and **D**[x]=**D**[y]. ∎

Theorem 3.2. *If* **D** *is an integral domain, then* **D**[x] *is also an integral domain.*

Proof: It follows from the preceding theorem that if f(x), g(X) are non-zero, then deg(f(x) g(X)) ≥ 0, and hence f(X) g(X) $\neq 0$. ∎

The field of rational numbers **Q** can be defined as the set of quotients of integers. We will generalize this method from integers to any integral domain as follows,

Definition 3.5. *Let* **D** *be an integral domain. The quotient field* **K(S)** *of* **D** *is defined as*

$$\mathbf{K(S)} = \left\{\frac{a}{b} : a, b \in \mathbf{D}, b \neq 0, \frac{a}{b} = \frac{c}{d} \iff a \cdot d = b \cdot c\right\}$$

with the summation + and multiplication · defined as

$$\frac{a}{b} + \frac{c}{d} = \frac{a \cdot d + b \cdot c}{b \cdot d}$$
$$\frac{a}{b} \cdot \frac{c}{d} = \frac{a \cdot c}{b \cdot d}$$

∎

Discussion

(1) The set $\mathbf{K(S)}$ may be defined as follows: let us consider the direct product $\mathbf{T} = \mathbf{D} \times \mathbf{D}^\times = \{(a,b) : a, b \in \mathbf{D}, b \neq 0\}$. In $\mathbf{T}$, we define a relation $\sim$ as follows,

$$(a,b) \sim (c,d) \Longleftrightarrow a \cdot d - b \cdot c = 0$$

It is easy to see that $\sim$ is an equivalence relation. Then we *define* $\mathbf{K(S)}$ as the quotient set of $\mathbf{T}$ with respect to $\sim$.

(2) it is not hard to see that the quotient field $\mathbf{K(S)}$ of $\mathbf{D}$ is indeed a field. Note that the unit for summation is

$$\frac{0}{1} = \frac{0}{b}, \qquad \text{for } b \neq 0$$

which may be written as 0. The multiplicative unit is

$$\frac{1}{1} = \frac{b}{b}, \qquad \text{for } b \neq 0$$

which may be written as 1. If $\frac{a}{b}$ is not 0, then a is not zero and the inverse of $\frac{a}{b}$ is $\frac{b}{a}$. ∎

Example 6. (Affine Geometry). *Given the affine space* R^n, *we may consider the ring* $\mathbf{A}$ *of all polynomial functions on* R^n. *We may write* $\mathbf{A} = \mathsf{R}[x_1, x_2, \cdots, x_n]$. *The quotient field* $\mathbf{K(A)}$ *of* $\mathbf{A}$ *is written as* $\mathsf{R}(x_1, x_2, \cdots, x_n)$, *the rational function field of* R^n. *It is easy to see*

$$\mathsf{R}(x_1, \cdots, x_n) = \left\{ \frac{f(x_1, \cdots, x_n)}{g(x_1, \cdots, x_n)} : f, g \in \mathsf{R}[x_1, \cdots, x_n], g \neq 0, \frac{f}{g} = \frac{h}{q} \Longleftrightarrow f \cdot q = g \cdot h \right\}$$

The concept of rational functions is very interesting. (1) Note that a rational function $f(x_1, x_2, \cdots, x_n)/g(x_1, x_2, \cdots, x_n)$ *is not a function on* R^n *in the sense of Analysis. Usually it is not defined over the whole* R^n. *For instance, the rational function* $1/(x_1 - a)$ *is not defined at all points with* $x_1 = a$. *In fact, there is no common* **domain of definition** *for the whole set of rational functions as indicated by the preceding example. On the other hand, rational functions with distinct domains of definition satisfy precisely the requirements of* **sheaf theory** *which produces a rich structure for the rational function theory of geometry.*

(2) Let us consider the n*-dimensional real affine space* R^n. *We may fix a subset* $\mathbf{V}$ *of* R^n *and consider all rational functions* $f(x_1, x_2, \cdot, x_n)/g(x_1, x_2, \cdot, x_n)$ *which are defined over* $\mathbf{V}$. *In this way, we introduce the* **local** *version of rational function theory. Let us consider an extreme case,* $\mathbf{V} = \{a \text{ point}\}$. *Let the point be* $(a_1, a_2, \cdots, a_n)$. *Then the set of rational functions which are defined at this point are those with denominator* $g(x_1, x_2, \cdots, x_n)$ *satisfying* $g(a_1, a_2, \cdots, a_n) \neq 0$. *Abstractly speaking, the set* $\mathbf{M}$ *of all possible denominators*

satisfies $g_1, g_2 \in \mathbf{M} \Longrightarrow g_1 g_2 \in \mathbf{M}$. It turns out this is the right definition for the local theory. See the following definition. ∎

Definition 3.6. *Let $\mathbf{D}$ be an integral domain with quotient field $\mathbf{K(S)}$. A non-empty subset $\mathbf{M}$ of $\mathbf{D}$ is said to be a multiplicative set if the following conditions are satisfied,*

(1) $0 \notin \mathbf{M}$.
(2) $d_1, d_2 \in \mathbf{M} \Longrightarrow d_1 \cdot d_2 \in \mathbf{M}$.

Let $\mathbf{M}$ be a multiplicative set of $\mathbf{D}$. Then the localization of $\mathbf{D}$ with respect to $\mathbf{M}$, $\mathbf{D}_M$, is defined to be $\{s/d : s/d \in \mathbf{K(S)}, d \in \mathbf{M}\}$. ∎

Discussion

(1) Since $\mathbf{D}_M$ is a subring of the quotient field, then it must be an integral domain.

(2) Let $\mathbf{M} = \{f(x_1) : f(a_1) \neq 0\} \in \mathbb{R}[x_1]$. Then $\mathbf{M}$ is a multiplicative set of $\mathbb{R}[x_1]$. It is easy to see that $\mathbb{R}[x_1]_M$ is the set of all rational functions in one variable whose domain of definition contains the point (a_1), namely those rational functions which are defined around the point (a_1).

(3) Let $\mathbf{M} = \{m^n : n \text{ a positive integer}\} \subset \mathbb{Z}$. Then we have

$$\mathbb{Z}_M = \{\frac{a}{m^n} : a, n \in \mathbb{Z}, n \geq 0\}$$

(4) The discussion of the localization of a general ring $\mathbf{R}$ which is not an integral domain will be postponed to the ring theory. In there we will first generalize the quotient fields to the quotient rings. ∎

Exercises

(1) Let $\mathbf{R}$ be a ring. We define the *formal power series ring* of one variable over $\mathbf{R}$ to be

$$\mathbf{R}[[x_1]] = \{\sum_{i=0}^{\infty} a_i x_1^i : a_i \in \mathbf{R}\}$$

with summation and multiplication as follows,

$$(\sum a_i x_1^i) + (\sum b_i x_1^i) = \sum (a_i + b_i) x_1^i$$
$$(\sum a_i x_1^i) \cdot (\sum b_i x_1^i) = \sum (\sum_{j+k=i} a_j \cdot b_k) x_1^i$$

Prove $\mathbf{R}[[x_1]]$ is a ring.

(2) Let $\mathbf{R}$ be a ring and $\mathbf{R}[[x_1]]$ be defined as before. For any $f(x_1) = \sum a_i x_1^i \in \mathbf{R}[[x_1]]$, we define $\operatorname{ord}(f(x_1)) = \min\{i : a_1 \neq 0\}$. Prove that if $\mathbf{R}$ is an integral domain, then we have

$$\operatorname{ord}(f(x_1) \cdot g(x_1)) = \operatorname{ord}(f(x_1)) + \operatorname{ord}(g(x_1))$$

moreover, $\mathbf{R}[[x_1]]$ is an integral domain.

(3) Prove that $f(x_1) = \sum a_i x_1^i \in \mathbf{R}[[x_1]]$ is invertible $\Longleftrightarrow$ a_0 is invertible in $\mathbf{R}$. [hint: use the identity $1/(1-y) = 1 + y + y^2 + \cdots + y^n + \cdots$].

(4) Let us use the notations of the preceding exercise. Furthermore let $\mathbf{R}$ be a field. Show that $f(x_1)$ is invertible $\Longleftrightarrow$ $\operatorname{ord}(f(x_1))$ is zero.

(5) Let $\mathbf{R}$ be a field. Show that the quotient field of the formal power series ring $\mathbf{R}[[x_1]]$ is the formal meromorphic function field

$$\mathbf{R}((x_1)) = \{\sum_{-m}^{\infty} a_i x_1^i : a_i \in \mathbf{R}, m \text{ non-negative integer}\}$$

(6) Define $\mathbf{R}[[x_1, x_2]] = \mathbf{R}[[x_1]][[x_2]]$. In general, define $\mathbf{R}[[x_1, x_2, \cdots, x_n]] = \mathbf{R}[[x_1, \cdots, x_{n-1}]][[x_n]]$. Show that if $\mathbf{R}$ is an integral domain, then $\mathbf{R}[[x_1, \cdots, x_n]]$ is an integral domain. Furthermore, if $\mathbf{R}$ is a field, then $f \in \mathbf{R}[[x_1, \cdots, x_n]]$ is a unit $\Longleftrightarrow$ f has a non-zero constant term.

(7) Show that $\mathbf{Z}_6[x]$ is not an integral domain.

(8) Let p be a prime number and $M = \{p^n : n \text{ positive integer}\}$. Prove that every proper additive subgroup of $\mathbf{Z}_M/\mathbf{Z}$ is finite, while $\mathbf{Z}_M/\mathbf{Z}$ is not finite.

---

§3 Unique Factorization Theorem for Polynomials

There are many similarities between the integer ring $\mathbf{Z}$ and polynomial rings. One of the cornerstones of the theory of integers is the unique factorization theorem which can be proved for any polynomial ring over a field $\mathbf{k}$. This is the main topic of this section. Let us first establish the Euclidean algorithm for polynomial rings.

Theorem 3.3 (Euclidean algorithm). *Let $\mathbf{R}$ be a commutative ring and $f(x) \neq 0 \in \mathbf{R}[x]$ with*

$$f(x) = a_0 + a_1 x + a_2 x^2 + \cdots + a_n x^n, \qquad a_n \neq 0$$

For any polynomial $g(x) \in \mathbf{R}[x]$ with deg(g(x))=m, let $\ell = max\{0, m - n + 1\}, c = a_n^\ell$. Then there must be $q(x), r(x) \in \mathbf{R}[x]$ such that

$$cg(x) = q(x) \cdot f(x) + r(x), \qquad deg(r(x)) < deg(f(x))$$

Proof: (1) If deg(f(x))=$n = 0$, then $\ell = m + 1$. Our theorem is satisfied with $q(x) = a_0^{\ell-1} g(x), r(x) = 0$.

(2) Suppose that deg(f(x))> 0 and deg(g(x))= $m < n =$ deg(f(x)). Our theorem is satisfied with $q(x) = 0, r(x) = cg(x)$.

(3) Suppose that deg(f(x))> 0 and deg(g(x))= $m \geq n =$ deg(f(x)). We will use mathematical induction on deg(g(x)). Let the expressions of $f(x), g(x)$ be as follows,

$$f(x) = a_0 + a_1x + a_2x^2 + \cdots + a_nx^n, \qquad a_n \neq 0$$
$$g(x) = b_0 + b_1x + b_2x^2 + \cdots + b_mx^n, \qquad b_m \neq 0$$

Let

$$h(x) = a_n - b_m x^{m-n} f(x)$$

Then it is easy to see that deg(h(x))< deg(g(x)). It follows from mathematical induction that there are polynomials $q'(x)$ and $r(x)$ satisfying

$$a_n^{\ell-1}h(x) = q'(x) \cdot f(x) + r(x), \qquad deg(r(x)) < deg(f(x))$$

Therefore we have

$$\begin{aligned} cg(x) &= a_0^{\ell}g(x) = a_0^{\ell-1}(h(x) + b_m x^{m-1} f(x)) \\ &= (q'(x) + b_m x^{m-1})f(x) + r(x) = q(x)f(x) + r(x) \end{aligned}$$

∎

Corollary. *If the ring* **R** *is a field, then in the above theorem, we may take* $c = 1$. *Then the polynomials* $q(x)$ *and* $r(x)$ *are uniquely determined by* $f(x)$ *and* $g(x)$.

Proof: Note that in a field, every non-zero element is invertible. We shall multiply the equation in the theorem by c^{-1} and produce the following

$$g(x) = c^{-1}q(x) \cdot f(x) + c^{-1}r(x), \qquad deg(c^{-1}r(x)) < deg(f(x))$$

Let $q'(x)$ and $r'(x)$ be another pair of polynomials with

$$g(x) = q'(x) \cdot f(x) + r'(x)(= q(x) \cdot f(x) + r(x)), \qquad deg(r'(x)) < deg(f(x))$$

Then we have

$$(q(x) - q'(x)) \cdot f(x) = r'(x) - r(x)$$

If $q(x) - q'(x) \neq 0$, then we get a contradiction as follows,

$$\begin{aligned} deg((q(x) - q'(x)) \cdot f(x)) &= deg(q(x) - q'(x)) + deg(f(x)) \geq deg(f(x)) \\ &> deg(r(x) - r'(x)) = deg((q(x) - q'(x)) \cdot f(x)) \end{aligned}$$

∎

Discussion

(1) Usually, a good theorem indicates a common important property of a category of objects. Therefore, we may enlarge the category to include *all* objects sharing this important property. In other words, we turn this property, this theorem, into a definition. It is usually said that ***a good theorem becomes a good definition*** in mathematics. The Euclidean algorithm is very important, and we turn it into a definition: an integral domain $\mathbf{D}$ is said to be a ***Euclidean domain*** if there is a non-negative integer valued function δ on $\mathbf{D}^* = \mathbf{D}\backslash\{0\}$, we assign $\delta(0) = -\infty$, such that (1) if $a \mid b$, then $\delta(a) \leq \delta(b)$, (2) for any pair of elements $a, b \neq 0$, there exists elements q, r such that $a = bq + r$ and $\delta(r) < \delta(b)$. For instance, we let $\delta(a) = |a|$ if $a \neq 0$ in the ring of integers $\mathbf{Z}$, then it is easy to see that $\mathbf{Z}$ is a Euclidean domain. In the ring of polynomials $\mathbf{K}[x]$ of one variable over a field $\mathbf{K}$, we let $\delta(f(x)) = \deg(f(x))$. Then $\mathbf{K}[x]$ is a Euclidean domain. ∎

Example 7. *(1) Let us consider the polynomial ring $\mathbf{K}[x]$ of one variable over a field $\mathbf{K}$. Using Euclidean algorithm, for a fixed polynomial $f(x)$, we may define the **f(x)-adic expansion** of all polynomials g(x) as follows,*

$$g(x) = \sum_{i=0}^{n} h_i(x)f(x)^i, \qquad deg(h_i(x)) < deg(f(x)).$$

Note that this is just like the decimal expansions of integers and can be proved in the same way.

(2) **Finite difference:**[4] *Let us consider the polynomial ring $\mathbf{R}[x]$ over the real numbers $\mathbf{R}$. Let polynomials $d_n(x)$ be defined as*

$$d_n(x) = \frac{1}{n!}x(x-1)\cdots(x-n+1), \qquad \text{for } n = 1, 2, \cdots$$

Using the Euclidean algorithm, any polynomial $g(x)$ can be expressed as

$$(1) \qquad g(x) = \sum_{i=0}^{n} a_i d_i(x)$$

Furthermore, let us define

$$D^0(g(i)) = g(i)$$
$$D^1(g(i)) = D^0(g(i+1)) - D^0(g(i))$$
$$\cdots$$
$$D^k(g(i)) = D^{k-1}(g(i+1)) - D^{k-1}(g(i))$$

[4] "Ssu Yuan Yu Chien" by Chu Shih-Chieh 1303 A.D.. Finite difference was known as "Chao chha Ssu".

Then the coefficients in the equation (1) $a_i = D^i(g(0))$. *For instance, let* $g(x)$ *be a polynomial of degree 2 with* $g(0) = 2, g(1) = 5, g(2) = 9, g(3) = 14, \cdots$. *Then* $D^1(g(0) = 3, D^1(g(1) = 4, D^1(g(2) = 5, \cdots ; D^2(g(0) = 1, D^2(g(1) = 1, D^2(g(2) = 1, \cdots$; *and*

$$g(x) = 2 + 3x + \frac{1}{2}x(x-1)$$

It is not hard to see that if all coefficients of the equation (1) are integers, then $g(i)$ *will be an integer for any integer* i. *On the other hand if* $g(x) \in \mathbf{Z}[x]$ *is a polynomial with integer values for all* $x \in \mathbf{Z}$, *then all* $D^i(g(0))$ *are integers, hence all* a_i *will be integers.* ∎

Another application of the Euclidean algorithm is the unique factorization property of the integers $\mathbf{K}[x_1, x_2, \cdots, x_n]$. For this purpose, let us introduce,

Definition 3.7. *Let* $\mathbf{D}$ *be an integral domain. Let* α, β, γ *be elements in* $\mathbf{D}$. *If* $\alpha = \beta \cdot \gamma$. *then we say that* α *is a* ***multiple*** *of* β *and* β *is a* ***divisor*** *of* α, *in symbol,* $\beta \mid \alpha$. *If* $\beta \mid \alpha_1, \beta \mid \alpha_2, \cdots, \beta \mid \alpha_n$, *then we say that* β *is a* ***common divisor*** *of* $\alpha_1, \alpha_2, \cdots, \alpha_n$. ∎

Definition 3.8. *Let* $\mathbf{D}$ *be an integral domain. Let* $\delta, \alpha_1, \alpha_2, \cdots, \alpha_n$ *be elements* $\in \mathbf{D}$. *If the following conditions are satisfied,*

(1) $\delta \mid \alpha_1, \delta \mid a_2, \cdots, \delta \mid \alpha_n$.
(2) $\delta_1 \mid \alpha_1, \delta_1 \mid a_2, \cdots, \delta_1 \mid \alpha_n \Longrightarrow \delta_1 \mid \delta$.

then δ *is called a* ***greatest common divisor***, *in symbol, g.c.d.*$(\alpha_1, \alpha_2, \cdots, \alpha_n)$. ∎

Theorem 3.4. *Let* $\mathbf{K}$ *be a field and* $\mathbf{K}[x]$ *be the polynomial ring of one variable over* $\mathbf{K}$. *Let* $\alpha_1, \alpha_2 \in \mathbf{K}[x]$ *and the set* $(\alpha_1, \alpha_2) = \{\beta_1 \cdot \alpha_1 + \beta_2 \cdot \alpha_2 : \beta_i \in \mathbf{K}[x]\}$. *Then there is a polynomial* $\alpha \in \mathbf{K}[x]$ *such that* $(\alpha_1, \alpha_2) = (\alpha) = \{\beta\alpha : \beta \in \mathbf{K}[x]\}$. *Suppose that one of* $\alpha_1, \alpha_2 \in \mathbf{K}[x]$ *is non-zero. Then* α *is a greatest common divisor of* a_1, a_2.

Proof. (1) If $\alpha_1 = 0$, then we have $(\alpha_1, \alpha_2) = (\alpha_2)$ and our theorem is established.

(2) If $\alpha_1 \neq 0$, then let α be an element in (α_1, α_2) with the smallest non-negative degree. It is easy to see $(\alpha) \subset (\alpha_1, \alpha_2)$. We claim $(\alpha) \supset (\alpha_1, \alpha_2)$. Applying the Euclidean algorithm to the pair α, α_1, there exist η_1 and γ_1 with

$$\alpha_1 = \eta_1 \cdot \alpha + \gamma_1, \qquad deg(\gamma_1) < deg(\alpha)$$

If $\gamma_1 \neq 0$, then we get

$$\gamma_1 = \alpha_1 - \eta_1 \cdot \alpha = (1 - \beta_1 \cdot \eta_1)\alpha_1 + (-\beta_2 \cdot \eta_1)\alpha_2 \in (\alpha_1, \alpha_2)$$

Note that then γ_1 is in (α_1, α_2) with a degree smaller than deg(α). A contradiction! We conclude that $\gamma_1 = 0$, namely,

$$\alpha \mid \alpha_1$$

Similarly, we can prove

$$\alpha \mid \alpha_2$$

(3) Let α' be another common divisor of α_1, α_2. Then we have

$$\alpha' \mid \alpha_1, \quad \alpha' \mid \alpha_2 \Longrightarrow \alpha' \mid \beta_1 \cdot \alpha_1 + \beta_2 \cdot \alpha_2 = \alpha$$

Therefore α a greatest common divisor of α_1, α_2. ∎

Corollary. *Taking any number of polynomials $\{\alpha_i\} \subset \mathbf{K}[x]$ where $\mathbf{K}$ is a field, let $(\{\alpha_i\})=\{\sum_{finite} \beta_i\alpha_i : \beta_i \in \mathbf{K}[x]\}$. Then there exists $\alpha \in \mathbf{K}[x]$ such that $(\alpha)=\{\beta\alpha : \beta \in \mathbf{K}[x]\}= (\{\alpha_i\})$.*

Proof: Same as before. ∎

Discussion

(1) Let $b \in \mathbf{R}$ a ring. For any polynomial $f(x) \in \mathbf{R}[x]$. Then we have

$$f(x) = q(x)(x-b) + r, \qquad r \in \mathbf{R}$$

After replacing x by b in the above equation, we have

$$f(b) = r$$

In other words $f(b) = 0 \Longleftrightarrow (x-b) \mid f(x)$.

(2) Given any ring $\mathbf{R}$ and $\{\alpha_i\} \subset \mathbf{R}$, later on (§5) we will call $(\{\alpha_i\})=\{\sum_{finite} \beta_i\alpha_i : \beta_i \in \mathbf{R}\}$ the *ideal* generated by $\{\alpha_i\}$. In fact, every ideal will be generated by a subset of $\mathbf{R}$. If every ideal is generated by one element, then the ring is said to be a *principal ideal ring* or P.I.R.. Furthermore, if it is an integral domain, then it will be called a *principal ideal domain* or P.I.D.. It is not hard to reformulate the above theorem and its corollary as "Every Euclidean domain is a P.I.D." (see below Theorem 3.25.). ∎

Example 8. (Lagrange Interpolation). *Let us consider the Chinese Remainder Theorem for the polynomial ring $\mathbf{K}[x]$ of one variable over a field $\mathbf{K}$. Let a finite subset of n elements, $\{b_i\}$, of $\mathbf{K}$ be given with $b_i \neq b_j$ if $i \neq j$ and another finite set of n elements, $\{a_i\}$, of $\mathbf{K}$ be given. We want to find a polynomial $f(x) \in \mathbf{K}[x]$ such that*

$$f(b_i) = a_i, \qquad \forall\, i$$

Note that all b_i are distinct. Therefore all $(x-b_i)$ are distinct irreducible elements in $\mathbf{K}[x]$ and 1 is a greatest common divisor of $(x-b_i)$ and $\prod_{j\neq i}(x-b_j)$, namely

$$1 = g(x)(x-b_i) + h(x)\prod_{j\neq i}(x-b_j)$$

$$1 - g(x)(x-b_i) = h(x)\prod_{j\neq i}(x-b_j) = \alpha_i(x)$$

Then the polynomial $\alpha_i(x)$ *satisfies*

$$\alpha_i(x) \equiv 1 \ (mod\,(x_i - b_i)), \qquad \alpha_i(x) \equiv 0 \ (mod\,(x_j - b_j)) \quad \forall\, j \neq i$$

In fact, we may take polynomial $\alpha_i(x) = f_i(x)$ *defined as,*

$$f_i(x) = \frac{\prod_{j\neq i}(x - b_j)}{\prod_{j\neq i}(b_i - b_j)}$$

Moreover, the polynomial $f(x)$ *defined as*

$$f(x) = \sum_j a_j f_j(x) = \sum_{j=1}^{n} \frac{a_j}{\prod_{i\neq j}(b_j - b_i)} \frac{\prod_i (x - b_i)}{(x - b_j)}$$

satisfies the required system of congruence relations

$$f(x) \equiv a_i \ (mod\,(x_i - b_i)) \quad \forall\, i \quad i.e, \quad f(b_i) = a_i$$

∎

We will introduce the concepts of the *irreducible elements* and the *prime elements* as follows. Note that in an integral domain **D**, there may be many invertible elements with respect to multiplication.

Definition 3.9. *Let* **D** *be an integral domain,* $\alpha \neq 0$ *and* α *non-invertible. If in any factorization of* $\alpha = \beta \cdot \gamma$, *we must have* β *or* γ *invertible, then* α *is said to be an irreducible element. For any two elements* $\alpha_1, \alpha_2 \in$ **D**, *if there are two elements* $\delta_1, \delta_2 \in$ **D** *with* $\alpha_1 = \delta_2\alpha_2, \alpha_2 = \delta_1\alpha_1$, *then* α_1, α_2 *are said to be associated elements, in symbol,* $\alpha_1 \overset{a}{\sim} \alpha_2$.

∎

Discussion

(1) An invertible element is also called a *unit*. The set of all invertible elements of **D** are usually denoted by $\mathbf{D}^*$. If $\mathbf{D}^*$ is non-empty, then it is a group.

(2) Let **D** be an integral domain. Suppose that $\alpha \in \mathbf{D}[x]$ is invertible. Then there exists a β such that $\alpha\beta = 1$. Considering the degrees on both sides of the preceding equation, we have

$$deg(\alpha\beta) = 0 = deg(\alpha) + deg(\beta)$$

Therefore, we get deg(α)=deg(β)=0 and $\alpha, \beta \in$ **D**. We conclude that all invertible elements $\mathbf{D}[x]^*$ of $\mathbf{D}[x]$ are invertible elements $\mathbf{D}^*$ of **D**. Let $\mathbf{D}=\mathbf{K}[x_1, x_2, \cdots, x_{n-1}]$ where **K** is a field. Then it follows that the set of invertible elements $\mathbf{K}[x_1, x_2, \cdots, x_n]^*$ of $\mathbf{K}[x_1, x_2, \cdots, x_n]$ is $\mathbf{K}^* = \mathbf{K}\backslash\{0\} = \mathbf{K}^\times$.

(3) Let **K** be a field. Then using degree formula, it is easy to see that any linear polynomial $ax + b$ with $a \neq 0$ is irreducible.

(4) The relation $\stackrel{\bullet}{\sim}$ is clearly an equivalence relation.

(5) Let $\alpha_1 \stackrel{\bullet}{\sim} \alpha_2$. If $\alpha_1 = 0$, then it is clear that $\alpha_2 = 0$. If $\alpha_1 = 1$, then $\{\alpha_2 : \alpha_2 \stackrel{\bullet}{\sim} \alpha_1\} = \mathbf{D}^*$. If $\alpha_1 \neq 0$, then the relation $\alpha_1 = \delta_2\alpha_2 = \delta_2\delta_1\alpha_1$ implies $(1-\delta_2\delta_1)\alpha_1$ implies $\delta_2\delta_1 = 1$. Therefore δ_1, δ_2 are invertible.

(6) Let $\alpha_1 \stackrel{\bullet}{\sim} \alpha_2$. Then α_1 is irreducible $\Longleftrightarrow$ α_2 is irreducible. ∎

Definition 3.10. *Let* $\mathbf{D}$ *be an integral domain and* $\alpha \in \mathbf{D}$ *with* $\alpha \neq 0$ *and* α *not invertible. If* $\alpha \mid \beta \cdot \gamma \Longrightarrow \alpha \mid \beta$ *or* $\alpha \mid \gamma$, *then* α *is called a prime element.* ∎

Theorem 3.5. *In every integral domain* $\mathbf{D}$, *every prime element* π *is irreducible.*

Proof: let π be an element which can be written as $\pi = \alpha\beta$. Therefore, we have $\pi \mid \alpha\beta$ and $\pi \mid \alpha$ or $\pi \mid \beta$. We may assume that $\pi \mid \alpha$. Let

$$\alpha = \pi\delta, \qquad \pi = \pi\delta\beta$$
$$0 = (1-\delta\beta)\pi, \qquad \delta\beta = 1$$

Therefore, β is a unit and we are done. ∎

To simplify our notations, let us define

Definition 3.11. *Let* $\mathbf{D}$ *be an integral domain. If* $\mathbf{D}$ *satisfies the following conditions, then it is called a unique factorization domain, or simply called a U.F.D.,*

(1) *Any* $\alpha \neq 0 \in \mathbf{D}$ *can be factored as a product of irreducible elements,* $\alpha = \delta\prod_{i=1}^{n} p_i^{m_i}$ *where* δ *is a unit,* m_i *positive integer,* p_i*'s are non-associative irreducible elements.*

(2) *Let* $\alpha = \delta\prod_{i=1}^{n} p_i^{m_i} = \epsilon\prod_{i=1}^{s} q_i^{r_i}$ *be two factorizations of* α *as in (1). Then there is a reordering of* q_i *such that* $p_i \stackrel{\bullet}{\sim} q_i$ *and* $m_i = r_i$ *for all* i. ∎

Discussion

(1) To prove that unique factorization theorem exists in an integral domain $\mathbf{D}$ is nothing but to prove $\mathbf{D}$ is a unique factorization domain.

(2) In a field $\mathbf{K}$, every non-zero element is invertible, hence $\mathbf{K}$ is a unique factorization domain.

(3) The integral domains $\mathbf{Z}$ and $\mathbf{Z}[i]$ are unique factorization domains.

(4) The following is an integral domain which is not a unique factorization domain. Let $\mathbf{D}=\mathbf{R}[x^2,x^3] = \{\sum a_i x^i : \sum a_i x^i \in \mathbf{R}[x], a_1 = 0\}$. Note that $x \notin \mathbf{D}$, x^2 and x^3 are non-associative. We have the following distinct factorizations of x^6,

$$x^6 = (x^2)^3 = (x^3)^2$$

Therefore, $\mathbf{D}$ is not a unique factorization domain. ∎

The essential factorization properties of an integral domain $\mathbf{D}$ we have to examine are (1) the existence of factorization, (2) the relation between the concept of irreducible elements and prime elements. Our aim is to establish that $\mathbf{K}[x_1, x_2, \ldots, x_n]$ is a U.F.D. where $\mathbf{K}$ is a field. We shall denote $\mathbf{K}[x_1, \cdots, x_{n-1}]$ by $\mathbf{D}$. Inductively we know that $\mathbf{D}$ is a U.F.D.. Therefore it suffices to prove that $\mathbf{D}[x]$ is a U.F.D. as long as $\mathbf{D}$ is a U.F.D.. We have the following theorems.

Theorem 3.6. *Let* $\mathbf{D}$ *be a U.F.D. and* $\alpha_1, \alpha_2, \cdots, \alpha_n \in \mathbf{D}$. *Then there is a greatest common divisor* δ *of* $\alpha_1, \alpha_2, \cdots, \alpha_n$.

Proof: Allowing non-negative exponents n_{ij}, we may take the same set of irreducible factors in the following expressions,

$$\alpha_i = \delta_i \prod_j p_j^{n_{ij}}, \qquad \delta_i \text{ unit, distinct } p_i \text{ non-associative.}$$

Let $m_j = \min\{n_{ij} : \text{ for all } i\}$ and $\delta = \prod_j p_j^{m_j}$. Then it is routine to prove that δ is a greatest common divisor of $\alpha_1, \alpha_2, \cdots, \alpha_n$. ∎

Theorem 3.7. *Let* $\mathbf{D}$ *be an integral domain. We have*

(1) $\mathbf{D}$ *is a U.F.D..* $\Longrightarrow$ *every irreducible element is prime.*

(2) *Every element in* $\mathbf{D}$ *can be expressed by a product of irreducible elements and every irreducible element is prime* $\Longrightarrow$ $\mathbf{D}$ *is a U.F.D..*

Proof: (1) Let π be irreducible with $\pi \mid \alpha\beta$. Then there is an element γ and the following decompositions as products of irreducible elements,

$$\pi\gamma = \alpha\beta, \qquad \alpha = \epsilon_1 \prod \pi_i^{\ell_i}$$
$$\beta = \epsilon_2 \prod \pi_i^{m_i}, \qquad \gamma = \epsilon_3 \prod \pi^{n_i}$$
$$\pi\gamma = \gamma = \epsilon_3 \pi \prod \pi^{n_i} = \alpha\beta = \epsilon_1\epsilon_2 \prod \pi_i^{\ell_i} \prod \pi_i^{m_i}$$

where ϵ_i are units and π, π_i are irreducible. It is easy to conclude that π must be associated with one of π_i which has a non-zero exponent ℓ_i or m_i. Therefore π is a factor of either α or β. Hence π is prime.

(2) Let α have two expressions as products of irreducible elements as follows,

$$(1) \qquad \alpha = \delta \prod \pi_i^{m_i} = \epsilon \prod \chi_i^{n_i}, \qquad m_i, n_i \text{ positive integers.}$$

Then we have

$$\pi_1 \mid a = \epsilon\chi_1(\chi^{n_1 - 1} \prod_{j>1} \chi_j^{n_j})$$

which implies, with every irreducible element prime, the following

$$\pi_1 \mid \chi_1 \text{ or } \pi_1 \mid (\chi^{n_1 - 1} \prod_{j>1} \chi_j^{n_j})$$

If $\pi_1 \nmid \chi_1$, then we have $\pi_1 \mid \chi_2(\chi_2^{n_2-1}\prod_{j>2}\chi_j)$, namely

$$\pi_1 \mid \chi_2 \text{ or } \pi_1 \mid (\prod_{j>2}\chi_j)$$

Step by step, there must be a χ_s with

$$\pi_1 \mid \chi_s$$

while χ_s is irreducible. Thus we have

$$\delta'\pi_1 = \chi_s$$

Now let us consider $\alpha/(\delta'\pi_1 = \alpha/\chi_s = \delta\delta'^{-1}\pi_1^{m_1-1}\prod_{j>1}\pi_j^{m_j} = \epsilon\chi_s^{n_s-1}\prod_{j\neq s}\chi_j^{n_j-1}$. By mathematical induction on the total sum of the exponents of both sides of the expression (1), our theorem is proved.

The following definition and theorem are due to Gauss. They are the essential steps to establish that $\mathbf{K}[x_1, x_2, \cdots, x_n]$ is a U.F.D..

Definition 3.12. *Let* $\mathbf{D}$ *be a U.F.D. and* $0 \neq f(x) = \sum_{i=0}^{n}\alpha_i x^i \in \mathbf{D}[x]$. *Let the content of* $f(x)$, $c(f(x))$, *be defined as the set of greatest common divisors of the coefficients* $\alpha_1, \alpha_2, \cdots, \alpha_n$ *of* $f(x)$. *If* deg$(f(x)) \geq 1$ *and* $c(f(x))$ *is the set of all units, then* $f(x)$ *is said to be a primitive polynomial.* ∎

Lemma. *If* $f(x), g(x)$ *are primitive polynomials, then* $f(x)g(x)$ *is a primitive polynomial.*

Proof: Let

$$f(x) = \alpha_0 + \alpha_1 x^1 + \cdots + \alpha_m x^m, \qquad \alpha_m \neq 0$$
$$g(x) = \beta_0 + \beta_1 x^1 + \cdots + \beta_n x^n, \qquad \beta_n \neq 0$$

Suppose that $f(x)g(x)$ is not a primitive polynomial. Let π be a prime factor of all coefficients of $f(x)g(x)$. Let the integers i, j be defined as follows,

$$i = \min\{\ell : \pi \mid \alpha_r \ \ \forall r > \ell.\}$$
$$j = \min\{\ell : \pi \mid \beta_r \ \ \forall r > \ell.\}$$

Let

$$f(x)g(x) = \gamma_0 + \gamma_1 x^1 + \cdots + \gamma_{m+n} x^{m+n}$$

Then we have the following

$$\begin{aligned}
\gamma_0 &= \alpha_0\beta_0 \\
\gamma_1 &= \alpha_0\beta_1 + \alpha_1\beta_0 \\
&\cdots \\
&\cdots \\
\gamma_{i+j} &= \alpha_i\beta_j + \sum_{r=i+1}^{i+j} \alpha_r\beta_{i+j-r} + \sum_{r=j+1}^{i+j} \beta_r\alpha_{i+j-r} \\
&\cdots \\
&\cdots \\
\gamma_{m+n} &= \alpha_m\beta_n
\end{aligned}$$

Since we already have the following,

$$\pi \mid \gamma_{i+j}, \quad \pi \mid \sum_{r=i+1}^{i+j} \alpha_r\beta_{i+j-r}, \quad \pi \mid \sum_{r=j+1}^{i+j} \beta_r\alpha_{i+j-r}$$

then it follows from the expansion of γ_{i+j} that

$$\pi \mid \alpha_i\beta_j \Longrightarrow \pi \mid \alpha_i \text{ or } \pi \mid \beta_j$$

This contradicts the definition of i or j. ∎

Theorem 3.8. (Gauss Lemma). *Let* $\mathbf{D}$ *be a U.F.D. and* $0 \neq f(x), 0 \neq g(x) \in \mathbf{D}[x]$. *Then we have*

$$c(f(x)g(x)) = c(f(x))c(g(x))$$

Therefore, the product of two primitive polynomials is a primitive polynomial.

Proof: Let $\delta \in c(f(x)g(x))$, $\delta_1 \in c(f(x))$ and $\delta_2 \in c(g(x))$. Let

$$\begin{aligned}
f(x) &= \delta_1 f^*(x) \\
g(x) &= \delta_2 g^*(x) \\
f(x)g(x) &= \delta_1\delta_2 f^*(x)g^*(x)
\end{aligned}$$

It is clear that both $f^*(x), g^*(x)$ are primitive polynomials. It follows from the preceding lemma that $f^*(x)g^*(x)$ is a primitive polynomial. Therefore, $\delta_1\delta_2 \in c(f(x)g(x))$ and $c(f(x)g(x)) \supset c(f(x))c(g(x))$. It is clear that $\delta = \delta_1\delta_2\epsilon$ where ϵ is a unit and $\delta_2\epsilon \in c(g(x))$. Therefore $c(f(x)g(x)) \subset c(f(x))c(g(x))$. ∎

One of the applications of Gauss' lemma is the following existence theorem of factorizations,

Theorem 3.9. *Let* $\mathbf{D}$ *be an integral domain. Then every non-zero polynomial* $f(x) \in \mathbf{D}[x]$ *can be written as a product of irreducible elements in* $\mathbf{D}[x]$.

Proof: (1) We claim that every irreducible element α in $\mathbf{D}$ is an irreducible element in $\mathbf{D}[x]$ and every non-zero element in $\mathbf{D}$ can be written as a product of irreducible elements in $\mathbf{D}[x]$. Let $\alpha = g(x)h(x)$ in $\mathbf{D}[x]$. Comparing the degrees on both sides, we have

$$0 = deg(\alpha) = deg(g(x)) + deg(h(x))$$

Therefore, both $g(x), h(x)$ are of degree 0 and are elements of $\mathbf{D}$. Since α is irreducible in $\mathbf{D}$, then one of $g(x), h(x)$ is a unit in $\mathbf{D}$ and unit in $\mathbf{D}[x]$.

(2) We claim that every primitive polynomial can be written as a product of irreducible elements in $\mathbf{D}[x]$. Let $f(x)$ be a primitive polynomial. If $f(x)$ is irreducible, then we are done. If not, then $f(x) = g(x)h(x)$ with $g(x), h(x)$ non-invertible. It follows from the preceding theorem that

$$c(f(x)) = c(g(x))c(h(x))$$

Therefore, $\deg(g(x)) \geq 1$ and $\deg(h(x)) \geq 1$, and both $g(x), h(x)$ are primitive polynomials. By induction on the degree of the primitive polynomials, we conclude that both $g(x), h(x)$ can be written as product of irreducible elements in $\mathbf{D}[x]$. Henceforth, $f(x)$ can be written as product of irreducible elements of $\mathbf{D}[x]$.

(3) Let $f(x)$ be any non-zero element in $\mathbf{D}[x]$. Then $f(x) = \delta f^*(x)$ where $\delta \in \mathbf{D}$ and $f^*(x)$ is primitive. It follows from (1) and (2) that both δ and $f^*(x)$ can be written as products of irreducible elements in $\mathbf{D}[x]$, therefore $f(x)$ can be written as product of irreducible elements in $\mathbf{D}[x]$. ∎

Recall that our main objective of this section is to establish for any field $\mathbf{K}$, the polynomial ring $\mathbf{K}[x_1, x_2, \cdots, x_n]$ is a U.F.D.. To start our induction process, we will prove the following theorems.

Theorem 3.10. *Let* $\mathbf{K}$ *be a field. Then every irreducible element of* $\mathbf{K}[x]$ *is prime.*

Proof: Let $f(x)$ be an irreducible element of $\mathbf{K}[x]$ and $f(x) \mid g(x)h(x)$. Let us use the notations of Theorem 3.4. and set $(f(x), g(x)) = (\ell(x))$. There are two cases: (1) $\ell(x)$ is a unit, (2) $\ell(x)$ is a non-unit.

(1) In this case, we have $\ell(x) \in \mathbf{K}^* = \mathbf{K} \backslash \{0\}$. Let $\ell = \ell(x)$. Then we have

$$\ell = \alpha(x)f(x) + \beta(x)g(x)$$
$$h(x) = \ell^{-1}\alpha(x)f(x) + \ell^{-1}\beta(x)g(x)h(x)$$

Since $f(x)$ is a divisor of the right hamd side, then $f(x) \mid h(x)$.

(2) We have $f(x) = 1f(x) + 0g(x), g(x) = 0f(x) + 1g(x) \in (f(x), g(x)) = (\ell(x))$ and

$$f(x) = \alpha(x)\ell(x), \qquad g(x) = \beta(x)\ell(x)$$

Since $f(x)$ is irreducible and $\ell(x)$ non-unit, then we must have $\alpha(x)$ a unit, namely $\alpha = \alpha(x) \in \mathbf{K}^*=\mathbf{K}\backslash\{0\}$. Therefore, we have

$$g(x) = \alpha^{-1}\beta(x)f(x)$$

Henceforth, $f(x) \mid g(x)$. ∎

Theorem 3.11. *Let* **K** *be a field. Then* $\mathbf{K}[x]$ *is a U.F.D..*

Proof: We know that **K** is a U.F.D.. It follows from Theorem 3.8. that every non-zero element in $\mathbf{K}[x]$ can be factored into product of irreducible elements. Moreover, it follows from Theorem 3.10. that every irreducible element in $\mathbf{K}[x]$ is prime. Therefore, $\mathbf{K}[x]$ satisfies the two conditions of (2) of Theorem 3.6. and is a U.F.D.. ∎

Discussion

(1) It can be shown directly that Euclidean domain $\Longrightarrow$ P.I.D. $\Longrightarrow$ U.F.D. (see below Theorems 3.25. & 3.29). Therefore, the U.F.D. property of $\mathbf{K}[x]$ follows from the Euclidean algorithm. ∎

Example 9. *Let us prove that a complex affine curve* **C** *defined by the following equation*

(1) $$X^n - Y^n = 1, \qquad n \geq 3$$

can not be parametrized by $X = f(t), Y = g(t)$ *for non-constant rational functions* $f(t), g(t) \in \mathbb{C}(t)$. *In* ***algebraic geometry*** *a complex curve which can be parametrized as above will be called a* ***genus zero*** *curve, otherwise it is called a positive genus curve. In other words, we want to show the existence of positive genus curves.*

Let us use ω *to denote the following complex number*

(2) $$\omega = e^{\frac{2\pi i}{n}}$$

and call ω *a* ***primitive n-th root of unity***. *Suppose that there is a parametrization* $X = f(t), Y = g(t)$. *Let*

$$f(t) = \frac{\alpha(t)}{\gamma(t)}, \qquad g(t) = \frac{\beta(t)}{\gamma(t)}$$

where $\alpha(t), \beta(t), \gamma(t) \in \mathbb{C}[t]$ *and with 1 as their greatest common divisor and* $\deg(\alpha(t)) + \deg(\beta(t)) + \deg(\gamma(t)) > 0$. *We claim that for any such triple* $\alpha(t), \beta(t), \gamma(t)$, *we can produce another triple* $\alpha'(t), \beta'(t), \gamma'(t)$ *with* $\deg(\alpha(t)) + \deg(\beta(t)) + \deg(\gamma(t)) > \deg(\alpha'(t)) + \deg(\beta'(t)) + \deg(\gamma'(t)) > 0$. *Then, by* ***infinite descent***, *namely mathematical induction on the total degree, we get a contradiction.*

After we multiply the equation (1) by $\gamma(t)^n$, *it becomes*

(3) $$\alpha(t)^n - \beta(t)^n = \prod_{i=0}^{n-1}(\alpha(t) - \omega^i\beta(t)) = \gamma(t)^n$$

Note that any two factors of $\prod_{i=0}^{n-1}(\alpha(t)-\omega^i\beta(t))$ can not have a common divisor. Otherwise, let $\pi(t)$ be a common divisor of $(\alpha(t)-\omega^i\beta(t))$ and $(\alpha(t)-\omega^j\beta(t))$. Then it is easy to see that $\pi(t)$ is a divisor of $\alpha(t),\beta(t)$ and hence of $\gamma(t)$. This contradicts our assumption on $\alpha(t),\beta(t),\gamma(t)$.

Since the right hand side of the equation (3) is an n-th power and $\mathbf{C}[x]$ is a U.F.D., we conclude that each factor is an n-th power as follows,

$$\alpha(t)-\beta(t)=\overline{\alpha(t)}^n$$
$$\alpha(t)-\omega\beta(t)=\overline{\gamma(t)}^n$$
$$\alpha(t)-\omega^2\beta(t)=\overline{\beta(t)}^n$$

where $\overline{\alpha(t)\beta(t)\gamma(t)}\mid\gamma(t)$. Multiplying the above equations by $\omega,-1-\omega,1$ and then adding them together, we have

$$\omega\overline{\alpha(t)}^n+\overline{\beta(t)}^n=(1+\omega)\overline{\gamma(t)}^n \tag{4}$$

Form high school we assume that the n-th roots of a complex number exist. We define

$$\alpha'(t)^n=\omega\overline{\alpha(t)}^n,\quad \beta'(t)^n=-\overline{\beta(t)}^n,\quad \gamma'(t)^n=(1+\omega)\overline{\gamma(t)}^n \tag{5}$$

Replacing the equation (4) by the new polynomials in the equation (5), we get

$$\alpha'(t)^n-\beta'(t)^n=\gamma'(t)^n \tag{6}$$

We have shown that $\alpha'(t),\beta'(t),\gamma'(t)$ have 1 as their greatest common divisor and have smaller total degree. A contradiction.

∎

Let us consider the problem of establishing the U.F.D. property of a polynomial ring $\mathbf{K}[x_1,\cdots,x_n]$ of n variables over a field $\mathbf{K}$. We let $\mathbf{D}=\mathbf{K}[x_1,\cdots,x_{n-1}]$, $\mathbf{D}[x_n]=\mathbf{K}[x_1,\cdots,x_n]$. Inductively we know that $\mathbf{D}$ is a U.F.D.. It has been proved in Theorem 3.8. that every element in $\mathbf{D}[x_n]$ can be written as a product of irreducible elements. In the light of Theorem 3.6., it suffices to show that every irreducible element in $\mathbf{D}[x_n]$ is prime. This is what we will achieve in the next theorems.

Theorem 3.12. *Let $\mathbf{D}$ be a U.F.D. with $\mathbf{K}$ as its quotient field. Let $f(x)\in\mathbf{D}[x]$ be a primitive polynomial. Then $f(x)$ can be factored as a product of irreducible polynomials in $\mathbf{K}[x]$ such that every one of them is primitive in $\mathbf{D}[x]$. Especially, $f(x)$ is irreducible in $\mathbf{D}[x]\iff f(x)$ is irreducible in $\mathbf{K}[x]$.*

Proof: ($\Longrightarrow$) Let a primitive polynomial $f(x)\in\mathbf{D}[x]$ be decomposed as follows in $\mathbf{K}[x]$,

$$f(x)=\alpha(x)\beta(x),\qquad deg(\alpha(x))\geq 1,\ deg(\beta(x))\geq 1$$

Note that all coefficients of $\alpha(x), \beta(x)$ are of the form $a/b, b \neq 0, a, b \in \mathbf{D}$. Select two elements $d_1, d_2 \in \mathbf{D}$ with $d_1\alpha, \delta_2\beta \in \mathbf{D}[x]$. Let $\alpha_1(x), \beta_1(x)$ be two primitive polynomials in $\mathbf{D}[x]$ satisfying

$$d_1\alpha(x) = e_1\alpha_1(x), \qquad d_2\beta(x) = e_2\beta_1(x), \qquad e_1, e_2 \in \mathbf{D}$$

Then we have from the above that

$$d_1d_2f(x) = d_1\alpha(x)d_2\beta(x) = e_1e_2\alpha_1(x)\beta_1(x)$$

Using Gauss lemma to take the contents on both sides, we have

$$d_1d_2 \overset{a}{\sim} e_1e_2, \qquad \frac{e_1e_2}{d_1d_2} \in \mathbf{D}$$

Therefore, we have

$$f(x) = \frac{e_1e_2}{d_1d_2}\alpha_1(x)\beta_1(x)$$

It is easy to see $\alpha_1(x), \beta_1(x)$ are primitive, and $(e_1e_2)/(d_1d_2)$ is a unit in $\mathbf{D}$. Therefore, $f(x)$ can be decomposed in $\mathbf{D}[x]$.

($\Longleftarrow$) Suppose that $f(x) = g(x)h(x)$ in $\mathbf{D}[x]$ with $g(x), h(x)$ non-invertible. Since $f(x)$ is primitive, then it is easy to deduce from Gauss' lemma that deg($g(x)$)≥ 1 and deg($h(x)$)≥ 1, and both are primitive. Therefore, it is a decomposition of $f(x)$ in $\mathbf{K}[x]$ as product of irreducible elements. ∎

Theorem 3.13. *Let* $\mathbf{D}$ *be a U.F.D.. Then every irreducible element* $f(x)$ *in* $\mathbf{D}[x]$ *is prime.*

Proof: (1) Suppose that $f(x) = f \in \mathbf{D}$. We claim f is an irreducible element, and hence a prime element, in $\mathbf{D}$. Assume that $f = \alpha\beta$ in $\mathbf{D}$ with α, β non-invertible in $\mathbf{D}$. On the other hand, $f = f(x)$ is irreducible in $\mathbf{D}[x]$, therefore, one of α, β, say α, is invertible in $\mathbf{D}[x]$. Let $\gamma(x)$ be the inverse of α in $\mathbf{D}[x]$. It is easy to see that deg($\gamma(x)$)=0, and $\gamma = \gamma(x) \in \mathbf{D}$. Therefore, α is invertible in $\mathbf{D}$. A contradiction.

(2) Suppose that $f(x) = f \in \mathbf{D}$. We claim f is a prime element in $\mathbf{D}[x]$. Assume that $f \mid g(x)h(x)$ where $g(x), h(x) \in \mathbf{D}[x]$. Let $g^*(x), h^*(x), \ell^*(x)$ be primitive polynomials and $d_1, d_2, d_3 \in \mathbf{D}$ satisfying the following equations

$$g(x) = d_1g^*(x), \qquad h(x) = d_2h^*(x)$$
$$f\ell(x) = g(x)h(x), \qquad \ell(x) = d_3\ell^*(x)$$

It follows from Gauss lemma that

$$c(f\ell(x)) = c(f)c(\ell(x)) = c(g(x))c(h(x))$$

Therefore, $fd_3 \overset{a}{\sim} d_1d_2$ and there exists an invertible element $\delta \in \mathbf{D}$ such that $\delta fd_3 = d_1d_2$, namely, $f \mid d_1$ or $f \mid d_2$. In other words, $f \mid g(x)$ or $f \mid h(x)$. We conclude that $f = f(x)$ is a prime element in $\mathbf{D}[x]$.

(3) Suppose that the irreducible element $f(x)$ of $\mathbf{D}[x]$ is not in $\mathbf{D}$. Then we have $\deg(f(x)) \geq 1$ and $f(x)$ must be primitive. We claim $f(x)$ is a prime element in $\mathbf{D}[x]$. Let $\mathbf{K}$ be the quotient field of $\mathbf{D}$. It follows from Theorem 3.12. that $f(x)$ is an irreducible element in $\mathbf{K}[x]$ and henceforth, by Theorem 3.10., a prime element in $\mathbf{K}[x]$.

Assume that $f(x) \mid g(x)h(x)$ where $g(x), h(x) \in \mathbf{D}[x] \subset \mathbf{K}[x]$. Let us consider the preceding division relation in $\mathbf{K}[x]$. Since $f(x)$ is prime in $\mathbf{K}[x]$, then we have $f(x) \mid g(x)$ or $f(x) \mid h(x)$ in $\mathbf{K}[x]$, say $f(x) \mid g(x)$. Therefore, there exists $\gamma(x)$ with $\gamma(x)f(x) = g(x)$. Let d be a common multiple of all denominators of the coefficients of $\gamma(x)$. Then we have

$$(1) \qquad (d\gamma(x))f(x) = dg(x)$$

Let $\gamma^*(x)$ be a primitive polynomial in $\mathbf{D}[x]$ and $e \in \mathbf{D}$ satisfying the following equation

$$d\gamma(x) = e\gamma^*(x)$$

Then the equation (1) can be rewritten as

$$(2) \qquad (d\gamma(x))f(x) = e(\gamma^*(x)f(x)) = dg(x)$$

Since $\gamma^*(x), f(x)$ are both primitive, then their product is primitive. Therefore, e is a greatest common divisor of the coefficients of the above polynomial and d is a common divisor of the coefficients of the above polynomial. Henceforth, we have

$$d \mid e, \qquad \frac{e}{d}\gamma^*(x) \in \mathbf{D}[x]$$

Factoring the equation by d, we conclude that $f(x) \mid g(x)$. Therefore, $f(x)$ is a prime in $\mathbf{D}[x]$. ∎

Theorem 3.14. *Let* $\mathbf{D}$ *be a U.F.D.. Then* $\mathbf{D}[x]$ *is a U.F.D..*

Proof. It follows from Theorem 3.8. that every non-zero element in $\mathbf{D}[x]$ can be factored into product of irreducible elements. Moreover, it follows from Theorem 3.13. that every irreducible element in $\mathbf{D}[x]$ is prime. Therefore, $\mathbf{D}[x]$ satisfies the two conditions of (2) of Theorem 3.6. and is a U.F.D.. ∎

Corollary 1. *Let* $\mathbf{D}$ *be a U.F.D.. Then* $\mathbf{D}[x_1, x_2, \cdots, x_n]$ *is a U.F.D..*

Proof. Note that $\mathbf{D}[x_1, \cdots, x_n] = \mathbf{D}[x_1, \cdots, x_{n-1}][x_n]$. By induction, we know that $\mathbf{D}[x_1, \cdots, x_{n-1}]$ is a U.F.D., therefore our Corollary follows. ∎

Corollary 2. *Let* $\mathbf{K}$ *be a field. Then* $\mathbf{K}[x_1, x_2, \cdots, x_n]$ *is a U.F.D..*

Proof. Replace $\mathbf{D}$ in the preceding lemma by the field $\mathbf{K}$. ∎

Exercises

(1) Find a greatest common divisor of $x^6+2x^4+2x^3+x^2+2x+1$ and $x^5+2x^3+x^2+x+1$ in $\mathbf{Z}[x]$.
(2) Are there infinitely many primes in $\mathbf{Z}[\sqrt{2}]$?
(3) Let $\mathbf{R}$ be a ring and $f(x)=\sum a_i x^i \in \mathbf{R}[x]$. Show that $f(x)$ is invertible $\iff a_0$ is invertible and a_i is *nilpotent*, namely there is m_i such that $a_i^{m_i}=0$, for all i.
(4) Construct an example to show that for some ring $\mathbf{R}$ and $y=\sum a_i x^i \in \mathbf{R}[x]$, one has (1) deg(y)$\geq 2$ and (2) $\mathbf{R}[x]=\mathbf{R}[y]$.
(5) Let $\mathbf{D}$ be a U.F.D. with quotient field $\mathbf{K}$. Let $f(x)=\sum_{i=0}^{n} a_i x^i \in \mathbf{D}[x]$ be a *monic polynomial*, namely, $a_n=1$. Let $g(x)$ be a *monic* divisor of $f(x)$ in $\mathbf{K}[x]$. Show that $g(x)\in \mathbf{D}[x]$.
(6) Let $a_1,a_2,\cdots,a_n$ be distinct positive integers. Prove the following polynomial $f(x)$ is irreducible in $\mathbf{Q}[x]$,

$$f(x)=\prod_{i=1}^{n}(x-a_i)+1$$

(7) Prove that $(x^2+x+1)\mid(x^{3m}+x^{3n+1}+x^{3p+2})$.
(8) Find a greatest common divisor of x^m-1, x^n-1.
(9) Prove that $x^2-y^2+2x-y+3$ is not a divisor of $x^4y^2z^2+2x^3-3y^3+1$.
(10) Decompose $x^3+y^3+z^3-3xyz$ into a product of irreducible factors in $\mathbf{Z}[x,y,z]$ and $\mathbf{C}[x,y,z]$.
(11) Find all monic irreducible cubic polynomials in $\mathbf{Z}_3[x]$.
(12) Let $f(x)$ be a monic polynomial in $\mathbf{Z}[x]\subset\mathbf{Q}[x]$. Prove that if $f(x)$ is reducible in $\mathbf{Q}[x]$, then $\overline{f(x)}=f(x)\bmod(p)$ in $\mathbf{Z}_p[x]$ is reducible for any prime p.
(13) Prove that $f(x)=x^4-10x^2+1$ is irreducible in $\mathbf{Q}[x]$, while $\overline{f(x)}=f(x)\bmod(p)$ in $\mathbf{Z}_p[x]$ is reducible for any prime p.
(14) Let $f(x)=x^p-x+a\in\mathbf{Z}_p[x]$ where p is a prime number. Prove that $f(x)$ is irreducible $\iff a\neq 0$.
(15) Prove that $x^{89}+89x^{88}+178x^{50}-90x+11$ is irreducible in $\mathbf{Q}[x]$.
(16) Prove that x^6+x^3+1 is irreducible in $\mathbf{Q}[x]$.
(17) Prove that $f(x)=x^3+x^2+1$ is irreducible in $\mathbf{Z}_2[x]$
(18) Find a decomposition of $f(x)=x^5-3x^3+3x^2+4x+2$ as product of irreducible polynomials in $\mathbf{Z}_5[x]$.
(19) Find an irreducible polynomial $f(x)\in\mathbf{Z}[x]$ and a prime number p, such that $\overline{f(x)}=f(x)\bmod(p)$ in $\mathbf{Z}_p[x]$ is irreducible.
(20) Let $\mathbf{R}$ be a ring and $f(x)$ be a zero-divisor in $\mathbf{R}[x]$. Prove that there is an element $a\in\mathbf{R}$ such that $af(x)=0$.
(21) We say two elements α,β are *coprime* if 1 is a greatest common divisor of α,β. Let $f(x),g(x)\in\mathbf{Z}[x]$. Then $f(x),g(x)$ are coprime in $\mathbf{Q}[x]\iff$ for all but finitely many prime numbers p, $\overline{f(x)}=f(x)\bmod(p)$, $\overline{g(x)}=g(x)\bmod(p)$ are coprime in $\mathbf{Z}_p[x]$.

(22) Decompose the following polynomial $f(x,y,z)$ into product of irreducible polynomials in $\mathbf{C}[x,y,z]$,

$$f(x,y,z) = -x^3 - y^3 - z^3 + x^2(y+z) + y^2(x+z) + z^2(x+y) - 2xyz$$

§4 Symmetric Polynomial, Resultant and Discriminant

Given any polynomial $f(x) = \sum a_i x^i$, an interesting problem is to study the relation between the *roots* and the coefficients a_i of $f(x)$. Let us define,

Definition 3.13.. *Let* $\mathbf{R}$ *be a commutative ring and* $f(x) = \sum a_i x^i \in \mathbf{R}[x]$. *The value* $f(b)$ *of* $f(x)$ *at* $x = b$ *is defined to be* $f(b) = \sum a_i b^i$. *If* $f(b) = 0$, *then we say* $x = b$ *is a root of* $f(x)$. *A root is also called a zero point. For several variables, if* $f(b_1, b_2, \cdots, b_n) = 0$, *then we say* $(b_1, b_2, \cdots, b_n)$ *is a zero point of* $f(x_1, x_2, \cdots, x_n) \in \mathbf{R}[x_1, x_2, \cdots, x_n]$. ∎

Let us restate Discussion (1) of Theorem 3.4. as the following theorem for reference.

Theorem 3.15. *Let* $\mathbf{R}$ *be a commutative ring and* $f(x) \in \mathbf{R}[x]$. *Then* $x = b$ *is a root of* $f(x) \Longleftrightarrow (x-b) \mid f(x) \Longleftrightarrow f(x) \in ((x-b)) = \{g(x)(x-b) : g(x) \in \mathbf{R}[x]\}$. ∎

Corollary 1. *If* $\mathbf{R}$ *is a field and deg(*$f(x)$*)*$= n$, *then* $f(x)$ *has at most* n *distinct roots.*

Proof. We have then $\mathbf{R}[x]$ a U.F.D.. Let

$$f(x) = \delta \prod p_i(x)^{m_i}, \qquad \delta \text{ unit, and } m_i \geq 1$$

Then it follows from comparing the degrees of both sides that there are at most n distinct prime factors $p_i(x)$ and hence at most n distinct prime factors of the form $x - b_i$. ∎

Corollary 2. *If* $\mathbf{R}$ *is an integral domain and deg(*$f(x)$*)*$= n$, *then* $f(x)$ *has at most* n *distinct roots.*

Proof. Let $\mathbf{K}$ be the quotient field of $\mathbf{R}$. Then we have $f(x) \in \mathbf{K}[x]$. ∎

Discussion

(1) If $\mathbf{R}$ is not an integral domain, then $f(x)$ may have more than n roots. For instance, let $\mathbf{R}$ be $\mathbf{Z}_8$ and $f(x) = x^3 \in \mathbf{R}[x]$. Then $x = [0]_8, [2]_8, [4]_8, [6]_8$ are all roots of $f(x)$.

(2) If $(x-b) \mid f(x), (x-b)^2 \nmid f(x)$, then $x = b$ is said to be a *simple root* of $f(x)$. If $(x-b)^2 \mid f(x)$, then $x = b$ is said to be a *multiple root* of $f(x)$. ∎

Let us assume that the ring is an integral domain $\mathbf{D}$. Sometimes a polynomial $f(x)$ will have all roots in $\mathbf{D}$. Later on (*Chapter V*) we will see that this is indeed the case for any polynomial if **D** is an *algebraically closed field*, especially if **D**=**C**. Let we *re-index* the coefficients of $f(x)$ as

$$\begin{aligned} f(x) &= a_0x^n + a_1x^{n-1} + \cdots + a_ix^{n-i} + \cdots + a_{n-1}x + a_n \\ &= a_0\prod_{i+1}^{n}(x - b_i) \end{aligned}$$

For the time being, let us consider the following *generic polynomial*,

$$\begin{aligned} F(y_1, y_2, \cdots, y_n, x) &= \prod_{i=1}^{n}(x - y_i) \\ &= x^n + \sum_{i=1}^{n}(-1)^i\theta_i(y_1, y_2, \cdots, y_n)x^{n-i} \end{aligned}$$

where

$$\begin{aligned} \theta_1(y_1, y_2, \cdots, y_n) &= \sum_i y_i \\ \theta_2(y_1, y_2, \cdots, y_n) &= \sum_{i_1<i_2} y_{i_1}y_{i_2} \\ &\cdots\cdots \\ \theta_j(y_1, y_2, \cdots, y_n) &= \sum_{i_1<i_2<\cdots<i_j}\prod_{k=1}^{j} y_{i_k} \\ &\cdots\cdots \\ \theta_n(y_1, y_2, \cdots, y_n) &= \prod_{k=1}^{n} y_k \end{aligned}$$

Definition 3.14. *The above defined polynomials θ_i are called the i-th elementary symmetric polynomials of $y_1, y_2, \cdots, y_n$.* ∎

Discussion

(1) Since the polynomial $F(y_1, y_2, \cdots, y_n, x)$ is invariant under all permutations of the variables $y_1, y_2, \cdots, y_n$, then the coefficients of its expansion, the elementary symmetric polynomials θ_i, are invariant under all permutations of the variables $y_1, y_2, \cdots, y_n$. In other words, let $\sigma \in \mathbf{S}_n$ and define $\sigma(\theta(y_1, y_2, \cdots, y_n)) = \theta(y_{\sigma(1)}, y_{\theta(2)}, \cdots, y_{\theta(n)})$, then we have

$$\sigma(\theta(y_1, y_2, \cdots, y_n)) = \theta(y_1, y_2, \cdots, y_n)$$

for all $\sigma \in \mathbf{S}_n$.

(2) Instead of the generic polynomial $F(y_1, y_2, \cdots, y_n, x)$, we may consider any *special* polynomial as follows,

$$\begin{aligned} f(b_1, b_2, \cdots, b_n, x) &= \prod(x - b_i) \\ &= x^n + \sum_{i=1}^{n}(-1)^i a_i x^{n-i} \end{aligned}$$

Then we have,

$$\begin{aligned} a_1 &= \sum_i b_i \\ a_2 &= \sum_{i_1<i_2} b_{i_1} b_{i_2} \\ &\cdots\cdots \\ a_j &= \sum_{i_1<i_2<\cdots<i_j} \prod_{k=1}^{j} b_{i_k} \\ &\cdots\cdots \\ a_n &= \prod_{k=1}^{n} b_k \end{aligned}$$

(3) Let the ring **R** be the complex field **C**. Since the coefficients a_i are polynomials of the roots b_j, then the coefficients a_i are *continuous function* of the roots b_j. In fact, the converse is also true, namely, *the roots b_j are continuous functions of the coefficients a_i*. However, this assertion will not be used and will not be proved in this book. ∎

Definition 3.15. *Let $g(y_1, y_2, \cdots, y_n) \in \mathbf{R}[y_1, y_2, \cdots, y_n]$. If, for any $\sigma \in \mathbf{S}_n$, defining $\sigma(g(y_1, y_2, \cdots, y_n)) = g(y_{\sigma(1)}, y_{\theta(2)}, \cdots, y_{\theta(n)})$, we always have*

$$\sigma(g(y_1, y_2, \cdots, y_n)) = g(y_1, y_2, \cdots, y_n)$$

then $g(y_1, y_2, \cdots, y_n)$ is said to be a symmetric polynomial. ∎

Definition 3.16. *The weight of θ_i is defined to be i. For any polynomial with coefficients in* **R**, *$g(\theta_1, \theta_2, \cdots, \theta_n) = \sum a_{j_1\cdots j_n} \prod \theta_i^{j_i}$, the weight of g is defined to be $\max\{\sum i j_i : a_{j_1\cdots j_n} \neq 0\}$.* ∎

Theorem 3.16 (Newton's Theorem)[5]. *Let $f(y_1, y_2, \cdots, y_n) \in \mathbf{R}[y_1, y_2, \cdots, y_n]$ be a symmetric polynomial of degree m. Then there is a polynomial $g(\theta_1, \theta_2, \cdots, \theta_n) \in \mathbf{R}[\theta_1, \theta_2, \cdots, \theta_n]$ with weight m such that*

$$f(y_1, y_2, \cdots, y_n) = g(\theta_1, \theta_2, \cdots, \theta_n).$$

[5] English Mathematician and Scientist 1642-1727.

Proof: (1) Let us make a double induction on n and m. If $n = 1$, then $y_1 = \theta_1$ and our theorem is evident. Let assume that $n > 1$ and assume that our theorem is true if the number of variables $= 1, 2, \cdots, (n-1)$.

(2) Let us assume the number of variables is n. If $m = 0$, then $f(y_1, \cdots, y_n)$ is a constant and our theorem is evident. Let us assume that $m > 0$ and our theorem is true for all polynomials of weight less than m.

(3) Let us assume the number of variables is n and the weight of the polynomial f is m. Let

$$\overline{\theta}_i(y_1, y_2, \cdots, y_{n-1}) = \theta(y_1, y_2, \cdots, y_{n-1}, 0), \quad \text{for } i = 1, 2, \cdots, n-1.$$

It is easy to see that $\overline{\theta}_i$ is the i-th elementary symmetric polynomial of the variables, $y_1, y_2, \cdots, y_{n-1}$. Let us consider the polynomial

$$\overline{f}(y_1, y_2, \cdots, y_{n-1}) = f(y_1, y_2, \cdots, y_{n-1}, 0)$$
$$deg(\overline{f}(y_1, y_2, \cdots, y_{n-1})) \leq deg(f(y_1, y_2, \cdots, y_{n-1}, 0)) \leq m$$

Then $\overline{f}$ is a symmetric polynomial in $y_1, y_2, \cdots, y_{n-1}$. It follows from the induction hypothesis that there is a polynomial $\overline{g}$ such that

(i) $\overline{g}(\overline{\theta}_1, \overline{\theta}_2, \cdots, \overline{\theta}_{n-1}) = \overline{f}(y_1, y_2, \cdots, y_{n-1}) = f(y_1, y_2, \cdots, y_{n-1}, 0)$.

(ii) The weight of $\overline{g}(\overline{\theta}_1, \overline{\theta}_2, \cdots, \overline{\theta}_{n-1})$ with respect to $\overline{\theta}_1, \overline{\theta}_2, \cdots, \overline{\theta}_{n-1}$ is less than or equal to m.

Let

$$h(y_1, y_2, \cdots, y_n) = f(y_1, y_2, \cdots, y_n) - \overline{g}(\theta_1, \theta_2, \cdots, \theta_{n-1})$$

Then we have

(iii) The polynomial $h(y_1, \cdots, y_n)$ is a symmetric polynomial in $y_1, \cdots, y_n$.

(iv) $\overline{g}(\theta_1, \cdots, \theta_{n-1})$ has weight $\leq m$.

(v) $\deg_{y_1, \cdots, y_n} \overline{g}(\theta_1, \ldots, \theta_{n-1}) \leq m$.

(vi) $\deg_{y_1, \cdots, y_n} h(y_1, \cdots, y_n) \leq m$.

(vii) $h(y_1, \cdots, y_{n-1}, 0) = f(y_1, \cdots, y_{n-1}, 0) - \overline{g}(\overline{\theta}_1, \cdots, \overline{\theta}_{n-1}) = 0$.

Therefore, it follows from Theorem 3.15. that $y_n \mid h(y_1, \cdots, y_n)$. On the other hand, h is symmetric in $y_1, \cdots, y_n$. It is easy to see that $y_i \mid h(y_1, \cdots, y_n)$ for all i. We conclude routinely that

$$h(y_1, \cdots, y_n) = (\prod y_i)\overline{h}(y_1, \cdots, y_n) = \theta_n \overline{h}(y_1, \cdots, y_n)$$

Henceforth, we have

(viii) $\overline{h}(y_1, \cdots, y_n)$ is a symmetric polynomial in $y_1, \cdots, y_n$.

(ix) The weight of $\overline{h}$ is less than or equal to $m - n$.

It follows from the hypothesis of induction that there is a polynomial g^* such that

$$\overline{h}(y_1, \cdots, y_n) = g^*(\theta_1, \cdots, \theta_n)$$
$$\text{Weight of } g^* \leq m - n.$$
$$f(y_1, \cdots, y_n) = \overline{g}(\theta_1, \cdots, \theta_n) + g^*(\theta_1, \cdots, \theta_n) = g(\theta_1, \cdots, \theta_n)$$

It is easy to see that the polynomial g defined above has weight $\leq m$. Comparing the degrees on both sides of the last equation, we conclude that the weight of g must be m. ∎

Example 10. (Newton's Formula).
Let the symmetric polynomials s_m be defined as follows,

$$s_m = x_1^m + x_2^m + \cdots + x_n^m, \qquad \text{for } m = 1, 2, \cdots$$

Then we have

$$s_0 = n, \qquad s_1 = \theta_1$$

In general, we have the following relations,

$$\begin{aligned}\theta_j s_{m-j} &= \Big(\sum \prod_{i_1 < i_2 < \cdots < i_j} x_{i_1} x_{i_2} \cdots x_{i_j}\Big)\Big(\sum x_k^{m-j}\Big) \\ &= \sum \prod_{i_1 < \cdots \widehat{i_\ell} \cdots < i_j, i_\ell \neq i_r} x_{i_1} \cdots \widehat{x_{i_\ell}} \cdots x_{i_j} x_{i_\ell}^{m-j+1}\end{aligned}$$

From the above relations, we deduce Newton's formula as follows,

$$\begin{aligned}&s_m - s_{m-1}\theta_1 + s_{m-2}\theta_2 - \cdots + (-1)^{m-1} s_1 \theta_{m-1} + (-1)^m m\theta_m = 0, \quad \text{for } m \leq n \\ &s_m - s_{m-1}\theta_1 + s_{m-2}\theta_2 - \cdots + (-1)^n s_{m-n}\theta_m = 0, \quad \text{for } m > n\end{aligned}$$

From Newton's formula, we deduce for a field K of characteristic zero the following

$$K[s_1, s_2, \cdots, s_n] = K[\theta_1, \theta_2, \cdots, \theta_n]$$

∎

The other important topic is the elimination of variables from several polynomials. Let us consider the simplest case of two polynomials $f(x,y), g(x,y) \in F[x,y]$ where we wish to eliminate y to produce a single polynomial $h(x)$. The criterion for $h(x)$ is that if there is an α with $h(\alpha) = 0$, then there must be an β with $f(\alpha, \beta) = 0, g(\alpha, \beta) = 0$. In other words, $f(\alpha, y), g(\alpha, y)$ must have a common factor (i.e., $y - \beta$). Let us take any field K (in the preceding discussion, we let $K = F(x)$). Then we wish to find the condition on the coefficients of $f(y), g(y) \in K[y]$ such that $f(y), g(y)$ have a common factor. We have the following

Definition 3.17. (Resultant).
Let K be a field, $f(y), g(y) \in K[y]$ with the following expressions,

$$f(y) = a_n y^n + a_{n-1} y^{n-1} + \cdots + a_1 y + a_0$$
$$g(y) = b_m y^m + b_{m-1} y^{m-1} + \cdots + b_1 y + b_0$$

We define the resultant[6] *of $f(y), g(y)$ with respect to y as the determinant of the following $(n+m) \times (n+m)$ matrix,*

$$Res_y(f(y), g(y)) =$$

$$\det \begin{pmatrix} a_n, & a_{n-1}, & \cdots & a_1, & a_0, & 0, & \cdots & \cdots & \cdots & 0 \\ 0, & a_n, & a_{n-1}, & \cdots & a_1, & a_0, & 0, & \cdots & \cdots & 0 \\ . & . & \cdots & \cdots & . & . & . & \cdots & \cdots & . \\ . & . & \cdots & \cdots & . & . & . & \cdots & \cdots & . \\ . & . & \cdots & \cdots & . & . & . & \cdots & \cdots & . \\ 0, & 0, & \cdots & \cdots & 0, & a_n, & a_{n-1}, & \cdots & a_1, & a_0 \\ b_m, & b_{m-1}, & \cdots & \cdots & b_1, & b_0, & 0, & \cdots & \cdots & 0 \\ 0, & b_m, & b_{m-1}, & \cdots & . & b_1, & b_0, & 0, & \cdots & 0 \\ . & . & \cdots & \cdots & . & . & . & \cdots & \cdots & . \\ . & . & \cdots & \cdots & . & . & . & \cdots & \cdots & . \\ 0, & \cdots & \cdots & 0, & b_m, & b_{m-1}, & \cdots & \cdots & b_1, & b_0 \end{pmatrix}$$

∎

Theorem 3.17. *Let K be a field and $f(y), g(y) \in K[y]$ with*

$$f(y) = a_n y^n + a_{n-1} y^{n-1} + \cdots + a_1 y + a_0$$
$$g(y) = b_m y^m + b_{m-1} y^{m-1} + \cdots + b_1 y + b_0$$

Then $Res_y(f(y), g(y)) = 0 \Leftrightarrow a_n = 0 = b_m$ or $f(y), g(y)$ have a common factor of degree ≥ 1.

Proof: ($\Leftarrow$) If $a_n = 0 = b_m$, then it is clear from the definition of the resultant that $Res_y(f(y), g(y)) = 0$. If $a_n \neq 0$ or $b_m \neq 0$, say $a_n \neq 0$, and $f(y), g(y)$ have a common factor $h(y)$ of degree ≥ 1, then we have

$$\begin{aligned} & f(y) = h(y)\alpha(y), \qquad g(y) = -h(y)\beta(y) \\ (1) \qquad & \deg \alpha(y) < \deg f(y) = n, \quad \deg \beta(y) < \deg g(y) \\ & \beta(y) f(y) + \alpha(y) g(y) = 0 \end{aligned}$$

[6] Due to English Mathematician Sylvester 1814-1897.

Let us consider the above equations in the light of indeterminate coefficients. Let $u_1, u_2, \cdots, u_n, v_1, v_2, \cdots, v_m$ be indeterminates and

$$A(y) = u_n y^{n-1} + u_{n-1} y^{n-2} + \cdots + u_2 y + u_1$$
$$B(y) = v_m y^{m-1} + v_{m-1} y^{m-2} + \cdots + v_2 y + v_1$$

We may assign values to the indeterminates $u_1, u_2, \cdots, u_n, v_1, \cdots, v_m$ such that $A(y)$ becomes $\alpha(y)$ and $B(y)$ becomes $\beta(y)$. In other words, after assigning values which are not all zeroes to the indeterminates $u_1, \cdots, u_n, v_1, \cdots, v_m$, the following equation becomes an identity in y,

$$A(y)f(y) + B(y)g(y) = 0$$

We shall collect the coefficients of y^i in the above equation to get a system of equations in the indeterminates $u_1, \cdots, u_n, v_1, \cdots, v_m$ as follows,

$$\begin{array}{lllllllll}
y^{n+m-1} : & a_n v_m & & & +b_m u_n & & & & =0 \\
y^{n+m-2} : & a_{n-1} v_m & +a_n v_{m-1} & & +b_{m-1} u_n & +b_m u_{n-1} & & & =0 \\
\cdot & \cdots & \cdots & \cdots & \cdots & \cdots & & \cdots & \cdots \\
\cdot & \cdots & \cdots & \cdots & \cdots & \cdots & & \cdots & \cdots \\
y^m & : a_1 v_m & +a_2 v_{m-1} & + \cdots & +b_{m-n-1} u_n & +b_{m-n-2} u_{n-1} & & + \cdots & =0 \\
y^{m-1} & : a_0 v_m & +a_1 v_{m-1} & + \cdots & +b_{m-n} u_n & +b_{m-n-1} u_{n-1} & & + \cdots & =0 \\
y^{m+1} & : 0 v_m & +a_0 v_{m-1} & + \cdots & +b_{m-n+1} u_n & +b_{m-n} u_{n-1} & & + \cdots & =0 \\
\cdot & \cdots & \cdots & \cdots & \cdots & \cdots & & \cdots & \cdots \\
\cdot & \cdots & \cdots & \cdots & \cdots & \cdots & & \cdots & \cdots \\
1 & : 0 v_m & +0 v_{m-1} & \cdots & +a_0 v_1 & & & +b_0 u_1 & =0
\end{array}$$

Note that the coefficient matrix of the above system of equations, with $u_1, \cdots, u_n, v_1, \cdots, v_m$ as variables, is the transpose of the matrix of the $Res_y(f(y), g(y))$. The existence of non-zero polynomials $\alpha(y), \beta(y)$ implies that the above system of equations has non-zero solutions for the $(n+m)$-tuple $(u_1, \cdots, u_n, v_1 \cdots, v_m)$. Therefore the determinant of the coefficient matrix is zero. Henceforth, $Res_y(f(y), g(y)) = 0$.

($\Rightarrow$) If $a_n = 0 = b_m$, then we are done. Otherwise we may assume that $a_n \neq 0$ and deg $(f(y)) = n$. Since $Res_y(f(y), g(y)) = 0$, then the above system of linear equations will has non-zero solutions for $(v_m, \cdots, v_1, u_n, \cdots, u_1)$ (cf Example 4). Therefore there are non-zero polynomials $\alpha(y), \beta(y)$ with

$$\beta(y)f(y) + \alpha(y)g(y) = 0$$
$$\deg \alpha(y) < \deg f(y) = n$$

If $f(y), g(y)$ have no common factor of degree ≥ 1, then the g.c.d of $f(y), g(y)$ must be unit, i.e., there are polynomials $\delta(y), \epsilon(y)$ with

$$1 = \delta(y)f(y) + \epsilon(y)g(y)$$
$$\alpha(y) = \alpha(y)\delta(y)f(y) + \epsilon(y)\alpha(y)g(y)$$
$$= (\alpha(y)\delta(y) - \beta(y)\epsilon(y))f(y)$$
$$\deg \alpha(y) \geq \deg f(y)$$

a contradiction. ∎

We shall find the relation between $Res_y(f(y), g(y))$ and the roots of $f(y), g(y)$. For the time being, let us assume that $a_n = b_m = 1$ and consider the following *generic polynomials* $f(y), g(y)$ with x_i, z_j symbols,

$$f(y) = \prod_{i=1}^{n}(y - x_i) = \sum_{i=1}^{n} a_i y^i$$

$$g(y) = \prod_{j=1}^{m}(y - z_j) = \sum_{j=1}^{m} b_j y^j$$

Then we have,

Theorem 3.18. *Given the above assumptions and notations. We have,*

$$\begin{aligned} Res_y(f(y), g(y)) &= \prod_{i,j}(x_i - z_j) \\ &= \prod_{i} g(x_i) \\ &= (-1)^{mn} \prod_{j} f(z_j) \end{aligned}$$

Proof. Let us separate the proof into three steps.

(1) We claim that $Res_y(f(y), g(y))$ is a homogeneous polynomial of degree mn in x_i, z_j. Let us consider the following matrix for $Res_y(f(y), g(y))$,

$$Res_y(f(y), g(y)) = det \begin{pmatrix} 1, & a_{n-1}, & \cdots & a_1, & a_0, & 0, & \cdots & \cdots & \cdots & 0 \\ 0, & 1, & a_{n-1}, & \cdots & a_1, & a_0, & 0, & \cdots & \cdots & 0 \\ \cdot & \cdot & \cdots & \cdots & \cdot & \cdot & \cdot & \cdots & \cdots & \cdot \\ \cdot & \cdot & \cdots & \cdots & \cdot & \cdot & \cdot & \cdots & \cdots & \cdot \\ \cdot & \cdot & \cdots & \cdots & \cdot & \cdot & \cdot & \cdots & \cdots & \cdot \\ 0, & 0, & \cdots & \cdots & 0, & 1, & a_{n-1}, & \cdots & a_1, & a_0 \\ 1, & b_{m-1}, & \cdots & \cdots & b_1, & b_0, & 0, & \cdots & \cdots & 0 \\ 0, & 1, & b_{m-1}, & \cdots & \cdot & b_1, & b_0, & 0, & \cdots & 0 \\ \cdot & \cdot & \cdots & \cdots & \cdot & \cdot & \cdot & \cdots & \cdots & \cdot \\ \cdot & \cdot & \cdots & \cdots & \cdot & \cdot & \cdot & \cdots & \cdots & \cdot \\ 0, & \cdots & \cdots & 0, & 1, & b_{m-1}, & \cdots & \cdots & b_1, & b_0 \end{pmatrix}$$

Note that a_i is $(-1)^i\times$(the i-th elementary symmetric polynomial) of $x_1, \cdots, x_n$ and b_j is $(-1)^j\times$(the j-th elementary symmetric polynomial) of $z_1, \cdots, z_m$. Let us multiply the i-th row by x_1^i for $1 \le i \le n$, and multiply the (n+j)-th row by x_1^j for $1 \le j \le m$. Then it is easy to see that the k-th column consists of a homogeneous polynomial of degree k.

Therefore, we conclude that its determinant, which is $Res_y(f(y), g(y))$, is homogeneous of degree

$$\begin{aligned} &1+2+\cdots+(m+n)-(1+2+\cdots m)-(1+2+\cdots n) \\ &= \frac{(m+n)(m+n+1)}{2} - \frac{m(m+1)}{2} - \frac{n(n+1)}{2} \\ &= mn \end{aligned}$$

(2) We claim that $\prod_{i,j}(x_i - z_j) = \prod_i g(x_i) \mid Res_y(f(y), g(y))$.

It is obvious that in the polynomial ring of $n+m$ variables $K[x_1, \cdots, x_n, z_1, \cdots, z_m]$, if $i \neq \ell$, then $g(x_i) = \prod_j(x_i - z_j)$ and $g(x_\ell) = \prod_j(x_\ell - z_j)$ have no common factor. Therefore it suffices to prove $g(x_i) \mid Res_y(f(y), g(y))$ for all i.

Let us do the following operations to the matrix in (1): (i) multiplying the first column by x_i^{m+n}, (ii) adding to the first column the result of multiplying the ℓ-th column by $x_i^{m+n-\ell+1}$ for all $\ell = 2, \cdots, m+n$. Then the first column of the resulting matrix will be as follows,

$$\begin{pmatrix} x_i^m f(x_i) \\ x_i^{m-1} f(x_i) \\ \cdot \\ \cdot \\ x_i f(x_i) \\ x_i^n g(x_i) \\ x_i^{n-1} g(x_i) \\ \cdot \\ \cdot \\ x_i g(x_i) \end{pmatrix} = \begin{pmatrix} 0 \\ 0 \\ \cdot \\ \cdot \\ 0 \\ x_i^n g(x_i) \\ x_i^{n-1} g(x_i) \\ \cdot \\ \cdot \\ x_i g(x_i) \end{pmatrix}$$

Therefore the determinant of the new matrix is divisible by $g(x_i)$, namely,

$$g(x_i) \mid x_i^{m+n} Res_y(f(y), g(y))$$

Moreover, there is no common factor between $g(x_i) = \prod_j(x_i - z_j)$ and x_i^{m+n}. We conclude that $g(x_i) \mid Res_y(f(y), g(y))$.

(3) Since the degrees of $\prod_{i,j}(x_i - z_j)$ and $Res_y(f(y), g(y))$ are both $m+n$ in the variables $x_1, \cdots, x_n, z_1, \cdots, z_m$, then we must have $Res_y(f(y), g(y)) = c\prod_{i,j}(x_i - z_j)$, where $c \in K^*$. We claim that $c = 1$.

Since c appears in a polynomial identity, we may specify the variables $x_1, \cdots, x_n$, $z_1, \cdots, z_m$ to any set of numbers to compute the value of c. Let $x_1 = x_2 = \cdots = x_n = 0$, $z_1 = z_2 = \cdots = z_m = 1$. Then we have $a_0 = a_1 = \cdots = a_{n-1} = 0$, $b_0 = (-1)^m$. Therefore we have, from the matrix in (1), $Res_y(f(y), g(y)) = (-1)^{mn}$. It is obvious that $\prod_{i,j}(0-1) = (-1)^{mn}$. Henceforth we conclude $c = 1$. ∎

Corollary 1. *If* $f(y) = a_n \prod_i (y - x_i), g(y) = b_m \prod_j (y - z_j)$, *then we have*

$$\begin{aligned} Res_y(f(y), g(y)) &= a_n^m b_m^n \prod_{i,j} (x_i - z_j) \\ &= a_n^m \prod_i g(x_i) \\ &= (-1)^{mn} b_m^n \prod_j f(z_j) \end{aligned}$$

∎

Corollary 2. *If* $f(y) = a_n \prod_i (y - \alpha_i), g(y) = b_m \prod_j (y - \beta_j)$ *for* $\alpha_i, \beta_j \in K$, *then we have*

$$\begin{aligned} Res_y(f(y), g(y)) &= a_n^m b_m^n \prod_{i,j} (\alpha_i - \beta_j) \\ &= a_n^m \prod_i g(\alpha_i) \\ &= (-1)^{mn} b_m^n \prod_j f(\beta_j) \end{aligned}$$

∎

Example 11. (Projection). *The algebraic operation of eliminating variables is equivalent to the geometric operation of projection. Let us consider an example. Let* $f(x, y)$, $g(x, y)$ *be two polynomials in* $\mathbb{R}[x, y]$. *After suitable selection of variables* x, y, *say by a suitable linear transformation, we may assume that*

$$\begin{aligned} deg_{x,y} f(x, y) = deg_y f(x, y) = n \\ deg_{x,y} g(x, y) = deg_y g(x, y) = m \end{aligned}$$

Namely,

$$\begin{aligned} f(x, y) &= a_n y^n + \sum_{i=1}^{n} f_i(x) y^{n-i} \\ g(x, y) &= b_m y^m + \sum_{j=1}^{m} g_j(x) y^{m-j} \\ deg_x f_i(x) &\le i, \quad deg_x g_j(x) \le j \end{aligned}$$

Let $L = \mathbf{R}(x)$ = the field of rational functions. Let us consider $Res_y(f(y), g(y))$ in the ring $L[y]$. Then we have

$$Res_y(f(y), g(y)) =$$

$$\det \begin{pmatrix} a_n, & f_1(x), & \cdots & f_n(x), & 0, & \cdots & \cdots & \cdots & 0 \\ 0, & a_n, & f_1(x), & \cdots & f_n(x), & 0, & \cdots & \cdots & 0 \\ \cdot & \cdot & \cdots & \cdots & \cdot & \cdot & \cdots & \cdots & \cdot \\ \cdot & \cdot & \cdots & \cdots & \cdot & \cdot & \cdots & \cdots & \cdot \\ \cdot & \cdot & \cdots & \cdots & \cdot & \cdot & \cdots & \cdots & \cdot \\ 0, & 0, & \cdots & \cdots & 0, & a_n, & f_1(x), & \cdots & f_n(x) \\ b_m, & g_1(x), & \cdots & \cdots & g_m(x), & 0, & \cdots & \cdots & 0 \\ 0, & b_m, & g_1(x), & \cdots & \cdot & g_m(x), & 0, & \cdots & 0 \\ \cdot & \cdot & \cdots & \cdots & \cdot & \cdot & \cdots & \cdots & \cdot \\ \cdot & \cdot & \cdots & \cdots & \cdot & \cdot & \cdots & \cdots & \cdot \\ 0, & \cdots & \cdots & 0, & b_m, & g_1(x), & \cdots & \cdots & g_m(x) \end{pmatrix}$$

We may imitate the proof of (1) of Theorem 3.18. to prove if $Res_y(f(x,y), g(x,y)) \neq 0$, then we have the following formula,

$$(1) \qquad deg_x(Res_y(f(x,y), g(x,y)) \leq mn$$

The above inequality is meaningful in geometry: let us explain it as follows. If $x = \alpha, y = \beta$ is a common solution of the following system of equations,

$$f(x,y) = 0, \qquad g(x,y) = 0$$

then the following system of equations will have a common solution,

$$f(\alpha,y) = 0, \qquad g(\alpha,y) = 0$$

Namely,

$$Res_y(f(\alpha,y), g(\alpha,y)) = 0$$

Therefore, $Res_y(f(x,y), g(x,y))$ defines the projection to the x-axis of the common intersections of the curves defined by $f(x,y) = 0$ and $g(x,y) = 0$. Similarly, $Res_z(f(x,y,z), g(x,y,z)) = 0$ defines the projection to the (x,y)-plane of the common intersections of the surfaces in the three dimensional space defined by $f(x,y,z) = 0$ and $g(x,y,z) = 0$. The above equation (1) states that the projection to the x-axis of the common intersections of two curves defined by two polynomials $f(x,y) = 0$ and $g(x,y) = 0$ of degrees m, n respectively, such that they do not have common components, consists of mn points at most. In fact, we have the following theorem due to Bezout. ∎

Example 12. (Bezout's Theorem). *Let $f(x,y), g(x,y)$ be two polynomials without non-constant factors. The the two curves defined by $f(x,y) = 0, g(x,y) = 0$ have at most mn common intersections where $m = deg f(x,y), n = deg g(x,y)$.*

Proof: Suppose that there are more than mn common intersections. Select any $mn+1$ of them. Connect those points by straight lines. There are finitely many lines. Let the y-axis be chosen such that it is not parallel to any of those lines. Then the projections of the $mn+1$ points to the x-axis will be all different and the degree of $Res_y(f(x,y), g(x,y))$ will have at least $mn+1$. A contradiction. ∎

Another important application of the ***resultant*** is to find the ***multiple roots*** (cf Discussion of Theorem 3.15.) of a polynomial $f(x, y)$. For this purpose, we will introduce the concept of ***derivatives***.

Definition 3.18. *Let* K *be a field and $f(y) = \sum a_i y^i \in K[y]$. Then the* ***derivative*** *of $f(y)$, $f'(y)$, is defined to be $f'(y) = \sum i a_i y^{i-1}$.* ∎

Theorem 3.19. *The derivative defined above satisfies the usual rules in Calculus as follows,*

(1) $c' = 0$ *for all* $c \in K$.
(2) $(af(y) + bg(y))' = af'(y) + bg'(y)$ *for all* $a, b \in K$.
(3) $(f(y)g(y))' = f'(y)g(y) + f(y)g'(y)$.

Proof: They are left to the reader as exercises. ∎

We define the following important notion of ***discriminant***.

Definition 3.19. *We define the discriminant of $f(y)$, $Dis_y(f(y))$, as follows,*

$$Dis_y(f(y)) = Res_y(f(y), f'(y))$$

∎

We have the following theorem.

Theorem 3.20. *We have*

(1) *If α is a root of $f(y)$, then α is a multiple root of $f(y) \iff \alpha$ is a root of $f'(y)$.*
(2) *An element α in* K *is a multiple root of $f(y) \iff \alpha$ is a root of $Dis_y(f(y))$.*
(3) *Let $f(y) = a_n \prod_{i=1}^{n} (y - \alpha_i)$. Then we have*

$$Dis_y(f(y)) = a_n^{2n-1} \prod_{i \neq j} (\alpha_i - \alpha_j)$$

Proof: (1) Let $f(y) = (y-\alpha)^r h(y)$ where $r \geq 1$ and $h(\alpha) \neq 0$. Then it follows from Theorem 3.19. that

$$\begin{aligned} f'(\alpha) &= r(\alpha-\alpha)^{r-1} h(\alpha) + (\alpha-\alpha)^r h'(\alpha) \\ &= r(\alpha-\alpha)^{r-1} h(\alpha) \end{aligned}$$

which is zero if and only if $r \geq 2$.

(2) It follows from (1).

(3) It is routine to check

$$f'(y) = a_n(\sum_i \prod_{j \neq i}(y - \alpha_j))$$

Therefore, our statement follows from Theorem 3.18.. ∎

Example 13. *(1) Let $f(y) = y^2 - by + c$. Then we have*

$$\begin{aligned} Dis_y(f(y)) &= det \begin{pmatrix} 1, & -b, & c \\ 2, & -b, & 0 \\ 0, & 2, & -b \end{pmatrix} \\ &= b^2 - 2b^2 + 4c = -b^2 + 4c \\ &= -\Delta \end{aligned}$$

where Δ is the discriminant of a quadratic equation as taught in high school. Let $f(y) = (y - \alpha_1)(y - \alpha_2)$. Then we have

$$\begin{aligned} b = \alpha_1 + \alpha_2&, \qquad c = \alpha_1\alpha_2 \\ Dis_y(f(y)) &= (\alpha_1 - \alpha_2)(\alpha_2 - \alpha_1) \\ &= -b^2 + 4c \end{aligned}$$

(2) Let us consider cubic polynomials. Let $f(y) = y^3 - ay^2 + by - c = (y - \alpha_1)(y - \alpha_2)(y - \alpha_3)$. Then we have

$$a = \alpha_1 + \alpha_2 + \alpha_3, \quad b = \alpha_1\alpha_2 + \alpha_2\alpha_3 + \alpha_3\alpha_1, \quad c = \alpha_1\alpha_2\alpha_3$$

$$\begin{aligned} Dis_y(f(y)) &= det \begin{pmatrix} 1, & -a, & b, & -c, & 0 \\ 0, & 1, & -a, & b, & -c \\ 3, & -2a, & b, & 0, & 0 \\ 0, & 3, & -2a, & b, & 0 \\ 0, & 0, & 3, & -2a, & b \end{pmatrix} \\ &= 4a^3c - a^2b^2 - 18abc + 4b^3 + 27c^2 \end{aligned}$$

In many applications, we assume that $a = 0$. Then we have

$$Dis_y(f(y)) = 4b^3 + 27c^2$$

∎

Exercises

(1) Let $\mathbf{K}$ be a field of characteristic 0. Given $f(x) \in K[x]$, prove that $f'(x) = 0 \iff f(x) \in K$.

(2) Find n such that x^2+x+1 is a factor of $x^{13}+2x^{12}+10x^9+x^8+x^5-x^2+6x+1$ in $\mathbf{Z}/n\mathbf{Z}$.

(3) Show that if the characteristic of a field $\mathbf{K}$ is non-zero, then there exists $f(x) \notin K$ with $f'(x) = 0$.

(4) In the formal power series ring $K[[x]]$, we may define the derivative $f'(x)$ for $f(x) = \sum a_i x^i$ as follows,

$$f'(x) = \sum i a_i x^{i-1}$$

Show that the derivative defined above have the three properties of Theorem 3.19..

(5) Let us define the derivative of $g(x)^{-1}$ as $(g(x)^{-1})' = -g(x)^{-2}g'(x)$. Show that the definition of derivative can be generalized to $K(x), K((x))$.

(6) Let $\mathbf{K}$ be a field of infinitely many elements. Let $f(x_1, \cdots, x_n) \in K[x_1, \cdots, x_n] \backslash K$. Show that there exists $a_1, \cdots, x_n \in K$ such that $f(a_1, \cdots, a_n) \neq 0$.

(7) Prove that in the following expression (cf Example 10),

$$s_n = \sum a_{\ell_1 \cdots \ell_n} \theta_1^{\ell_1} \cdots \theta_n^{\ell_n}$$

we always have

$$a_{\ell_1 \cdots \ell_n} = (-1)^{\ell_2+\ell_4+\cdots+\ell_2[n/2]} \frac{m(\ell_1 + \cdots + \ell_n - 1)!}{\ell_1! \cdots \ell_n!}$$

(8) Use the elementary symmetric polynomials $\theta_1, \cdots, \theta_n$ to express the following polynomial,

$$\sum_{i<j<k} (x_i - x_j)^2 (x_j - x_k)^2 (x_k - x_i)^2$$

(9) Let $f(x), g(x)$ be polynomials of degrees m, n respectively. Show that,

$$Res_x(f(x), g(x)) = (-1)^{mn} Res_x(g(x), f(x))$$

(10) Show that

$$Res_x(f(x), g(x)h(x)) = Res_x(f(x), g(x)) Res_x(f(x), h(x))$$

(11) Solve the following system of equations over $\mathbf{Z}$,

$$\begin{cases} 5y^2 - 6xy + 5x^2 - 16 = 0 \\ y^2 - xy + 2x^2 - y - x - 4 = 0 \end{cases}$$

(12) Find the projections of the common solutions of the following system of equations to the x, y-plane,

$$\begin{cases} x^2 + xy + z^2 = 0 \\ x^2 - y - z^3 = 0 \end{cases}$$

(13) Let $f(x) = x^3 + 4x^2 - x - 4$. Find the discriminant $Dis_x(f(x))$. Determine all prime numbers p, such that $f(x) mod\ p$ has multiple roots.

(14) Let $f(x) = x^5 - 5x - 5$. Determine all prime numbers p, such that $f(x) mod\ p$ has multiple roots.

(15) Let $f(x) = x^{12} + 2x^{11} + x^{10} + x^3 + x^2 - x - 1$. Find $Dis_x(f(x))$.

(16) Prove that for any $n \geq 1$, the following polynomial $f(x)$ has no multiple root,

$$f(x) = 1 + x + \frac{x^2}{2!} + \cdots + \frac{x^n}{n!}$$

---

§5 Ideals

The concept of *ideal* was created by Kummer in 1846 to study the famous "Fermat's Last Problem". Now, *ideal* become one of the most important concepts in mathematics. We have the following definition,

Definition 3.20. *Let* **R** *be a commutative ring and* **A** *be a subset of* **R**. *Then we define the ideal generated by* **A**, (**A**), *as,*

$$(A) = \begin{cases} \{0\}, & if A = \emptyset \\ \{\sum_{finite} r_i a_i : r_i \in R, a_i \in A, & otherwise \end{cases}$$

If **A** *is a finite set* $\{a_1, \cdots, a_n\}$, *then we write* (**A**) *as* $(a_1, \cdots, a_n)$. *If the ideal generated by a set* I *is* I, $(I) = I$, *then we say* I *is an ideal.* ∎

Example 14. (Algebraic Geometry). *(1) Let* **K** *be a field. The reader may assume that* **K**= **Q**, *or* **R**, *or* **C**. *Let* **A** $= \{f_i(x_1, \cdots, x_n)\} \in K[x_1, \cdots, x_n]$ *be the set of the polynomials in the following system of equations*

(I) $$f_i(x_1, \cdots, x_n) = 0 \quad \forall\, i$$

Let us consider the ideal (A) *generated by* **A**. *Note that* (A) *is the set of polynomials of the following system of equations,*

(II) $$\sum_{finite} h_i(x_1, \cdots, x_n) f_i(x_1, \cdots, x_n) = 0 \qquad \forall\, h_i \in K[x_1, \cdots, x_n], f_i \in A$$

It is easy to see that a common solution $(a_1, \cdots, a_n)$ of the system (I) is a common solution of the system (II). On the other hand, we always have $A \subset (A)$, therefore, every common solution of the system (II) is a common solution of the system (I). We conclude that the system (I) and system (II) have identical solution set.

(2) Let us use the above (1) for some problems. Let us give a simple argument to show that any two distinct circles have at most two intersection points. Let the equations of the two circles be as follows,

$$f(x,y) = x^2 + y^2 + a_1x + a_2y + a_3 = 0$$
$$g(x,y) = x^2 + y^2 + b_1x + b_2y + b_3 = 0$$

Subtracting one from the other, we get,

$$h(x,y) = (a_1 - b_1)x + (a_2 - b_2)y + (a_3 - b_3) = 0$$

It is easy to see that

$$(f(x,y), g(x,y)) = (f(x,y), h(x,y))$$

It follows from Bezout's Theorem that $f(x,y) = 0, h(x,y) = 0$ have at most two intersection points. Therefore, $f(x,y) = 0, g(x,y) = 0$ have at most two intersection points.

(3) Another application of (1) is to find a third circle which passes the intersection points of two given circles and a third point. Let the equations of the given two circles be given as $f(x,y) = 0, g(x,y) = 0$ as in the above (2). Let the third point be (c_1, c_2). Let

$$h(x,y) = \alpha f(x,y) + \beta g(x,y), \qquad \alpha, \beta \in K$$

Then $h(x,y) \in (f(x,y), g(x,y))$ and passes through the common intersection points of the two given circles. We select α, β satisfying the following equation

$$h(c_1, c_2) = \alpha f(c_1, c_2) + \beta g(c_1, c_2) = 0$$

Then we have the equation of the third circle required. This example is significant in understanding the so-called Noether's $A\phi + B\psi$ Theorem in Algebraic Geometry. ∎

Example 15. (Algebraic Geometry). *(1) In the preceding example, we start with polynomials and then study the set of their common solutions. In the present example, we start with any subset **T** in $K^n = K \times K \times \cdots \times K$. Let an ideal I be defined as follows,*

$$I = \{f(x_1, \cdots, x_n) : f(a_1, \cdots, a_n) = 0 \ \forall \ (a_1, \cdots, a_n) \in T\}$$

*It is easy to see that I is indeed an ideal. It follows from the definition of I that the set **S** of common solutions of the system of equations, namely the **algebraic closure** of **T**,*

$$f(x_1, \cdots, x_n) = 0, \forall \ f(x_1, \cdots, x_n) \in I$$

contains **T**. *If* **T**=**S**, *then* **T** *is said to be an* ***affine algebraic variety***.

(2) In fact, the set of common solutions of a system of equations, finite or infinite, is an ***affine algebraic variety***. *However, the proof, which requires Hilbert's Nullstellensatz which will not be discussed in this book, will not be presented.*

(3) Let us give a set which is not an affine algebraic variety. Let **K** =**R** *and* $n = 1$. *Let* **T**=$\{r : r \geq 0\}$. *Then there are infinitely many distinct elements in* **T**. *Any polynomial* $f(x_1) \neq 0$ *has only finitely many roots. Therefore, the ideal* I *of all polynomials vanishing on* **T** *must be the zero ideal* (0). *Then the set* **S** *of solutions of the equation* $0 = 0$ *is* **S**= $\{r : r \in \mathbf{R}\}$ *which is not* **T**. *Therefore,* **T** *is not an affine algebraic variety.* ∎

Let us turn our attentions to the mapping theory. We will realize that ideals play the same role in ring theory as normal subgroups in group theory. We have the following sequence of definitions.

Definition 3.21. *Let* **R** *and* **R'** *be rings. A map* ρ: **R** → **R'** *is said to be a* ***homomorphism*** *if* ρ *keeps the relations of operations, namely,*

(1) $\rho(r_1 + r_2) = \rho(r_1) + \rho(r_2)\ \forall\, r_1, r_2 \in R$.

(2) $\rho(r_1 \cdot r_2) = \rho(r_1) \cdot \rho(r_2)\ \forall\, r_1, r_2 \in R$.

(3) $\rho(1) = 1'$, *where* 1 *and* $1'$ *are units of* **R** *and* **R'** *respectively.*

∎

Definition 3.22. *Let* **R** *and* **R'** *be rings and map* ρ: **R** → **R'** *be a* ***homomorphism***. *If* ρ *is surjective, then we say* **R** *is homomorphic to* **R'**. *If* ρ *is a bijective map, then we say* **R** *is isomorphic to* **R'**, *in symbol* **R** ≈ **R'**. *If furthermore* **R**=**R'**, *then we say* ρ *is an* automorphism. ∎

Definition 3.23. *Let* **R** *and* **R'** *be rings and map* ρ: **R** → **R'** *be a* ***homomorphism***. *We define the image of* ρ, *im*(ρ), *as*

$$im(\rho) = \{r' : r' \in R',\ \text{there exist } r \in R \text{ with } \rho(r) = r'\}$$

We define the kernel of ρ, *ker*(ρ), *as*

$$ker(\rho) = \{r : r \in R, \rho(r) = 0' \text{ where } 0' \text{ is the zero of } \mathbf{R'}\}$$

∎

The following theorem gives us a relation between the ideals and maps.

Theorem 3.21. *(1) Let* I *be an ideal of a ring* **R**. *Then the relation* $\sim$ *defined by*

$$r_1 \sim r_1 \iff r_1 - r_2 \in I$$

is an equivalence relation. Let the quotient set be $\mathbf{R}/\mathbf{I}$ *with the equivalence set of* r *denoted by* $[r]$*. Then* $\mathbf{R}/\mathbf{I}$ *forms a ring, the* ***quotient ring*** *of* $\mathbf{R}$ *over* $\mathbf{I}$*, under the following natural summation* $+$ *and multiplication* $\cdot$ *,*

$$[r_1] + [r_2] = [r_1 + r_2]$$
$$[r_1] \cdot [r_2] = [r_1 \cdot r_2]$$

(2) Let the map χ*:* $\mathbf{R} \to \mathbf{R}/\mathbf{I}$ *be defined as* $\kappa(r) = [r]$*. Then* χ *is called the canonical map from* $\mathbf{R}$ *to* $\mathbf{R}/\mathbf{I}$*. The canonical map is surjective.*

Proof: The theorem follows routinely from the definition of ideal. ∎

The following ***isomorphism theorem*** about rings is similar to the one about groups.

Theorem 3.22 (Isomorphism Theorem). *Let* ρ *be a surjective homomorphism from ring* $\mathbf{R}$ *to ring* $\mathbf{R'}$*. Let* $\mathbf{I'}$ *be an ideal of* $\mathbf{R'}$ *and* $\mathbf{I}$ *the pre-image of* $\mathbf{I'}$*. The induced map* $\overline{\rho}$ *:* $\mathbf{R}/\mathbf{I} \to \mathbf{R'}/\mathbf{I'}$*, defined as follows, is an isomorphism,*

$$\overline{\rho}([r]) = [\rho(r)] \in R'/I'$$

Therefore there is a bijective map, which keeps the inclusion relation, between the set of all ideals of $\mathbf{R'}$ *which contains the ideal* I' *and the set of all ideals of* $\mathbf{R}$ *which contains the ideal* I*.*

Proof: (1) Let us prove the map $\overline{\rho}$ is well-defined. In other words, the map $\overline{\rho}$ is independent of the representative of every coset. Let r_2 be other representatives of the coset $[r_1]$, namely

$$r_1 - r_2 \in I$$

Then we have

$$\rho(r_1 - r_2) = \rho(r_1) - \rho(r_2) \in I'$$
$$[\rho(r_1)] = [\rho(r_2)]$$

(2) We claim that $\overline{\rho}$ is a homomorphism. Let r_1, r_2 be elements of $\mathbf{G}$. Then we have

$$\begin{aligned}\overline{\rho}([r_1] + [r_2]) &= \overline{\rho}([r_1 + r_2]) \\ &= [\rho(r_1 + r_2)] = [\rho(r_1) + \rho(g_2)] \\ &= [\rho(r_1)] + [\rho(g_2)] \\ &= \overline{\rho}([r_1]) + \overline{\rho}([r_2])\end{aligned}$$

Similarly, we have

$$\overline{\rho}([r_1][r_2]) = \overline{\rho}([r_1])\overline{\rho}([r_2])$$

(3) We claim that $\overline{\rho}$ is surjective. Let $[r']$ be any coset in the quotient ring $\mathbf{R'}/\mathbf{I'}$. Since ρ is surjective, then there is $r \in \mathbf{R}$ with $\rho(r) = r'$. Therefore, we have

$$\overline{\rho}([r]) = [\rho(r)] = [r']$$

(4) We claim that $\overline{\rho}$ is injective. We have

$$\begin{aligned}\overline{\rho}([r_1]) &= \overline{\rho}([r_2]) \\ &\Longrightarrow [\rho(r_1)] = [\rho(r_2)] \\ &\Longrightarrow [\rho(r_1)] - [\rho(r_2)] = 0 \\ &\Longrightarrow [\rho(r_1 - r_2)] = 0 \\ &\Longrightarrow r_1 - r_2 \in I\end{aligned}$$

∎

Corollary. *Let ρ be a homomorphism from a ring* **R** *to a ring* **R'**. *Then* **R**/*ker*(ρ) *and* *im*(ρ) *are isomorphic rings.* ∎

Discussion

(1) Any homomorphism ρ from a ring **R** to a ring **R'** can be written as the composition of three homomorphisms as in the following diagram,

$$\begin{array}{ccc} \mathbf{R} & \xrightarrow{\ \rho\ } & \mathbf{R}' \\ {\scriptstyle \kappa}\downarrow & & \uparrow{\scriptstyle i} \\ \mathbf{R}/\mathrm{ker}(\rho) & \xrightarrow{\ \overline{\rho}\ } & \mathrm{im}(\rho) \end{array}$$

where $\rho = i * \overline{\rho} * \kappa$ with i injective, $\overline{\rho}$ an isomorphism and κ canonical. ∎

Example 16. *(1) Let us use the notations of Example 15.. Let* **T** *be a subset of* $\mathbf{K}^n$ *and* I *be the ideals of all polynomials in* $K[x_1, \cdots, x_n]$ *which vanish identically on* **T**. *We define the* ***ring of polynomials*** *of* **T**, **K**[**T**]*, as the ring of induced polynomials from* $\mathbf{K}^n$ *to* **T**, *namely,*

$$f(x_1, \cdots, x_n) \sim g(x_1, \cdots, x_n) \Longleftrightarrow f(x_1, \cdots, x_n) - g(x_1, \cdots, x_n) \in I$$

It then follows that the ring of polynomials of **T**, **K**[**T**]*, is canonically isomorphic to* $K[x_1, \cdots, x_n]/I$. *If the ring of polynomials of* **T**, **K**[**T**]*, is an integral domain, then we define the* ***field of rational functions***, **K**(**T**)*, as the quotient field of the ring of the polynomials of* **T**, **K**[**T**].

(2) Let us consider a special case with **T**$=\{(a_1, \cdots, a_n)\}$ *a singleton. Then a polynomial* $h(x_1, \cdots, x_n) \in I \Longleftrightarrow h(a_1, \cdots, a_n) = 0$. *Let us expand the polynomial* $h(x_1, \cdots, x_n)$ *around the point* $(a_1, \cdots, a_n)$ *as follows,*

$$h = \sum a_{j_1 \dots j_n} (x_1 - a_1)^{j_1} \cdots (x_n - a_n)^{j_n}$$

Then we have $h \in I \iff a_{0\cdots 0} = 0$. It is easy to see that

$$I = (x_1 - a_1, \cdots, x_n - a_n)$$
$$K[x_1, \cdots, x_n]/I \sim K$$

The canonical map κ is given by $\kappa : f(x_1, \cdots, x_n) \to f(a_1, \cdots, a_n)$. ∎

We will classify ideals in a ring **R** and use them to classify rings. We have the following definition.

Definition 3.24. *(1) Let **R** be a commutative ring and I an ideal in **R**. If there is no other ideals between **R** and I, then I is said to be a maximal ideal. In other words, if $I \subset J \subset \mathbf{R}$ where J is an ideal in **R**, implies $J = I$ or $J = \mathbf{R}$, then I is a maximal ideal.*

*(2) Let **R** be a commutative ring and I an ideal in **R**. If $I \neq \mathbf{R}$ and $ab \in I \Longrightarrow a \in I$ or $b \in I$, then I is said to be a prime ideal.* ∎

Example 17. *(1) Let $I = (a)$ be an ideal in $\mathbb{Z}$. Then I is a maximal ideal if and only if a is a prime number. Moreover, I is a prime ideal if and only if a is either 0 or a prime number. In this case, we see that every maximal ideal is prime, while not every prime ideal is maximal.*

(2) Let us use the notations of Example 15.. The ideal $I = (x_1 - a_1, \cdots, x_n - a_n)$ is maximal in $K[x_1, \cdots, x_n]$. Moreover, the ideal I is prime. Let $f(x_1, \cdots, x_n)$ be a prime element in $K[x_1, \cdots, x_n]$. Then the ideal $J = (f(x_1, \cdots, x_n))$ is prime. For instance, (x_1) is a prime ideal. If there is more than one variable, then (x_1) is not maximal. ∎

The role played by maximal ideals in the ring theory is similar to the role played by points in *topology*. We wish to prove the existence of maximal ideals. We have the following theorem.

Theorem 3.23. *(1) Let **R** be a commutative ring and $a \in \mathbf{R}$ with $a^i \neq 0$ for $i = 1, 2, \cdots, n, \cdots$. Then there is a prime ideal I such that $a \notin I$.*

*(2) Let **R** be a commutative ring. Then there are maximal ideals.*

Proof. (1) Let us use Zorn's Lemma to construct an ideal I. Let

$$\mathsf{F} = \{J : J \text{ is ideal in } \mathbf{R}, \text{ and } a^i \notin J \ \forall\, i = 1, 2, \cdots\}$$

Since $(0) \in \mathsf{F}$, then F is not an empty family. Let us define a partial ordering in F by the set-theoretic inclusion. Let $\{I_j\}$ be any chain in F. Let $I = \cup I_j$. We claim that I is an upper bound of the chain $\{I_j\}$. It is easy to see that I is an ideal in **R**. We shall prove that $a^i \notin I \ \forall\, i = 1, 2, \cdots$. Suppose that $a^m \in I = \cup I_j$. Then we must have $a^m \in I_s$ for some s. This contradicts the assumption on I_s. Therefore the assumptions of Zorn's Lemma is satisfied. We conclude there is a maximal element I in the family F.

We claim that I is a prime ideal. Let $bc \in I$. Suppose that $b \notin I$ and $c \notin I$. Due to the maximal property of I, the two ideals, $b \in (b)+I = \{rb+i : r \in R, i \in I\} \neq I$ and $c \in (c)+I = \{rc+i : r \in R, i \in I\} \neq I$, are not in $\mathbf{F}$. Therefore there are positive integers m, ℓ such that

$$a^m = r_1 b + i_1, \qquad i_1 \in I$$
$$a^\ell = r_2 c + i_2, \qquad i_2 \in I$$

Then we have

$$a^{m+\ell} = a^m a^\ell = r_1 r_2 bc + r_1 b i_2 + r_2 c i_1 + i_1 i_2 \in I$$

a contradiction.

(1) In our definition of rings, we assume that there is a unit 1. Let $a = 1$ in (1). Then the ideal I in (1) is a maximal ideal. ∎

The following theorem establishes other criteria for prime ideals and maximal ideals.

Theorem 3.24. *(1) Let* $\mathbf{R}$ *be a commutative ring,* I *be an ideal. Then* I *is a prime ideal* $\Longleftrightarrow$ $\mathbf{R/I}$ *is an integral domain.*

(2) Let $\mathbf{R}$ *be a commutative ring,* I *be an ideal. Then* I *is a maximal ideal* $\Longleftrightarrow$ $\mathbf{R/I}$ *is a field.*

Proof: Let us consider the canonical map $\kappa : \mathbf{R} \to \mathbf{R/I}$. Let $\kappa(r) = [r]$.

(1)($\Longrightarrow$) We have the following,

$$\begin{aligned}[a][b] = 0 &\Longleftrightarrow [ab] = 0\\ &\Longleftrightarrow ab \in I\\ &\Longleftrightarrow a \in I \text{ or } b \in I\\ &\Longleftrightarrow [a] = 0 \text{ or } [b] = 0\end{aligned}$$

($\Longleftarrow$) Let $ab \in I$. Then we have $[ab] = [a][b] = [0]$. Since $\mathbf{R/I}$ is an integral domain, then we must have $[a] = 0$ or $[b] = 0$. Therefore, we have $a \in I$ or $b \in I$.

(2) We have the following,

$$\begin{aligned} R/I \text{ is a field} &\Longleftrightarrow \text{R/I has only two ideals (0) and R/I}\\ &\Longleftrightarrow \text{any ideal J, which contains I, must be either I or R}\\ &\Longleftrightarrow \text{I is a maximal ideal.}\end{aligned}$$

∎

Corollary. *Every maximal ideal is a prime ideal.*

∎

We will use the types of ideals to classify rings. We have the following definition.

Definition 3.25. *(1) If an ideal* I *is generated by one element, namely* $I = (r)$*, then* I *is said to be a* ***principal ideal****.*

(2) Let **R** *be a commutative ring. If every ideal I in* **R** *is principal, then* **R** *is said to be a principal ideal ring, P.I.R.. Furthermore, if* **R** *is an integral domain, then* **R** *is said to be a principal ideal domain, P.I.D..* ∎

We have the following theorem about the relation between Euclidean domain (cf the discussion after Theorem 3.4.) and *P.I.D.*.

Theorem 3.25. *Every Euclidean domain* **R** *is a P.I.D..*

Proof: Let δ be the non-negative integer valued function for **R**. Let I be an ideal. If $I = (0)$, then I is principal. Suppose that $I \neq (0)$. Let a be any element in $I \setminus \{0\}$ such that $\delta(a)$ is minimal. We claim $I = (a)$.

Let b be any element in I. Then it follows from the Euclidean algorithm that

$$b = qa + r, \qquad \delta(r) < \delta(a)$$

Since $r = b - qa \in I$, then we must have $r = 0$, $b = qa \in (a)$. Therefore, I is principal. ∎

Example 18. *(1) The rings* $\mathbf{Z}, \mathbf{Z}[i]$, $\mathbf{K}[x]$ *are Euclidean domains, and hence each is a P.I.D., where* **K** *is a field.*

(2) The following ring **R** *is a P.I.D while not Euclidean,*

$$R = \{a + b(1+\sqrt{-19})/2 : a, b \in \mathbf{Z}\}$$

The P.I.D. property of the above ring follows from the following theorem.

Theorem (Dedekind & Hassa) Let **R** *be a subring of* **C**, *and* $N(x)$ *be the usual complex norm for* $x \in \mathbf{R}$. *If for any pair of elements* $a, b \in \mathbf{R}$ *with* $N(a) \leq N(b)$, *either* $a \mid b$ *or there exist* $u, v \in \mathbf{R}$ *with* $0 < N(ua - vb) < N(a)$, *then* **R** *is a P.I.D..*

Proof: Let I be an ideal. If $I = (0)$, then I is principal. Suppose that $I \neq (0)$. Let a be any element in $I \setminus \{0\}$ such that $N(a)$ is minimal. We claim $I = (a)$. Let $0 \neq b \in I$. Then we have $N(a) \leq N(b)$. If $a \nmid b$, then we have $u, v \in \mathbf{R}$ with $0 < N(ua - vb) < N(a)$. Note that $ua - vb \in I$. Therefore, it is impossible and $a \mid b$. ∎

Let us apply the above theorem to our present situation. Let $I \neq (0)$ be an ideal in **R**. Let a be any element in $I \setminus \{0\}$ such that $\delta(a)$ is minimal. We claim $I = (a)$. Given any element $0 \neq b \in I$, we have $N(b) \geq N(a)$. If $a \nmid b$, then let $b/a = (x + y\sqrt{-19})/z \in \mathbf{C}$ where $x, y, z \in \mathbf{Z}, z > 1$, and $(x, y, z) = (1)$. We claim that there are $u, v \in R$ such that $N(ua - vb) < N(a)$. Note that the previous inequality is equivalence to $N(u(a/b) - v) < 1$.

Since $(x, y, z) = 1$, then we may select $r, s, t \in \mathbf{Z}$ such that $rx + sy + tz = 1$. Let $sx - 19ry = qz + \ell$ with $|\ell| < z/2$.

Let us consider the case $z \geq 5$. Let $u = s + r\sqrt{-19}, v = q - t\sqrt{-19}$. Then we have

$$\begin{aligned} u(a/b) - v &= (x + y\sqrt{-19})(s + r\sqrt{-19})/z - (q - t\sqrt{-19}) \\ &= \ell/z + \sqrt{-19}/z \end{aligned}$$

With the assumption $z \geq 5$, it is easy to check

$$N((a/b) - v) = (\ell^2 + 19)/z^2 < 1$$

Let us consider the case $z = 4$. It follows from $(x, y, z) = 1$ that x, y can not be both even. If one of x, y is even, then let $u = x - y\sqrt{-19}, v = q$. If x, y are both odd, then let $u = (x - y\sqrt{-19})/2, v = q$. It is easy to verify our claim.

Let us consider the case $z = 3$. Let $u = x - y\sqrt{-19}$ and $v = q$. It is to verify our claim.

Let us consider the case $z = 2$. Let $u = 1, v = [(x-1) + y\sqrt{-19}]/2$. It is easy to verify our claim. Therefore we have established that **R** is a *P.I.D.*.

Let us prove[7] **R** is non-Euclidean. Let $\widetilde{R}$ = the set of all units of **R** and 0. In our present case, it is not hard to check that

$$\widetilde{R} = \{\pm 1, 0\}$$

An element $u \in R \setminus \widetilde{R}$ is said to be a *universal side divisor* if for any $x \in \mathbf{R}$ there exists $z \in \widetilde{R}$ such that $u \mid x - z$.

(i) We claim that 2, 3 are irreducible in **R**. This is easy and left to the reader.

(ii) We claim that there is no universal side divisor u in **R**. Let $x_1 = 2, 3$. Then $u = \pm 2$ or ± 3. Now let $x_2 = \frac{1}{2}(1 + \sqrt{-19})$. It is easy to show that none of $\pm 2, \pm 3$ is a factor of $x_2, x_2 \pm 1$. Therefore, there is no universal side divisor.

(iii) We claim that **R** is non-Euclidean. Suppose **R** is Euclidean. Let d be the Euclidean function. Consider the non-empty set $\mathbf{S}=\{d(v) : v \in \mathbf{R} \setminus \widetilde{R}\}$. Let $u \in \mathbf{R}\setminus\widetilde{R}$ be with $d(u)$ minimal. We will show that u is a universal side divisor. Let $x \in \mathbf{R}$. By the Euclidean algorithm, there are y, z such that

$$x = yu + z, \qquad d(z) < d(u)$$

By the minimal property of $d(u)$, we must have $z \in \widetilde{R}$. Therefore, $u \mid x - z$ and u is a universal side divisor, contradicting (ii).

Remark: In fact, it is known that the ring of *algebraic integers* in $\mathbf{Q}(\sqrt{-D})$, **R**, is of the following form

$$R = \{a + b(1 + \sqrt{-D})/2 : a, b \in \mathbf{Z}\}$$

It is also known that the only cases where **R** is *P.I.D.* are for $D = 1, 2, 3, 7, 11, 19, 43, 67$, 163. The ring of algebraic integers is Euclidean in the first five cases. For the remaining cases, proofs similar to the above will show that they are not Euclidean. ∎

From the ideal-theoretic point of view, a field **K** is the simplest ring, because in it there is only one proper ideal $\{0\}$ which is generated by the empty set. The *P.I.R.* is the second simplest ring, because in it all proper ideals are generated by singletons. We shall study

[7] The treatment is due to K.S. Williams, Mathematics Magazine, 48 (1975).

he rings defined by the following definition which turn out to be the most important rings n mathematics.

Definition 3.26. *Let* **R** *be a commutative ring. If every ideal I of* **R** *can be finitely generated, namely every ideal I can be generated by a finite set, then* **R** *is said to be a Noetherian ring.* ∎

The following theorem establishes some equivalence definitions for Noetherian rings.

Theorem 3.26. *Let* **R** *be a commutative ring. The following three conditions are equivalence,*

(1) *Every ideal I is finitely generated.*
(2) *ascending chain condition, a.c.c..* **R** *satisfies the following ascending chain condition on ideals: every ascending chain of ideals must terminate, namely, let the following be a chain of ascending ideals,*

$$I_1 \subset I_2 \subset \cdots \subset I_n \subset I_{n+1} \subset \cdots$$

then there must be an m such that

$$I_m = I_{m+1} = \cdots$$

(3) *Maximal principle. Let F be a non-empty family of ideals in* **R**. *Then there must be a maximal element I in F, namely,*

$$I \subset J \in \mathsf{F} \Longrightarrow I = J$$

Proof: We shall establish (1) $\Longrightarrow$ (2) $\Longrightarrow$ (3) $\Longrightarrow$ (1).

((1)$\Longrightarrow$ (2)) Let the following be a chain of ascending ideals,

$$I_1 \subset I_2 \subset \cdots \subset I_n \subset I_{n+1} \subset \cdots$$

Let us define $I = \cup_i I_i$. We claim that I is an ideal. Let f_i be elements in I. It suffices to prove that

$$g = \sum_{finite} h_i f_i \in I, \qquad \text{where } h_i \in \mathbf{R}$$

It follows from the definition of I that $f_i \in I_{n_i}$ for some suitable n_i. Let $m > n_i$ for all i. Then we have $g \in I_m \subset I$. Therefore, I is an ideal.

It follows from condition (1) that I is finitely generated. Let

$$I = (g_1, g_2, \cdots, g_n)$$

A similar argument as above will produce an integer ℓ such that $g_i \in I_\ell$ for all i. Therefore we have

$$I = (g_1, g_2, \cdots, g_n) \subset I_\ell \subset I_{\ell+1} \subset \cdots \subset I$$

Henceforth, we have $I = I_\ell = I_{\ell+1} = \cdots = I$.

((2)$\Longrightarrow$ (3)) Let F be a non-empty family of ideals in $\mathbf{R}$. Select any element I_1 in F. If I_1 is a maximal element in F, then we are done. If I_1 is not a maximal element in F, then there is an $I_1 \subsetneq I_2 \in \mathsf{F}$. If I_2 is a maximal element in F, then we are done. Otherwise, there is an ideal I_3. In this way, we construct an ascending chain of ideals,

$$I_1 \subsetneq I_2 \subsetneq \cdots \subsetneq I_n \subsetneq \cdots$$

It follows from condition (2) that the above chain must terminate after finitely many steps. In this way we pick up a maximal element in F.

((3)$\Longrightarrow$ (1)) Let F be a non-empty family of finitely generated ideals which are subsets of I. Let J be a maximal element in F. We claim $J = I$. Suppose not. Let $f \in I \setminus J$. Then $J \subsetneq J + f\mathbf{R}$ is a finitely generated ideal $\subset I$. A contradiction. ∎

The following theorem is a cornerstone of Algebraic Geometry and Algebraic Number Theory.

Theorem 3.27. (Hilbert's Base Theorem). [8] *If a ring* $\mathbf{R}$ *is Noetherian, then* $\mathbf{R}[x]$ *is Noetherian.*

Proof: Let I be any ideal of $\mathbf{R}[x]$. We wish to prove that I is finitely generated. Let

$$I_n = \{a_n : \text{ there is } f(x) \in I \text{ with } f(x) = a_0 + a_1 x + \cdots + a_n x^n\}$$

(1) We claim I_n is an ideal of $\mathbf{R}$. Let $b = \sum c_i a_{ni}$ with $c_i \in \mathbf{R}$, $a_{ni} \in I_n$. Let $a_{0i} + a_{1i}x + \cdots + a_{ni}x^n \in I$. Then we have,

$$\sum c_i(a_{0i} + a_{1i}x + \cdots + a_{ni}x^n) \in I$$
$$\Longrightarrow \sum c_i a_{ni} \in I_n$$

(2) We claim that $I_n \subset I_{n+1}$ for $n = 0, 1, \cdots$. Let $a_n \in I_n$. Then there exists $f(x) = a_0 + a_1 x + \cdots + a_n x^n \in I$. Therefore, $x f(x) \in I$ and $a_n \in I_{n+1}$. We have the following ascending chain of ideals in $\mathbf{R}$,

$$I_0 \subset I_1 \subset \cdots \subset I_n \subset \cdots$$

Since $\mathbf{R}$ is a Noetherian ring, then the *a.c.c.* is satisfied and there is an integer m such that

$$I_m = I_{m+1} = \cdots$$

[8] German Mathematician 1862-1943.

(3) Since $\mathbf{R}$ is a Noetherian ring, then the ideals I_i are all finitely generated. Let

$$I_n = (a_{n1}, a_{n2}, \cdots, a_{n\ell_n})$$

and

$$f_{ni}(x) = a_{0ni} + a_{1ni}x + \cdots + a_{ni}x^n \in I$$

We claim $I = (f_{01}(x), \cdots, f_{0\ell_0}(x), f_{11}(x), \cdots, f_{1\ell_1}(x), \cdots, f_{m1}(x), \cdots, f_{m\ell_m}(x))$ where m is determined in (2).

Let $g(x) = b_0 + b_1 x + \cdots + b_s x^s \in I$ with $deg(g(x)) = s$. Then $b_s \in I_s$. If $s \leq m$, then we have $b_s = \sum c_i a_{si}$. Let $h(x) = g(x) - \sum c_i f_{si} \in I$. It is easy to see that $deg(h(x)) < s$. Our claims follow from mathematical induction. If $s > m$, then we have $I_m = I_s$ and $b_s = \sum c_i a_{mi}$. Let $h(x) = g(x) - \sum c_i x^{s-m} f_{mi} \in I$. It is easy to see that $deg(h(x)) < s$. Our claims follow from mathematical induction again. ∎

Corollary. *Let* $\mathbf{K}$ *be a field. Then* $\mathbf{K}[x_1, \cdots, x_n]$ *is Noetherian.*

Proof: It is easy to see that $\mathbf{K}$ is Noetherian. By induction, we may assume that $\mathbf{R}=\mathbf{K}[x_1, \cdots, x_{n-1}]$ is Noetherian. It follows from the above theorem that $\mathbf{R}[x_n]$ is Noetherian. ∎

Example 19. *(1) Let* $\mathbf{K}$ *be a field and a system of equations* $\{f_i(x_1, \cdots, x_n\} \in K[x_1, \cdots, x_n]$, *finite or infinite, be given. It follows from the above Corollary that the ideal* $I = (\{f_i\})$ *is finitely generated. Therefore, the given system of equations, even if infinite, is equivalent to a finite system of equations in the sense that an element* $(a_1, \cdots, a_n) \in \mathbf{K}^n$ *is a common solution of the given system of equations if and only if it is a common solution of a finite system of equations.*

(2) Recall from Example (7) (2) that the ring $\mathbf{R}$ *of all polynomials in* $\mathbb{R}[x]$ *which are integer valued on the integers, is generated by* $\{d_n(x)\}$ *defined as follows,*

$$d_n(x) = \frac{1}{n!}x(x-1)\cdots(x-n+1), \qquad \text{for } n = 1, 2, \cdots$$

In other words, $\mathbf{R}= \mathbb{Z}[\{d_n(x)\}]$. *Let* $I = (\{d_n(x)\})$. *We claim that* I *is not finitely generated.*

Suppose $I = (f_1(x), \cdots, f_n(x))$ *is finitely generated. Then we have*

$$f_i(x) = \sum a_{ij_1\cdots j_m} \prod_{k=1}^{m} d_k^{j_k}, \qquad a_{ij_1\cdots j_m} \in \mathbb{Z}$$

We may assume that $I = (d_1, \cdots, d_m)$. *Therefore, let* $m + \ell = p > m$ *be a prime number, we have*

$$(1) \qquad d_p(x) = \sum_{i=1}^{m} b_i c_i(x) d_i(x), \qquad b_i \in \mathbb{Z}, c_i(x) \in R$$

Note that

$$d_p(p) = 1, \qquad d_i(p) \equiv 0(\ mod\ p), \qquad b_i(p) \in \mathbb{Z}$$

The above equation (1) produces a contradiction. Henceforth, I is not finitely generated and **R** *is not Noetherian.*

(3) Let C_R be the ring of all continuous functions on **R**. *It is easy to see that C_R is commutative ring under the usual addition + and multiplication · . Let*

$$I_n = \{f(x) : f(x) \in C_R, f(i) = 0 \forall\ i \geq n \text{ and } i \in \mathbb{Z}\}$$

Then we have the following ascending chain of ideals which never terminates,

$$I_1 \subsetneq I_2 \subsetneq \cdots \subsetneq I_n \subsetneq I_{n+1} \subsetneq \cdots$$

Therefore, C_R is not Noetherian.

(4) Let **K** *be a field and the polynomial ring of infinite variables* **R**=**K**$[x_1, \cdots, x_n, \cdots]$ *be defined as follows,*

$$R = \{ \sum_{finite} a_{i_1 \cdots i_m} \prod_j x_j^{i_j} \}$$

Let $I_n = (x_1, \cdots, x_n)$. Then we have the following ascending chain of ideals which never terminates,

$$I_1 \subsetneq I_2 \subsetneq \cdots \subsetneq I_n \subsetneq I_{n+1} \subsetneq \cdots$$

Therefore, **R** *is not Noetherian.* ∎

Theorem 3.28. *Let* **R** *be a Noetherian ring and* **R'** *an image of* **R**, *namely, there is a surjective homomorphism ρ:* **R** → **R'**. *Then* **R'** *is a Noetherian ring.*

Proof: We will prove that **R'** satisfies *a.c.c.*, namely every ascending chain of ideals must terminate. Let the following be an ascending chain of ideals in **R'**,

$$I'_1 \subset I'_2 \subset \cdots \subset I'_n \subset I'_{n+1} \subset \cdots$$

Let $I_n = \rho^{-1}(I'_n)$ = the pre-image of I'_n. Then we have

$$I_1 \subset I_2 \subset \cdots \subset I_n \subset I_{n+1} \subset \cdots$$

Since **R** is Noetherian, then there exists an integer m such that $I_m = I_{m+1} = \cdots$. Therefore, $I'_m = I'_{m+1} = \cdots$ and **R'** is a Noetherian ring. ∎

Example 20. (Trigonometry). *Recall Example 16.. Consider the set* **T** *of the points of the unit circle in the real plane,* $\mathbb{R}^2$. *Then the following equation,*

$$f(x) = x^2 + y^2 - 1 = 0$$

defines the unit circle. Note that $f(x,y)$ can not be factored into a product of two linear polynomials. Therefore, it is irreducible. Moreover, given any $g(x,y) \in \mathbf{R}[x,y]$, if $f(x,y) \nmid g(x,y)$, then it follows from Bezout's Theorem that $g(x,y)$ only vanishes at finitely many points of the unit circle. Therefore, $g(x,y) \notin I$, the ideal of all polynomials vanishing on the unit circle $\mathbf{T}$. *Henceforth, we conclude $g(x,y) \in I \Longleftrightarrow f(x,y) \mid g(x,y)$, namely, $I = (f(x,y))$.*

The ring of polynomial functions on the unit circle $\mathbf{T}$, $\mathbf{R}[\mathbf{T}]$, *is by definition $\mathbf{R}[x,y]/I$. On the other hand, let us define a map σ: $\mathbf{R}[x,y] \to \mathbf{R}[\sin\theta, \cos\theta]$ with $\sigma(x) = \sin\theta, \sigma(y) = \cos\theta$. It is easy to see that σ is surjective and $ker(\sigma) = \sigma^{-1}(0) = I$. Therefore, we have*

$$\mathbf{R}[T] \sim \mathbf{R}[x,y]/I \sim \mathbf{R}[\sin\theta, \cos\theta]$$

We conclude that the ring of polynomials in $\sin\theta, \cos\theta$ is the ring of polynomial functions on the unit circle, $\mathbf{R}[\mathbf{T}]$. *It then follows that the ring of all trigonometric functions*[9] *is the field of rational functions on the unit circle,* $\mathbf{R}(\mathbf{T})$. ∎

We will finish this chapter by clarifying the relations between three kinds of integral domains: Euclidean domain, P.I.D. and U.F.D.. Let us prove the following lemma,

Lemma. *In a Noetherian ring* $\mathbf{R}$, *every element can be written as a product of irreducible elements.*

Proof: Let $0 \neq b_0 \in \mathbf{R}$ be non-invertible. If b_0 is irreducible, then we are done. Otherwise, $b_0 = b_{01}b_{02}$ with b_{01}, b_{02} non-invertible. Therefore, we have,

$$(b_0) \subsetneq (b_{01}), \qquad (b_0) \subsetneq (b_{02})$$

If both b_{01}, b_{02} are irreducible, then we are done. Otherwise, we may assume that b_{01} is reducible and $b_{01} = b_{011}b_{012}$ with b_{011}, b_{012} non-invertible. Then we have,

$$(b_0) \subsetneq (b_{01}) \subsetneq (b_{011})$$

It follows from the Noetherian property of $\mathbf{R}$ that the above chain must terminate. We may view the above in the following way. A person b_0 has two sons b_{01}, b_{02}. Successively, every person $b_{0i_1\cdots i_n}$ in this linear family has either two sons, $b_{0i_1\cdots i_n1}$ and $b_{0i_1\cdots i_n2}$, with both non-invertible, or the family line terminates. The Noetherian property of the ring $\mathbf{R}$ says that every line of inheritance terminates in finitely many steps. We claim the whole linear family has only finitely members. The above statement can be proved by induction and will be left to the reader as an exercise. ∎

Theorem 3.29. *Every P.I.D. is an U.F.D..*

Proof: Every P.I.D. is Noetherian. It follows from the preceding lemma that every element can be written as a product of irreducible elements. The proof of the uniqueness of this product is similar to the proof of Theorem 10. of §3 and left to the reader. ∎

---

[9] Defined by the anciant Indian book 'Siddhānta' 400.

We have the following relations,

$$\text{Euclidean domain} \Longrightarrow P.I.D. \Longrightarrow U.F.D.$$

While Example 18. shows that

$$P.I.D. \not\Longrightarrow \text{Euclidean domain}$$

Furthermore, let $\mathbf{K}$ be any field. Then $\mathbf{K}[x, y]$ is a U.F.D.. It is easy to show that the ideal (x, y) is not principal. Therefore, we have

$$U.F.D. \not\Longrightarrow P.I.D.$$

∎

Exercises

(1) Solve the following system of equations,

$$\begin{cases} x^7 - 2x^5 - 3x^2 + 6 = 0 \\ x^5 - 2x^3 + 5x^2 - 10 = 0 \end{cases}$$

(2) Let $I = \{a + bi : a, b \in 2\mathbf{Z}\}$. Find all zero-divisors of $\mathbf{Z}[i]/I$.
(3) Find the cardinality of $\mathbf{Z}[x]/(x^2, 4)$ and all invertible elements.
(4) Let $\mathbf{R}$ be a ring and $\mathbf{S}$ be a subring of $\mathbf{R}$. Let I be an ideal of $\mathbf{R}$. Then $I \cap \mathbf{S}$ is an ideal of $\mathbf{S}$.
(5) Let I be a prime ideal of $\mathbf{Z}[x]$ such that $I \cap \mathbf{Z} = (0)$. Show that I is principal.
(6) Let I be an ideal of $\mathbf{Z}[x]$ generated by $x^3 - 3x - 3, 3x^2 - 3$. What is $I \cap \mathbf{Z}$?
(7) Is the ideal $(3, x^2 + 1)$ prime in $\mathbf{Z}[x]$?
(8) Show that all maximal ideal in $\mathbf{Z}[x]$ can be generated by two elements.
(9) Find all ideals in $\mathbf{Z}[x]$.
(10) Find all ideals of $\mathbf{Q}[x]/(x(x + 1)(x + 2))$.
(11) Let $\mathbf{C}([a, b])$ be the ring of all continuous functions on the closed interval $[a, b] \subset \mathbf{R}$ under the usual addition and multiplication. Prove
   (i) The map ρ: $\mathbf{C}([a, b]) \to \mathbf{R}$ with $\rho(f(x)) = f(a)$ is a homomorphism and $\mathbf{C}([a, b])/ker(\rho) \sim \mathbf{R}$.
   (ii) Let I be an ideal in $\mathbf{C}([a, b])$. Then there is $c \in [a, b]$ such that $f(c) = 0 \; \forall f(x) \in I$.
   (iii) Every maximal ideal I is of the form $I = \{f(x) : f(c) = 0 \text{ for some } c \in [a, b]\}$
(12) Let $\rho : \mathbf{C}[x, y] \to \mathbf{C}[t]$ with $\rho(x) = t^2, \rho(y) = t^3$. Show that $ker(\rho) = (y^2 - x^3)$.

(13) Let $\mathbf{R} = \mathbb{Z}[\sqrt{3}]$ and $I = (2\sqrt{3}+4)$. Describe $\mathbf{R}/I$.
(14) Show that in any ring $\mathbf{R}$, there is a unique subring which is isomorphic to $\mathbb{Z}/m\mathbb{Z}$.
(15) Let D be a multiplicative system in an integral domain $\mathbf{R}$. Show that there is a ideal I, which is maximal with $I \cap D = \emptyset$, is prime.
(16) Let $\mathbf{R}$ be an integral domain and D a multiplicative system. Show that $\mathbf{R}_D$ is Noetherian.
(17) Show that any finite integral domain is a field.
(18) Show that $\mathbb{Z}[i]/(2+3i)$ is a field. What is its characteristic?
(19) Show that any non-zero prime ideal in $\mathbb{Z}[i]$ is maximal.
(20) Let $\mathbf{R}$ be an integral domain which is not a field. Show that $\mathbf{R}[x]$ is not a *P.I.D.*.
(21) Show that the following ring $\mathbf{R}$ is Euclidean,

$$R = \{a + b(1+\sqrt{-11})/2 : a, b \in \mathbb{Z}\}$$

(22) Show that $\mathbb{Z}[x]$ is not a *P.I.D.*.
(23) Let us use the notations of Exercise (4). Is $\mathbf{C}([0,1])$ Noetherian?
(24) Let $\mathbf{R} = \{a/b : a, b \in \mathbb{Z}, b \neq 0, 2 \nmid b, 3 \nmid b\} \subset \mathbf{Q}$. Is $\mathbf{R}$ Noetherian?
(25) Let $\mathbf{R}_i$ be Noetherian rings for $i = 1, 2, \cdots, n$. Show that the direct sum $\sum \oplus \mathbf{R}_i$ is Noetherian.
(26) Imitate the proof of Theorem 3.27. to show that if $\mathbf{R}$ is Noetherian, then the formal power series ring $\mathbf{R}[[x]]$ is Noetherian.
(27) Show that if $\mathbf{R}$ is Noetherian, then the formal power series ring in n variables, $\mathbf{R}[[x_1, \cdots, x_n]]$ is Noetherian.
(28) Show that $\mathbf{Q}[[x_1, \cdots, x_n, \cdots]] = \cup_{i=1}^{\infty} \mathbf{Q}[[x_1, \cdots, x_i]]$ is not Noetherian.

CHAPTER IV

Linear Algebra

§1 Vector Spaces

Let us give the definition of a *vector space*[1] as follows,

Definition 4.1. *Let* **K** *be a field and* **V** *a nonempty set. If* **V** *satisfies the following conditions, then* **V** *is said to be a vector space (over* **K***),*

(1) *There is a binary operation* ***addition***, $+$, *in* **V**. *With respect to* $+$, **V** *is a commutative additive group. The additive unit, denoted by 0, is called the* ***zero-vector***.

(2) *There is a* ***scalar multiplication***, *namely, for any* $a \in \mathbf{K}$ *and* $v \in \mathbf{V}$, *a new vector* $a \cdot v \in \mathbf{V}$ *is defined with the following* ***unitary*** *law,*

$$1 \cdot v = v, \quad \forall v \in \mathbf{K}$$

(3) *The four operations, the addition and multiplication of the field* **K**, *the addition and the scalar multiplication of the vector space* **V**, *satisfy the associative and distributive laws as follows,*

$$(a_1 a_2)v = a_1(a_2 v), \qquad (a_1 + a_2)v = a_1 v + a_2 v$$
$$a(v_1 + v_2) = av_1 + av_2$$

If **V** *is a vector space over* **K**, *then the elements in* **V** *are called* ***vectors*** *and the elements in* **K** *are called scalars. The field* **K** *is called* ***scalar field*** *or* ***ground field***. ∎

[1] Italian Mathematician Peano defined 'vector space' over **R** in 1888.

Discussion

(1) Suppose that a ring $\mathbf{R} \supseteq \mathbf{K}$ a field. Then $\mathbf{R}$ is a vector space over $\mathbf{K}$. For instance, $\mathbf{C}$ is a vector space over $\mathbf{R}$, $\mathbf{R}$ is a vector space over $\mathbf{Q}$, and any polynomial ring $\mathbf{K}[x_1, x_2, \cdots, x_n]$ over a field $\mathbf{K}$ is a vector space over the field $\mathbf{K}$.

(2) It is easy to see that $0v = 0$ and $(-a)v = -(av)$ from the following computations,

$$0v = (0+0)v = 0v + 0v$$
$$0v = 0v + 0v - 0v = 0v - 0v = 0$$
$$(-a)v + av = (-a+a)v = 0v = 0$$

∎

Example 1. *One of the basic theorems from plane geometry which we learned in high school is the following, "Given any two triangles $\Delta\, abc$ and $\Delta a'b'c'$, if the angles $\angle\, abc = \angle a'b'c'$ and the two sides are proportional, $\overline{ab} = r\overline{a'b'}$, $\overline{bc} = r\overline{b'c'}$, then the third sides are proportional, $\overline{ab} = r\overline{a'b'}$. This is the basic ingredient in the concept of vector spaces. We may establish the fundamental connections between plane geometry and vector spaces as follows.*

Let us fix a point in the plane as the origin. Let $\mathbf{V}$ be the set of all arrows starting from the origin. For our convenience, we may fix a cartesian coordinate system in the plane as follows,

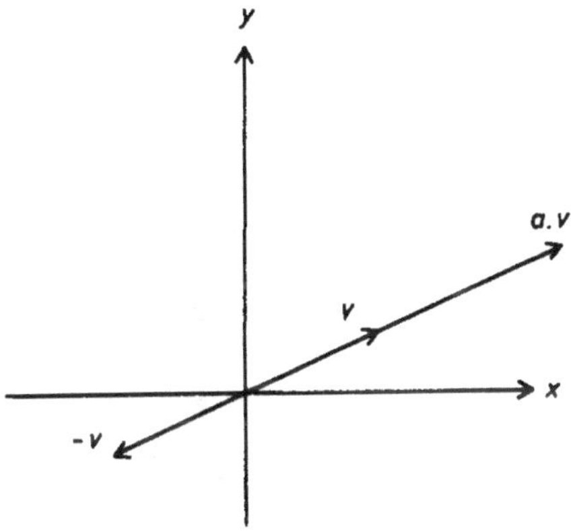

We shall use the tip (a_1, a_2) of an arrow v to represent the arrow. Let the field of real numbers $\mathbf{R}$ be the field $\mathbf{K}$. Let the scalar multiplication $a \cdot v$ be defined as follows,

$$a \cdot v = \begin{cases} \text{extend the vector } v \text{ by a proportionality of } a \text{ along the same direction} \\ \text{if } a \text{ is non-negative.} \\ \text{extend the vector } v \text{ by a proportionality of } |a| \text{ along the opposite direction} \\ \text{if } a \text{ is negative.} \end{cases}$$

Let the addition of two vectors v, u be defined as the diagonal arrow from the origin of the parallelogram spanned by the two vectors v, u. It is routine to see that **V** *is a vector space over* **R**.

Let us discuss the theorem of plane geometry mentioned at the beginning. Let us move the vertex b of the triangle Δabc to the origin. Let us move the vertex b' of the triangle $\Delta a'b'c'$ to the origin and the side $\overline{b'c'}$ to the line determined by the line segment $\overline{bc}$. Since $\angle abc = \angle a'b'c'$, if necessary by a further reflection, then we may assume that the side $\overline{a'b'}$ is on the line determined by the line segment $\overline{ab}$. It follows from the basic properties of parallelograms that the third side $\overline{ca}$ of the triangle Δaoc is parallel transformation of the vector $\overrightarrow{od}$ as in the following diagram,

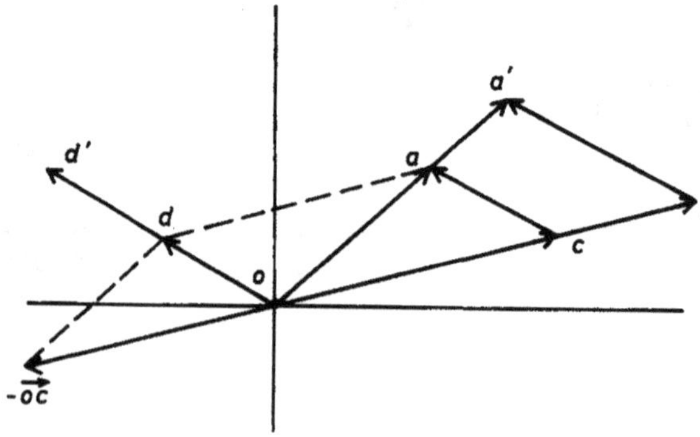

Furthermore, it follows from the basic properties of parallelograms that

$$\overrightarrow{od} = \overrightarrow{oa} - \overrightarrow{oc}$$

Thus our theorem in plane geometry follows from the basic rules of vector spaces as follows,

$$\begin{aligned}\overrightarrow{od'} &= \overrightarrow{oa'} - \overrightarrow{oc'} \\ &= r\overrightarrow{oa} - r\overrightarrow{oc} \\ &= r(\overrightarrow{oa} - \overrightarrow{oc}) \\ &= r\overrightarrow{od}\end{aligned}$$

∎

Example 2. *In the preceding Example, let us fix a coordinate system, we may then use the tip (a_1, a_2) of a vector $\overrightarrow{v}$ to represent the vector $\overrightarrow{v}$. In this way, we establish a* *representation of* $\mathbf{V}$ *by* $\mathsf{R} \oplus \mathsf{R} = \mathsf{R}^2$, *where the vector space operations on* R^2 *are defined as*

$$(a_1, a_2) + (b_1, b_2) = (a_1 + b_1, a_2 + b_2)$$
$$r(a_1, a_2) = (ra_1, ra_2), \qquad \text{where } r \in \mathsf{R}$$

Similarly, we may use $\mathbf{K}^n = \mathbf{K} \times \cdots \times \mathbf{K}$ *to denote the vector space over any field* $\mathbf{K}$ *of all n-tuples* $(a_1, \cdots, a_n)$ *of entries in* $\mathbf{K}$ *with the vector space operations defining as follows,*

$$(a_1, \cdots, a_n) + (b_1, \cdots, b_n) = (a_1 + b_1, \cdots, a_n + b_n)$$
$$r(a_1, \cdots, a_n) = (ra_1, \cdots, ra_n), \qquad \text{where } r \in \mathbf{K}$$

Furthermore, given any index set I, *we may use* $\prod_{i \in I} \mathbf{K}$ *to denote the vector space over any field* $\mathbf{K}$ *of the set* $\{(a_i)_{i \in I} : a_i \in \mathbf{K}.\}$ *with the vector space operations defining as follows,*

$$(a_i) + (b_i) = (a_i + b_i)$$
$$r(a_i) = (ra_i), \quad \text{where } r \in \mathbf{K}$$

∎

Example 3. *Let us consider a system of linear equations as follows,*

$$a_{11}x_1 + a_{12}x_2 + \cdots + a_{1n}x_n = 0$$
$$a_{21}x_1 + a_{22}x_2 + \cdots + a_{2n}x_n = 0$$
$$\cdots\cdots\cdots\cdots$$
$$\cdots\cdots\cdots\cdots$$
$$a_{m1}x_1 + a_{m2}x_2 + \cdots + a_{mn}x_n = 0$$

where $a_{ij} \in \mathbf{K}$ *a field. We may consider the set of all n-tuples* $\{(x_1, \cdots, x_n)\}$ *as the vector space* $\mathbf{K}^n$. *Let* $\mathbf{V} = \{(a_1, a_2, \cdots, a_n) : x_1 = a_1, x_2 = a_2, \cdots, x_n = a_n$ *is a solution for the system of equations.*$\}$. *Then* $\mathbf{V} \ni (0, 0, \cdots, 0)$, *and* $\mathbf{V}$ *is not empty. It is easy to see that*

$$\alpha = (a_1, a_2, \cdots, a_n),\ \beta = (b_1, b_2, \cdots, b_n) \in V$$
$$\Rightarrow \alpha + \beta \in V, c \cdot \alpha \in V$$

Therefore, $\mathbf{V}$ *is a subspace of* $\mathbf{K}^n$. ∎

Example 4. *Let the field* $\mathbf{K}$ *be the real field* R *or the complex field* $\mathbf{C}$. *Let us take a homogeneous ordinary differential equation with constant coefficients as follows,*

$$\sum_0^n a_i \frac{d^i f(x)}{d\, x^i} = (\sum_0^n a_i \frac{d^i}{d\, x^i}) f(x) = L(D) f(x) = 0, \qquad a_n \neq 0.$$

where $a_i \in \mathbf{K}$, $D = \frac{d}{dx}$, *and* $f(x)$ *is n-times differentiable. Let*

$$V = \{f(x) : L(D)f(x) = 0, f(x) \text{ n-times differentiable.}\}$$

Then we have the zero function $0 \in V$. *Let us use the usual addition and scalar multiplication of functions for* V. *Then it is easy to see that* V *forms a vector space.* ∎

Definition 4.2. *Let* **V** *be a vector space and* **U** *a subset of* **V**. *If* **U** *is a vector space with respect to the same operations, addition and scalar multiplication, of* **V**, *then* **U** *is said to be a* ***subspace*** *of* **V**. *Let* **S** *be a subset of* **V**. *Then the smallest subspace* [**S**] *of* **V** *which contains* **S** *is said to be the subspace* ***generated*** *or* ***spanned*** *by* **S**, *and* **S** *is said to be a* ***generating set*** *of* [**S**]. ∎

Discussion

(1) A vector space **V** is a subspace of itself. The zero set $\{0\}$ is a subspace of **V**.

(2) In fact, the subspace [**S**] generated by **S** is uniquely determined by **S**: If **S** is the empty set $\emptyset$, then it is clear that $[\emptyset]=\{0\}$. Assume that **S** is not empty. Let

$$U = \{\sum_{finite} a_i s_i : a_i \in K, s_i \in S\}$$

Then it is easy to see that **U** is a subspace of **V** which contains **S**. Let **W** be any subspace of **V** which contains **S**. Then we have

$$W \ni s_i \ \forall s_i \in S \quad \Rightarrow \sum_{finite} a_i s_i \in W$$

$$U \subset W$$

Therefore, **U** is the smallest subspace which contains **S**.

(3) It is easy to see [**U**]=**U** for any subspace **U** of **V**. Therefore, there is at least one generating set for any given subspace **U**.

(4) If **T**⊂**S**⊂**V**, then we have [**T**]⊂ [**S**].

(5) Let **U** and **W** be two subspaces of a vector space **V**. We use **U**+**W** to denote $\{u+w : u \in \mathbf{U}, w \in \mathbf{W}\}$. Then it is easy to see that **U**+**W** is a subspace of **V**. In fact, it is [**U**∪ **W**], the subspace generated by all elements in **U** ∪ **W**.

(6) Let **U** be a subspace of **V**. Then we may form the ***quotient space*** **V**/**U** as follows: Consider **V**, **U** as commutative groups, form the quotient commutative group **V**/**U** with elements denoted by $[v]$, and define the scalar multiplication $k[v] = [kv]$. It is routine to see that **V**/**U** is a vector space. ∎

Exercises

(1) Use vector space to show that the line passing through the middle points of two sides of a triangle is parallel to the third side.
(2) Let Δabc be any triangle. Let us mark down a point b' on the line $\overline{ab}$ such that the length $\overline{ab'}$ is r times the total length of $\overline{ab}$. Similarly, we mark down c' with the same r. Use vector space to show that the line passing through the points b', c' is parallel to the third side $\overline{bc}$.
(3) Show that the set of all real valued functions on a set **S** forms a vector space over the real numbers **R** under the usual addition and scalar multiplication.
(4) Show that the set of all real valued functions on the real space $\mathbf{R}^n$ forms a vector space over the real numbers **R** under the usual addition and scalar multiplication.
(5) Show that the set of all real valued continuous functions on the real space $\mathbf{R}^n$ forms a vector space over the real numbers **R** under the usual addition and scalar multiplication.
(6) Show that the set of all real valued differentiable functions on the real space $\mathbf{R}^n$ forms a vector space over the real numbers **R** under the usual addition and scalar multiplication.
(7) Show that the set of all complex valued functions on a set **S** forms a vector space over the complex numbers **C** under the usual addition and scalar multiplication.
(8) Show that the set of all polynomials $\mathbf{R}[x_1, x_2, \cdots x_n]$ on the real space $\mathbf{R}^n$ forms a vector space over the real numbers **R** under the usual addition and scalar multiplication.
(9) Show that the set of all polynomials of degree less than n, P_n, of one variable x with coefficients in a field **K** forms a vector space over **K** under the usual addition and scalar multiplication.
(10) Show that $\{1, x, \cdots, x^{n-1}\}$ is a generating set for P_n (cf Problem 9).
(11) Let **K** be a field. Show that $\prod_{i \in I} \mathbf{K}$ is a vector space over **K** with respect to the usual addition and multiplication.
(12) Let $\alpha_1, \alpha_2, \alpha_3$ be vectors in $\mathbf{R}^4$ as follows,

$$\alpha_1 = (1, 0, 2, 3), \quad \alpha_2 = (2, 1, 1, 1), \quad \alpha_3 = (0, -1, 3, 5)$$

Find a system of homogeneous linear equations for which the space of solutions is exactly the subspace of $\mathbf{R}^4$ spanned by these three vectors.

§2 Basis and Dimension

Definition 4.3. *Let* **V** *be a vector space over a field* **K**. *A subset* **S** *is said to be a* ***minimal generating set*** *if the following conditions are satisfied,*

(1) **S** *is a generating set of* **V**, *namely [S]=V.*

(2) *If any element s is deleted from* S, *then the remaining set is not a generating set of* V, *namely* $[S\backslash s] \neq V$.

A minimal generating set is said to be a ***basis*** *of* V. ∎

Definition 4.4. *Let* V *be a vector space over a field* K. *A collection of elements* $\{s_i\}_{i \in I}$ *is said to be* ***linearly independent*** *if the following condition is satisfied,*

(1) *Any equation* $\sum_{finite} a_i s_i = 0$ *where* $a_i \in K$, $i \in I \Rightarrow a_i = 0 \ \forall i$.

A collection of elements $\{s_i\}_{i \in I}$ *is said to be* ***linearly dependent*** *if it is not linearly independent. If a linearly independent (resp. dependent) collection is a set* S, *then it is said to be a* ***linearly independent (resp. dependent) set****. A linearly independent set* S *is said to be a* ***maximal linearly independent set*** *if the following condition is satisfied,*

(2) *Any subset* T *properly containing* S *is a linearly dependent set.*

∎

Discussion

(1) The empty set $\emptyset$ is a linearly independent set. The zero set $\{0\}$ is a linearly dependent set.

(2) If S is a linearly independent set, then any subset of S is a linearly independent set. If S is a linearly dependent set, then any set T between V and S is a linearly dependent set.

(3) Any collection with repeated elements is linearly dependent. Note that in this case we can not talk about sets.

(4) A linear relation as follows

$$\sum_{finite} a_i v_i = 0$$

is said to be *trivial* if all coefficients a_i are zeroes. Otherwise, the relation is *non-trivial.*

∎

Example 5. *Let us continue our discussion of Example 3. Given the following system of linear equations,*

$$\begin{aligned} a_{11}x_1 + a_{12}x_2 + \cdots + a_{1n}x_n &= 0 \\ a_{21}x_1 + a_{22}x_2 + \cdots + a_{2n}x_n &= 0 \\ &\cdots\cdots \\ &\cdots\cdots \\ a_{m1}x_1 + a_{m2}x_2 + \cdots + a_{mn}x_n &= 0 \end{aligned}$$

where $a_{ij} \in$ K *a field. We may use 'elimination theory'*[2] *to produce an equivalent system of linear equations, with reindexing* x_i *if necessary, as follows,*

[2]It was first invented in the Chinese book "Chiu Chang Suan Shu" of the first century A.D., later on rediscovered by Gauss, and customarily known as 'Gaussian elimination'

$$
\begin{array}{l}
x_1 \qquad -b_{1,s+1}x_{s+1} - \cdots - b_{1,n}x_n = 0 \\
\quad x_2 \qquad -b_{2,s+1}x_{s+1} - \cdots - b_{2,n}x_n = 0 \\
\qquad \cdots\cdots \\
\qquad \cdots\cdots \\
\qquad\qquad x_s - b_{s,s+1}x_{s+1} - \cdots - b_{s,n}x_n = 0
\end{array}
$$

where $s \le n$. If $s = n$, then clearly $(0, 0, \cdots, 0)$ is the only solution of the system of linear equations. Let assume $s < n$. Let t be an integer with $s < t \le n$, and $v_t = (b_{1,t}, b_{2,t}, \cdots, b_{s,t}, 0, \cdots, 1, \cdots, 0)$ where the last 1 appears at the $t-th$ place. Clearly, we have $v_t \in \mathbf{V}$ the solution subspace. It is not hard to see that $\{v_{s+1}, \cdots, v_n\}$ is a maximal linearly independent set of the vector space $\mathbf{V}$. ∎

Theorem 4.1. *Let $\mathbf{V}$ be a vector space over a field $\mathbf{K}$. Let $\mathbf{S}$ be a subset of $\mathbf{V}$. Then the following three conditions are equivalent, therefore any one of them can be used as the definition of a basis,*

(1) $\mathbf{S}$ *is a minimal generating set.*
(2) $\mathbf{S}$ *is a maximal linearly independent set.*
(3) $\mathbf{S}$ *is a linearly independent generating set.*

Proof:
((1)⇒ (2)).
(A) We claim that $\mathbf{S}$ is a linearly independent set. If not, then there is a linear equation of the following form,

$$a_1 s_1 + a_2 s_2 + \cdots + a_n s_n = 0$$

with at least one of a_i's non-zero. We may assume that $a_1 \neq 0$. Since $\mathbf{K}$ is a field, then the inverse a_1^{-1} exists. Multiplying the above equation by a_1^{-1}, and letting $b_i = -a_i a_1^{-1}$, we get the following equation,

$$s_1 = b_2 s_2 + \cdots + b_n s_n \tag{1}$$

Now we shall deduce that $\mathbf{S}\backslash\{s_1\}$ is a generating set for $\mathbf{V}$. Note that it contradicts the minimal property of $\mathbf{S}$ being a generating set for $\mathbf{V}$.

Given any element $v \in \mathbf{V}$, we have the following equation, since $\mathbf{S}$ is a generating set for $\mathbf{V}$,

$$v = c_1 s_1 + c_2 s_2 + \cdots + c_m s_m \tag{2}$$

where c_1 may be zero. Substituting Eq (1) into Eq (2) to eliminate s_1, we have a new expression as follows

$$v = (c_2 - c_1 b_2)s_2 + \cdots + (c_n - c_1 b_n)s_n + b_{n+1}s_{n+1} \cdots + c_m s_m \in [\mathbf{S}\backslash\{s_1\}]$$

Therefore, we deduce that $\mathbf{S}\backslash\{s_1\}$ is a generating set for $\mathbf{V}$. A contradiction.

(B) We claim that S is a maximal linearly independent set. Let $t \notin$ S. Since S is a generating set for V, then we must have the following equations,

$$t = d_1 s_1 + d_2 s_2 + \cdots + d_m s_m$$
$$0 = (-1)t + d_1 s_1 + d_2 s_2 + \cdots + d_m s_m$$

The last equation establishes that S$\cup\{t\}$ is a linearly dependent set.

((2)$\Rightarrow$(3)).

We want to show that [S]=V. Certainly we have S$\subset$ [S]. Let $t \notin$S. It follows from maximality of S that there must be a non-trivial relation of the following form,

$$bt + \sum_{finite} a_i s_i = 0$$

If $b = 0$, then we have a non-trivial relation among elements of S, impossible! Therefore, $b \neq 0$ and b^{-1} exists. After multiplying the above equation by b^{-1} and moving terms, we produce the following equation,

$$t = \sum_{finite} (-b^{-1} a_i) s_i \in [S]$$

Henceforth (3) is verified.

((3)$\Rightarrow$ (1)).

If S is not a minimal generating set, then there exists $s \in$ S such that S$\backslash\{s\}$ is a generating set. Therefore, we have the following equations,

$$s = \sum_{finite} a_i s_i, \qquad \text{where } s_i \neq s \ \forall\, i$$
$$0 = (-1)s + \sum_{finite} a_i s_i, \qquad \text{where } s_i \neq s \ \forall\, i$$

Clearly the last equation is a non-trivial linear equation among elements of S, which contradicts the linearly independent property of S. ∎

Example 6 Let us continue our discussion of Example 4. Let the field **K** be the real field **R** or the complex field **C**. Let us take a homogeneous ordinary differential equation with constant coefficients as follows,

$$\sum_0^n a_i \frac{d^i f(x)}{d\, x^i} = (\sum_0^n a_i \frac{d^i}{d\, x^i}) f(x) = L(D) f(x) = 0, \qquad a_n \neq 0.$$

where $a_i \in \mathbf{K}$, $D = \frac{d}{dx}$, and $f(x)$ is n-times differentiable. Let

$$V_{L(D)} = \{f(x) : L(D)f(x) = 0, f(x) \text{ n-times differentiable.}\}$$

Let us assume that

$$L(D) = \prod (D - a_j)^{m_j+1}, \qquad m_j \geq 0.$$

It will be obvious in Chapter V that the above assumption is not a restriction on $L(D)$ if $\mathbf{K}=$ C. We wish to find a basis of $\mathbf{V}_{L(D)}$. The following argument works for any characteristic zero field such that Lemma 1. holds and the exponential function e^x is defined. The usual method of using the *Wronskian* is not applied here. Let us prove the following lemmas,

Lemma 1. *If $Df(x) = 0$, then $f(x) = c$, a constant function.*

Proof: This is a place we have to use the completeness of the real field **R** or the complex field **C**. Suppose that $f(x)$ is not a constant function, say $f(a) \neq f(b)$. We may assume that the real parts of $f(a)$ and $f(b)$ are not equal. Connecting the points a, b by a real curve, it follows from the *mean value theorem* that there is a point c on the real curve such that $Df(c) \neq 0$. Contradiction. ∎

Lemma 2. *We have the following*

(1) $(D - a_1)(D - a_2) = (D - a_2)(D - a_1)$.
(2) *If $L(D) = L'(D)L''(D)$, then $L(D)f(x) = L'(D)(L''(D)f(x))$*

Proof: Direct checking. ∎

We shall use the following exponential functions

$$e^{ax} = 1 + ax + \frac{1}{2!}(ax)^2 + \cdots + \frac{1}{m!}(ax)^m + \cdots$$

The important property of the above functions is

$$D(e^{ax}) = ae^{ax}, \qquad \text{or} \qquad (D - a)e^{ax} = 0.$$

which can be verified by using power series expansions. Note that if the characteris... the field $\mathbf{K}$ is $p > 0$, then we always have

$$D^p(f(x)) = 0$$

for any power series $f(x)$. Therefore, the preceding differential equation $(D-a)(f(x)) = 0$ with $a \neq 0$ has no power series solution. We have the following lemma,

Lemma 3. *For any non-negative integer s, we have*

$$(D-c)^s(x^i e^{ax}) = [(a-c)^s x^i + \binom{s}{1}(a-c)^{s-1} i x^{i-1} + \cdots$$
$$+ \binom{s}{j}(a-c)^{s-j}(i(i-1)\cdots(i-j+1))x^{i-j} + \cdots$$
$$+ \binom{s}{s}(i(i-1)\cdots(i-s+1))x^{i-s}]e^{ax}$$

Proof: It is easy to see the above formulas are true for $s = 0, 1$. Inductively, it is easy to pass from s to $s+1$. ∎

Corollary 1. *For $a = c$ and $s > i$, we always have*

$$(D-a)^s(x^i e^{ax}) = 0$$

∎

Corollary 2. *For $a = c$ and $s = i$, we always have*

$$(D-a)^s(x^s e^{ax}) = s!\ e^{ax}$$

∎

Corollary 3. *For $a \neq c$, and $g(x) \in \mathbf{K}[x]$, we always have*

$$(D-c)^s(g(x)e^{ax}) = g^*(x)e^{ax}$$

where $g^*(x) \in \mathbf{K}[x]$, and $deg\ g(x) = deg\ g^*(x)$. ∎

Corollary 4. *For $a \neq c$, we always have*

$$(D-c)^s(e^{ax}) = (a-c)^s e^{ax}$$

∎

The following lemma is of interest, although not needed, for our arguments,

Lemma 4. *The set $\{e^{a_j x} : a_j \in \mathbf{K}$ and all distinct$\}$ is linearly independent.*

Proof: If it is linearly dependent, let the following be a non-trivial relation involving the smallest number of a_i's,

$$\sum_1^m c_j e^{a_j x} = 0$$

Note that $c_j \neq 0 \; \forall \, j$ and $m > 1$. Applying $(D - a_1)$ to the above equation, we produce the following non-trivial relation with a smaller number of $a_j's$,

$$(D - a_1)(\sum_1^m c_j e^{a_j x}) = \sum_2^m (a_j - a_1) c_j e^{a_j x} = 0$$

Therefore, we produce a contradiction. ∎

We have the following important lemma,

Lemma 5. *The set* $\{x^i e^{a_j x}$: *i non-negative integer,* $a_j \in \mathbf{K}$, *and all pairs* (i, a_j) *are distinct.* $\}$ *is linearly independent.*

Proof: We shall make an induction on the number s of distinct $a_j's$. Let us consider the case with $s = 1$. Let us consider all linear relations among $\{x^i e^{a_1 x}$: i non-negative integer.$\}$. Let the following, $g_1(x)$, be the one with the smallest degree, note that its degree ≥ 1,

$$\sum_0^{m_1} b_i x^i e^{a_1 x} = g_1(x) e^{a_1 x} = 0$$

Under our assumption, such a $g_1(x)$ exists. Applying $(D - a_1)$ to the above equation, we get

$$D(g_1(x)) e^{a_1 x} = 0$$

Since the degree of $D(g_1(x))$ is less than the degree of $g_1(x)$, we get a contradiction.

Let us assume that the lemma is true for s, and there are $s + 1$ distinct $a_j's$. If they are not linearly independent, then there must be a non-trivial linear relation of the following form

$$\begin{aligned} \sum_{finite} b_{ij} x^i e^{a_j x} &= \sum_j (\sum_i b_{ij} x^i) e^{a_j x} \\ &= \sum_j g_j(x) e^{a_j x}, \qquad g_j(x) \neq 0. \end{aligned}$$

where $g_j(x)$ is a polynomial of degree m_j.

Applying $L^*(D) = (D - a_1)^{m_1}$ to the above equation, it follows from the Corollaries of Lemma 3. that

$$L^*(D)(g_1(x) e^{a_1 x}) = 0$$
$$L^*(D)(g_j(x) e^{a_j x}) = g_j^*(x) e^{a_j x}, \qquad \forall \, j > 1$$
$$L^*(D)(\sum_j g_j(x) e^{a_j x}) = \sum_j g_j^*(x) e^{a_j x} = 0$$

Note that $deg\ g_j(x) = deg\ g_j^*(x)$, and $g_j^*(x) \neq 0$. Therefore, we produce a new non-trivial equation involving less number of $a_j's$, a contradiction. ∎

Corollary. *The functions* $\{1, x, x^2, \cdots, x^n, \cdots\}$ *are linearly independent.*

Let us assume as before that

$$L(D) = \prod_j (D - a_j)^{m_j+1}, \qquad m_j \geq 0, \qquad \sum_j (m_j + 1) = n.$$

It follows from Lemma 2. and Corollary 1. of Lemma 3. that for all $i \leq m_k$, we have

$$\begin{aligned} L(D)(x^i e^{a_k x}) &= \prod_{j \neq k} (D - a_j)^{m_j+1} (D - a_k)^{m_k+1} (x^i e^{a_k x}) \\ &= \prod_{j \neq k} (D - a_j)^{m_j+1} (0) \\ &= 0 \end{aligned}$$

Therefore, Lemma 5. implies that the following set $S_{L(D)}$ is a solution set of n linearly independent functions,

$$S_{L(D)} = \{x^i e^{a_j x} : 0 \leq i \leq m_j\}$$

We wish to prove that it is a basis for the vector space $V_{L(D)}$. Note that then a general solution of $L(D)(f(x)) = 0$ will be a linear combination of them. For this purpose, we will prove the following lemma,

Lemma 6. *We have*

$$V_{D-a} = \{f(x) : (D - a)(f(x)) = 0\} = \{ce^{ax} : c \in \mathbf{K}\}$$

Proof: Let $f(x) \in V_{D-a}$, and define

$$h(x) = \frac{f(x)}{e^{ax}}.$$

Then we have the following computations,

$$\begin{aligned} D(h(x)) &= \frac{e^{ax} D(f(x)) - f(x) D(e^{ax})}{e^{2ax}} \\ &= \frac{e^{ax} a(f(x)) - f(x) a(e^{ax})}{e^{2ax}} \\ &= 0 \end{aligned}$$

It follows from Lemma 1. that

$$h(x) = c, \qquad f(x) = ce^{ax}$$

Now we have the desired result,

Theorem. *The set $S_{L(D)}$ is a basis for $V_{L(D)}$.*

Proof. If the order n of the linear differential operator $L(D)$ is 1, then our Theorem follows from Lemma 6. Let us assume that $n > 1$, and

$$\begin{aligned} L(D) &= (D - a_1)L'(D) \\ L'(D) &= (D - a_1)^{m_1} \prod_{j>1} (D - a_j)^{m_j+1} \\ &= \prod_{j>1} (D - a_j)^{m_j+1} (D - a_1)^{m_1} \\ &= L''(D)(D - a_1)^{m_1} \end{aligned}$$

By induction on the order of the linear differential operator, we shall assume that the Theorem is true for $L'(D)$. It follows from Theorem 4.1. that it suffices to prove $S_{L(D)}$ generates $V_{L(D)}$. Given any $f(x) \in V_{L(D)}$, let

$$L'(D)(f(x)) = g(x)$$

Then we have

$$(D - a_1)(g(x)) = L(D)(f(x)) = 0$$

Therefore, it follows from Lemma 6. that we have

$$\begin{aligned} g(x) &= ce^{a_1 x} \\ L'(D)(f(x)) &= ce^{a_1 x} \end{aligned}$$

It follows from the Corollaries of Lemma 6. that

$$\begin{aligned} L'(D)(x^{m_1} e^{a_1 x}) &= L''(D)(D - a_1)^{m_1}(x^{m_1} e^{a_1 x}) \\ &= L''(D)(m_1! e^{a_1 x}) \\ &= m_1! \prod_{j>1} (D - a_j)^{m_j+1} (e^{a_1 x}) \\ &= m_1! \prod_{j>1} (a_1 - a_j)^{m_j+1} (e^{a_1 x}) \\ &= qe^{a_1 x} \end{aligned}$$

where $q \neq 0$ is defined by the above equation. Therefore, there exists r such that

$$L''(D)(f(x) - rx^{m_1} e^{a_1 x}) = 0$$

It follows from the induction hypothesis that

$$f(x) = rx^{m_1}e^{a_1x} + \sum_{i=0}^{m_1-1} b_{i1}x^ie^{a_1x} + \sum_{i=0,j>1}^{m_j} b_{ij}x^ie^{a_jx}$$

Henceforth, $S_{L(D)}$ is a generating set of $V_{L(D)}$. ∎

Example 6. is finished. ∎

Example 7. *Let us give a concrete discussion of Example 6.. Let the field* **K** *be the real field* **R** *or the complex field* **C** *as before. Let us take a homogeneous ordinary differential equation with constant coefficients as follows,*

$$\frac{d^3\,f(x)}{d\,x^3} - 3\frac{d^2\,f(x)}{d\,x^2} + 2\frac{d\,f(x)}{d\,x} = (\frac{d^3}{d\,x^3} - 3\frac{d^2}{d\,x^2} + 2\frac{d}{d\,x})f(x)$$
$$= D(D-1)(D-2)(f(x)) = L(D)(f(x)) = 0.$$

We are asked to find all solutions of the above equation which further satisfy the following initial conditions,

$$f(0) = 0, \qquad D(f(0)) = 0$$

It follows from the Theorem of Example 6. and the initial conditions that

$$f(x) = y_1 \cdot 1 + y_2 \cdot e^x + y_3 \cdot e^{2x}$$
$$f(0) = y_1 + y_2 + y_3 = 0$$
$$D(f(0)) = 0y_1 + y_2 + 2y_3 = 0$$

Solving the last two equations, we get

$$f(x) = y_3 - 2y_3e^x + y_3e^{2x} = y_3(1 - 2e^x + e^{2x})$$

where y_3 is arbitrary. ∎

The following theorem establishes the existence of a basis for a given vector space by Zorn's Lemma (cf Section 1, Chapter I),

Theorem 4.2. *Let* **V** *be a vector space over a field* **K**. *We have the following,*

1. *Let T be a linearly independent set. Then there is a basis $B \supset T$ for* **V**. *Especially, if $T = \emptyset$, then there is a basis for* **V**.
2. *Let T be a generating set for* **V**. *Then there is a basis $B \subset T$ for* **V**. *Especially, if $T =$* **V**, *then there is a basis for* **V**.

Proof: (1) By Theorem 4.1., it suffices to show the existence of a maximal linearly independent set $B \supset T$. We shall use Zorn's Lemma. Let

$$\mathsf{F} = \{S : S \supset T, S \text{ is a linearly independent set.}\}$$

Then we have $T \in \mathsf{F}$. Therefore, $\mathsf{F} \neq \emptyset$. Let us define a partial ordering $\leq$ in F by inclusion $\subset$,

$$S_1 \leq S_2 \Leftrightarrow S_1 \subset S_2$$

(A) We claim the hypothesis of Zorn's Lemma is satisfied. Let G be any chain of F. Let

$$S = \cup S_i, \qquad S_i \in \mathsf{G}$$

We shall show that $S \in \mathsf{F}$, then S is an upper bound of G.

Suppose not. Then there is a non-trivial linear relation among the elements of S. Let the following be one,

$$\sum_{finite} a_j v_j = 0, \qquad a_j \in \mathbf{K}, \quad v_j \in S = \cup S_i$$

Let $v_j \in S_{n_j}$. Since there are finitely many j and $\{S_i\}$ is a chain, then there must be one S_{n_k} which contains all others. Therefore, we have $v_j \in S_{n_k} \forall\, j$, and the above relation is among elements in S_{n_k}. This contradicts the assumption of S_{n_k} are linearly independent.

(B) It follows from Zorn's Lemma that there is a maximal linearly independent set which is a basis.

(2) We will show the existence of a basis $B \subset T$. We shall use Zorn's Lemma. Let

$$\mathsf{F} = \{S : S \subset T, S \text{ is a linearly independent set.}\}$$

Then we have $\emptyset \in \mathsf{F}$. Therefore, $\mathsf{F} \neq \emptyset$. Let us define a partial ordering $\leq$ in F by inclusion $\subset$,

$$S_1 \leq S_2 \Leftrightarrow S_1 \subset S_2$$

As in (1) it is routine to see that the hypothesis of Zorn's Lemma is satisfied. Let B be a maximal element in F. We claim that B is a basis for $\mathbf{V}$.

Let t be any element in T. It follows from the maximal property of B that there is a non-trivial linear relation for the set $\{t\} \cup B$,

$$at + \sum_{finite} b_i v_i = 0, \qquad a, b_i \in \mathbf{K}, v_i \in B$$

Note that $a \neq 0$. Otherwise, the above will be a relation among elements of B. Factoring out $(-a)$ from the above equation, we conclude that $t \in [B]$. Therefore, $T \subset [B]$, and $\mathbf{V} \subset [B]$. We conclude B is a linearly independent generating set. ∎

Usually there are many bases for a given vector space $\mathbf{V}$ over a field $\mathbf{K}$. We shall prove all bases have the same cardinality. We need the following lemma.

Lemma. *Given a vector space* $\mathbf{V}$ *over a field* $\mathbf{K}$. *Let* S *be a generating set of* $\mathbf{V}$ *and* S' *be a linearly independent set. Let* $W = S \cap S'$. *Then there exists a linearly independent subset* $R \subset S$ *and a bijective map* $\rho : R \to S'$ *with the restriction of* ρ *to* W *being identity. Therefore, we have,*

$$Card(S) \geq Card(S')$$

where as usual, Card is the cardinality.

Proof: Let $\mathsf{F} = \{(T, \rho, T')\}$, where

(i) $W \subset T \subset S, W \subset T' \subset S'$, and ρ is a bijective map: $T \to T'$, and the restriction of ρ to W being identity.

(ii) $T \cap (S' \setminus T') = \emptyset$, and $T \cup (S' \setminus T')$ is a linearly independent set.

The above conditions mean that the subset T' may be replaced by T which is with the same cardinality (as guaranteed by ρ), while after the replacement, the new set $T \cup (S' \setminus T')$ has the linearly independent property of S'. To prove our lemma, it suffices to show that $(T, \rho, S') \in \mathsf{F}$ for some T and ρ; it follows from elementary *set theory* that

$$Card(S) \geq Card(T) = Card(S')$$

(1) We claim that $\mathsf{F} \neq \emptyset$: since $(W, 1_W, W) \in \mathsf{F}$ where 1_W is the identity map of W.

(2) Let us define a partial ordering $\leq$ as follows, $(T_1, \rho_1, T_1') \leq (T_2, \rho_2, T_2')$ if and only if

(iii) $T_1 \subset T_2, T_1' \subset T_2'$.

(iv) The restriction of ρ_2 to T_1 equals ρ_1, namely,

$$\rho_2(t) = \rho_1(t), \qquad \forall\, t \in T_1$$

It is not hard to see that $\leq$ is a partial ordering.

(3) We want to show that the hypothesis of Zorn's Lemma is satisfied. Let $\mathsf{G} = \{(T_i, \rho_i, T_i')\}$ be a chain in F. Let us define

(v) $T = \cup T_i$, $T' = \cup T_i'$.

(vi) $\rho(t) = \rho_i(t)$ if $t \in T_i$

We will prove that the new element (T, ρ, T') thus defined is in F. It is then easy to see (T, ρ, T') is an upper bound of the chain G.

It is easy to check the condition (i),

$$W \subset T \subset S, W \subset T' \subset S', \text{ and } \rho \text{ is a bijective map: } T \to T'.$$

Let us check the condition (ii). We claim that $T \cap (S' \setminus T') = \emptyset$. Otherwise, let $t \in T \cap (S' \setminus T')$. Then we have $t \in T = \cup T_i$. Therefore, $t \in T_j$ for some j. Then we have

$$t \in T_j \cap (S' \setminus T') \subset T_j \cap (S' \setminus T_j') = \emptyset$$

A contradiction.

We claim that $T \cup (S' \setminus T')$ is a linearly independent set. Suppose not. Then there are finitely many elements $t_1, t_2, \cdots, t_n \in T \cup (S' \setminus T')$ which are linearly dependent. After reordering the indices of t_i if necessary, we may assume that

$$t_1 \in T_{m_1}, t_2 \in T_{m_2}, \cdots, t_s \in T_{m_s}, t_{s+1}, \cdots, t_n \in (S' \setminus T')$$

Since G is a chain, then there must be an r such that

$$t_1, t_2, \cdots, t_s \in T_r, \qquad (S' \setminus T') \subset (S' \setminus T'_r)$$

Therefore, we have

$$t_1, t_2, \cdots, t_n \in T_r \cup (S' \setminus T'_r)$$

This is a contradiction, because the last set is a linearly independent set.

(4) The hypothesis of Zorn's Lemma has been verified. It follows that there is a maximal element $(\overline{T}, \overline{\rho}, \overline{T}') \in$ F. We claim that $\overline{T}' = S'$. In other words, we will show that if $\overline{T}' \neq S'$, then $(\overline{T}, \overline{\rho}, \overline{T}')$ is not maximal. Note that our lemma will be established.

What we want to do is if $\overline{T}' \neq S'$, then we will find an element in F which is strictly greater than $(\overline{T}, \overline{\rho}, \overline{T}')$. We will construct the element as follows.

Assume $\overline{T}' \neq S'$. Let $s' \in S' \setminus \overline{T}'$. Since $\overline{T} \cup (S' \setminus \overline{T}')$ is a linearly independent set, then we have

$$s' \notin [\overline{T}'] = \text{ the subspace generated by } \overline{T}'.$$

We get

$$[\overline{T}'] \neq \mathbf{V} = [S]$$

Therefore, there must be an element $s \in S$ such that

$$s \notin [\overline{T}], \qquad \{s\} \cup \overline{T} \text{ is a linearly independent set.}$$

There are two possibilities: (A) $\{s\} \cup \overline{T} \cup (S' \setminus \overline{T}')$ is a linearly independent set. (B) $\{s\} \cup \overline{T} \cup (S' \setminus \overline{T}')$ is a linearly dependent set. In each case, we will construct an element in F which is strictly greater than $(\overline{T}, \overline{\rho}, \overline{T}')$.

Case (A), let $T^* = \{s\} \cup \overline{T}, T^{*'} = \{s'\} \cup \overline{T}'$, and the map ρ^* defined as

$$\rho^*(t) = \begin{cases} \rho(t) & if\ t \neq s \\ s' & if\ t = s \end{cases}$$

It is easy to see $(T^*, \rho^*, T^{*'}) \in$ F and

$$(\overline{T}, \overline{\rho}, \overline{T}') < (T^*, \rho^*, T^{*'})$$

Case (B), there is a non-trivial linear relation as follows,

$$as + \sum_{finite} a_i t_i + \sum_{finite} b_j t'_j, \qquad t_i \in \overline{T},\ t'_j \in \overline{T}' \tag{1}$$

Note that $a \neq 0$. Otherwise, the above will be a non-trivial relation among elements of the linearly independent set $\overline{T} \cup (S' \setminus \overline{T}')$. We may take $a = -1$. Then Eq (1) will be unique with $a = -1$. Otherwise, two distinct equations of the form Eq (1) will produce a third non-trivial equation after one equation is subtracted from the other. Note that the third equation will not involve s which is impossible. Furthermore, it is impossible for $b_j = 0 \forall j$. Otherwise, it will be a non-trivial relation among elements of the linearly independent set $\{s\} \cup \overline{T}$. Let us assume that $b_1 \neq 0$. Let $T^* = \{s\} \cup \overline{T}, T^{*'} = \{t'_1\} \cup \overline{T}'$, and the map ρ^* defined as

$$\rho^*(t) = \begin{cases} \rho(t) & if\ t \neq s \\ t'_1 & if\ t = s \end{cases}$$

We claim that $(T^*, \rho^*, T^{*'}) \in \mathsf{F}$. It will follow that

$$(\overline{T}, \overline{\rho}, \overline{T}') < (T^*, \rho^*, T^{*'})$$

which establishes our Lemma.

It is easy to verify condition (i). Let us concentrate on condition (ii). If $(T^* \cap (S' \setminus T^{*'})) \neq \emptyset$, then we must have $s \in (S' \setminus T^{*'}) \subset (S' \setminus \overline{T}')$, which means the unique equation (1) must be of the following form,

$$s - s = 0$$

Therefore, $s = t'_1 \notin (S' \setminus T^{*'}$, a contradiction. Henceforth, $(T^* \cap (S' \setminus T^{*'})) = \emptyset$.

We claim that $(T^* \cup (S' \setminus T^{*'}))$ is a linearly independent set. Otherwise, there will be a linear relation of the following form,

$$\alpha s + \sum_{finite} \alpha_i t_i + \sum_{j \neq 1} \beta_j t'_j, \qquad t_i \in \overline{T},\ t'_j \in \overline{T}' \tag{2}$$

Note that Eq (2) is in the form of Eq (1), therefore violating the uniqueness of Eq (1). We have finished the proof of $(T^*, \rho^*, T^{*'}) \in \mathsf{F}$, and constructed an element which is strictly greater than $(\overline{T}, \overline{\rho}, \overline{T}')$, a contradiction. Therefore, we must have $\overline{T}' = S'$. Our Lemma is thus proved. ∎

We have the following theorem and definition,

Theorem 4.3. *Let T be a linearly independent set of a vector space $\mathbf{V}$ over a field $\mathbf{K}$. For any two bases $B_1, B_2 \supset T$, there is a bijective map:$(B_1 \setminus T) \to (B_2 \setminus T)$. Especially,* *let $T = \emptyset$, then any two basis have the same cardinalities. The common cardinality of all bases is called the dimension of* **V**, *in notation, dim_K* **V** *or dim* **V**.

Proof: Let B_1 and B_2 be any two bases of **V** which contain T. It follows from our Lemma that there exists a subset $R \subset B_1$ and a bijective $\rho : R \to B_2$ with the restriction of ρ to T being identity. In other words, there is an injective map

$$\rho : B_2 \setminus T \to B_1 \setminus T$$

By the same argument, we have an injective map.

$$\phi : B_1 \setminus T \to B_2 \setminus T$$

Therefore, it follows from *set theory* that there is a bijective map $:B_1 \setminus T \to B_2 \setminus T$. ∎

Definition 4.5. *Let* **U**, **W** *be two subspaces of a vector space* **V**. *The vector space* **V** *is said to be a direct sum of* **U**, **W**, **V**=**U**⊕ **W**, *if* **V**= **U**+**W** (=[**U**∪**W**]) *and* **U**∩ **W**= $\{0\}$. ∎

Discussion

(1) **V**=**U**⊕ **W**, if and only if for any basis C of **U** and any basis D, **V**=$[C \cup D]$ and the collection of all elements of C and D are linearly independent.

(2) Given any subspace **U** of a vector space **V**, there are subspaces **W** such that **V**=**U**⊕ **W**: let C be any basis of **U**. It follows from Theorem 4.2. that C can be extended to a basis B of **V**. Let **W** be $[B \setminus C]$. Then it is trivial to see **V**=**U**⊕ **W**. ∎

Corollary. *Let* **U**, $\mathbf{W}_1$ *and* $\mathbf{W}_2$ *be subspaces of a vector space* **V** *such that* **V**=**U**⊕$\mathbf{W}_1$ =**U**⊕$\mathbf{W}_2$. *Then we have*

$$dim\ W_1 = dim\ W_2$$

The common cardinality of the above is said to be the codimension of **U** *in* **V**, $codim_V(U)$. *Especially, in case dim V is finite, we have*

$$codim_V(U) = dim\ V - dim\ U$$

Proof: Let C be a basis of **U** and D_i be a basis of $\mathbf{W}_i$, for $i = 1, 2$. Then we have $C \cap B_i = \emptyset$ and $C \cup B_i$ basis for **V**. Our corollary follows from Theorem 4.3.. ∎

Theorem 4.4. *Let* **U**, **W** *be subspaces of a vector space* **V**. *Then we have*

(1) *dim* **U**+*dim* **W**=*dim* (**U**+**W**)+*dim* (**U**∩ **W**).

(2) *codim* **U**+*codim* **W**=*codim* (**U**+**W**)+ *codim* (**U**∩ **W**).

Proof: Clearly (2) follows from (1). Let $\mathbf{R}=(\mathbf{U}\cap \mathbf{W})$, and $\mathbf{U}=\mathbf{R}\oplus \mathbf{U}_1$, $\mathbf{W}=\mathbf{R}\oplus \mathbf{W}_1$ where $\mathbf{U}_1$, $\mathbf{W}_1$ are subspaces. Let C, D, E be bases for $\mathbf{U}_1$, $\mathbf{W}_1$, $\mathbf{R}$ respectively. Then it is easy to see

(1) $\mathbf{U}+\mathbf{W}=\mathbf{U}_1\oplus\mathbf{R}\oplus\mathbf{W}_1$.
(2) $C\cap E=D\cap E=C\cap D=\emptyset$.
(3) $C\cup E$ is a basis for $\mathbf{U}$, $D\cup E$ is a basis for $\mathbf{W}$, and $C\cup D\cup E$ is a basis for $\mathbf{U}+\mathbf{W}$.

Conclusion (1) follows by direct counting. ∎

Exercises

(1) Let $\mathbf{K}$ be a field. Consider $\mathbf{K}[x]$ as a vector space over $\mathbf{K}$. Let $f_i(x)\in\mathbf{K}[x]$ for $i=1,2,\cdots,n$ with all distinct degrees. Show that $\{f_i(x)\}$ is a linearly independent set.

(2) In the preceding problem, replace 'degree' by 'order' and form the same conclusion.

(3) Let $\mathbf{R}^3=\mathbf{R}\oplus\mathbf{R}\oplus\mathbf{R}$ be the three dimensional real vector space. Let $e_1^{(3)}=(1,0,0), e_2^{(3)}=(0,1,0)$, $e_3^{(3)}=(0,0,1)$, $\xi_1=(-1,1,1)$, $\xi_2=(1,-1,1)$, $\xi_3=(1,1,-1)$. Show that

(i) the set $\{e_1^{(3)},e_2^{(3)},e_3^{(3)}\}$ is a basis.

(ii) express ξ_i as a linear combination of $\{e_1^{(3)},e_2^{(3)},e_3^{(3)}\}$.

(4) Let $\mathbf{R}^3=\mathbf{R}\oplus\mathbf{R}\oplus\mathbf{R}$ be the three dimensional real vector space. Find all maximal linearly independent subsets of the set of four vectors $\{(1,2,3), (4,5,6), (7,8,9), (10,11,12)\}$.

(5) Let $L(D)$ be the following linear differential operator,

$$L(D)=D(D-1)^2$$

Find all solutions of the homogeneous differential equation $L(D)f(x)=0$ over real numbers.

(6) Let $L(D)$ be defined as above. Find all solutions of the nonhomogeneous differential equation $L(D)f(x)=e^x$ over real numbers.

(7) Find the dimension of $FL(4,\mathbf{R})$ as vector space over $\mathbf{R}$. Let $\mathbf{U}$ be the subspace spanned by all matrices of the form $AB-BA$ with $A,B\in FL(4,\mathbf{R})$. Find the dimension of $\mathbf{U}$.

(8) Let $\mathbf{U}$ and $\mathbf{W}$ be subspaces of a vector space $\mathbf{V}$. Show that $\mathbf{V}=\mathbf{U}\oplus\mathbf{W}$ if and only if *dim* $\mathbf{V}$=*dim* $\mathbf{U}$+*dim* $\mathbf{W}$.

(9) Let $\mathbf{U}_i$ be a subspace of a vector space $\mathbf{V}$ for $i=1,2,\cdots,n$. Suppose that $\mathbf{V}=\mathbf{U}_1\oplus\mathbf{U}_2\oplus\cdots\oplus\mathbf{U}_n$. Show that the collection of bases, one from each $\mathbf{U}_i$, is a basis of $\mathbf{V}$.

(10) Suppose that $\mathbf{K}$ is an infinite field and $\mathbf{V}$ is a non-zero vector space over $\mathbf{K}$. Show that there are infinitely many bases for $\mathbf{V}$.

(11) Suppose that $\mathbf{K}$ is a finite field of $p^m = q$ elements and $\mathbf{V}$ is an n-dimensional vector space over $\mathbf{K}$. Show that the number of bases of $\mathbf{V}$=$(q^n - 1)(q^n - q)(q^n - q^2)\cdots(q^n - q^{n-1})$.

(12) Let a subspace $\mathbf{U}$ of the n-dimensional real vector space $\mathbf{R}^n$ be given by $\mathbf{U} = \{(a_1, a_2, \cdots, a_n) : \sum a_i = 0\}$. Find a basis for $\mathbf{U}$.

(13) (Cf Exercise (11), Sect 2, Chapter II) Let *Symm(n,R)* be the set of all real symmetric matrices, and $Symm^*(n, \mathbf{R})$ be the set of all real anti-symmetric matrices, namely

$$Symm(n, \mathbf{R}) = \{(a_{ij}) : a_{ij} = a_{ji}\}$$
$$Symm^*(n, \mathbf{R}) = \{(a_{ij}) : a_{ij} = -a_{ji}\}$$

Show that both *Symm(n,R)* and $Symm^*(n, \mathbf{R})$ are subspaces of the vector space $\mathbf{FL}$(n, $\mathbf{R}$). Furthermore, show that

$$\mathbf{FL}(n, \mathbf{R}) = Symm(n, \mathbf{R}) \oplus Symm^*(n, \mathbf{R}).$$

§3 Linear Transformation and Matrix

We have the following definition,

Definition 4.6. *Given two vector spaces* $\mathbf{V}$ *and* $\mathbf{W}$ *over a field* $\mathbf{K}$, *let* T *be a map from* $\mathbf{V}$ *to* $\mathbf{W}$. *If the following conditions are satisfied, then the map* T *is said to be a linear transformation* (or a $\mathbf{K}$-*linear transformation*) *from* $\mathbf{V}$ *to* $\mathbf{W}$,

(1) $T(v_1 + v_2) = T(v_1) + T(v_2), \quad \forall\, v_1, v_2 \in \mathbf{V}$.
(2) $T(av) = aT(v), \quad \forall\, a \in \mathbf{K}, v \in \mathbf{V}$.

The kernel, $ker(T)$, *and the image of* T, $im(T)$, *are defined as follows,*

$$ker(T) = \{v : T(v) = 0\}$$
$$im(T) = \{w : w = T(v) \text{ for some } v \in \mathbf{V}\}$$

The $ker(T)$ *is called the null space of* T, *and the* $im(T)$ *is called the range of* T. ∎

Discussion

(1) In the case that $\mathbf{W}$=$\mathbf{K}$, the linear transformation is called a *linear functional.*

(2) It is easy to see that both $ker(T)$ and $im(T)$ are vector spaces over $\mathbf{K}$. We define the *rank* of T as $dim\ im(T)$, the *nullity* of T as $dim\ ker(T)$.

(3) Let T_1 and T_2 be linear transformations from $\mathbf{V}$ to $\mathbf{W}$ and from $\mathbf{W}$ to $\mathbf{U}$ respectively. Let us define $(T_2T_1)(v) = T_2(T_1(v))$. Then it is to see that the composition T_2T_1 is a linear transformations from $\mathbf{V}$ to $\mathbf{U}$.

(4) Let T, T_1, T_2 be linear transformations from **V** to **W**. Let us define $(aT)(v) = a(T(v)), (T_1+T_2)(v) = T_1(v)+T_2(v)$. Then it is to see that aT and T_1+T_2 are linear transformations from **V** to **W**. Let us use $Hom(V,W)$ to denote the set of all linear transformations from **V** to **W**. It is routine to see that $Hom(V,W)$ is a vector space over **K**. Let **W**=**K**. Then $Hom(V,K)$ is called the ***dual space*** $\mathbf{V}^*$ of **V**.

(5) In the above if **V**=**W**, then we shall define $(T_1T_2)(v) = T_1(T_2(v))$. Then it is routine to see that $Hom(V,V)$ forms a (maybe non-commutative) ring with the identity transformation $I(v) = v$ as the identity 1 of the ring.

(6) Let $\{v_i\},\{w_j\}$ be bases for **V**, **W** respectively. Then a linear transformation T determines the following data,

$$T(v_i) = \sum_{finite} b_{ij}w_j, \qquad b_{ij} \in \mathbf{K}$$

Furthermore, the above data uniquely determine a linear transformation T by the following equation,

$$T(\sum_{finite} a_iv_i) = \sum_j(\sum_i a_ib_{ij})w_j$$

(7) A subspace **U** of **V** is said to be an ***invariant subspace*** of a linear transformation T from **V** to **V** if $T(\mathbf{U}) \subset \mathbf{U}$. Note that **U** is an invariant subspace of T if and only if T induces a linear transformation from **U** to **U**. ∎

Example 8. (Projection). *(1) Let* **V**=**W**=$\mathbf{K}^3$. *Let T be defined as $T(a_1,a_2,a_3) = (a_1,a_2,0)$. It is easy to see that T is a linear transformation. In fact, it is the* ***projection*** *from the 3-dimensional space $\mathbf{K}^3$ to a plane in it. Formally, we define a* ***projection*** *P as a linear transformation which satisfies the following condition,*

$$P^2(v) = PP(v) = P(v), \qquad \forall\, v \in V$$

It is trivial to see the above defined T is a projection.

(2) Let P be any projection. We claim **V**=$ker(P) \oplus im(P)$: *(A) For any $v \in$* **V**, *we have*

$$P(v - P(v)) = P(v) - PP(v) = P(v) - P(v) = 0$$

Therefore, $v = (v - P(v)) + P(v) \in ker(P) + im(P)$. (B) Let $w \in ker(P) \cap im(P)$. Then we have $P(w) = 0, w = P(u)$ which implies $w = P(u) = PP(u) = P(w) = 0$.

(3) Let **V**=**U**⊕ **W**. *For any $v \in$* **V**, *$v = u + w$ with $u \in$* **U** *and $w \in$* **W** *in a unique way. Let us define*

$$P(v) = u$$

Then it is routine to see that P is a projection. ∎

Example 9. *(1) Let* **V**=**W**=$\mathbf{R}^3$ *and R a rotation which fixes the origin. Then R is a linear transformation. If R is not the identity, then R is not a projection.*

(2) Let $\mathbf{V}=C^1(\mathbf{R})$ *the set of differentiable real functions and* $\mathbf{W}=C^0(\mathbf{R})$ *the set of continuous real functions as vector spaces over* $\mathbf{R}$. *Let the differential operator* D *be defined as*

$$D=\frac{d}{d\,x}, \qquad \text{with } Df(x)=\frac{d\,f(x)}{d\,x}$$

Then D *is a linear transformation from* $\mathbf{V}$ *to* $\mathbf{W}$.

(3) Let $\mathbf{V}$, $\mathbf{W}$ *be defined as in (2). Let the indefinite integral* I_0^x *from* 0 *to* x *be defined as*

$$I_0^x(f(x))=\int_0^x f(x)d\,x$$

Then I_0^x *is a linear transformation from* $\mathbf{W}$ *to* $\mathbf{V}$.

*(4) The **fundamental theorem of Calculus** states*

$$DI_0^x = \text{ identity.}$$

On the other hand, $I_0^x D(c)=0$ *for all constant functions* $f(x)=c$. *Therefore, we have*

$$I_0^x D \neq \text{ identity.}$$

(5) Let $\mathbf{W}$ *be defined as in (2). Let the definite integral* I_a^b *from* a *to* b *be defined as*

$$I_a^b(f(x))=\int_a^b f(x)d\,x$$

Then I_a^b *is a linear functional on* $\mathbf{W}$. ∎

Example 10. *Given the following system of linear equations,*

$$(*)\qquad \begin{aligned} a_{11}x_1+a_{12}x_2+\cdots+a_{1n}x_n&=b_1\\ a_{21}x_1+a_{22}x_2+\cdots+a_{2n}x_n&=b_2\\ &\cdots\cdots\\ &\cdots\cdots\\ a_{m1}x_1+a_{m2}x_2+\cdots+a_{mn}x_n&=b_m \end{aligned}$$

where $a_{ij}\in\mathbf{K}$ *a field, we may consider it as a linear transformation from* $\mathbf{K}^n=\{(x_1, x_2,\cdots,x_n): x_i\in\mathbf{K}.\}$ *to* $\mathbf{K}^m=\{(b_1,b_2,\cdots,b_m): b_i\in\mathbf{K}.\}$. *We shall use the following matrix*[3] *notation,*

$$(**)\qquad A=\begin{pmatrix} a_{11} & a_{12} & \cdots & a_{1n}\\ a_{21} & a_{22} & \cdots & a_{2n}\\ \cdots & \cdots & \cdots & \cdots\\ \cdots & \cdots & \cdots & \cdots\\ a_{m1} & a_{m2} & \cdots & a_{mn} \end{pmatrix} \qquad X=\begin{pmatrix} x_1\\ x_2\\ \cdots\\ \cdots\\ x_m \end{pmatrix} \qquad B=\begin{pmatrix} b_1\\ b_2\\ \cdots\\ \cdots\\ b_m \end{pmatrix}$$

$$AX=B$$

[3] The concept of matrix was in ancient Chinese book 'Chiu Chang Suan Shu' 100, the addition of matrices was defined by Chu Shih-Chieh, and the multiplication was defined by English Mathematicians Hamilton 1805-1865, and Cayley 1821-1895.

The system of linear equations (*) can be solved if and only if the vector B is in $im(A)$. Note that if the inverse A^{-1} of the matrix A exists, then the system of equations (**) can be solved easily by

$$X = A^{-1}B$$

Theorem 4.5. *(1) Let* **V**, **W** *be two finite dimensional vector spaces over* **K**. *Let* $\{v_1, v_2, \cdots, v_n\}$, $\{w_1, w_2, \cdots, w_m\}$ *be basis for* **V**, **W** *respectively. Let* Δ_i^j *be linear transformations defined as follows (cf Discussion 6, Definition 4.6.),*

$$\Delta_i^j(v_s) = \delta_s^j w_i, \qquad \text{where } \delta_s^j = \begin{cases} 1 & if\ j = s \\ 0 & if\ j \neq s \end{cases}$$

Then $\{\Delta_i^j\}$ *forms a basis of* $Hom(V, W)$. *Therefore, we have*

$$dim\ Hom(V, W) = (dim\ V)(dim\ W).$$

(2) Especially, if **W**=**K**, *then let* $\{1\}$ *be a basis for* **W**. we

(i) $Hom(V, K) = \mathbf{V}^*$ *the dual space of* **V**.

(ii) *Let* $v_i^* = \Delta_1^i$. *Then* $\{v_i^*\}$ *is the dual basis of the dual space* $\mathbf{V}^*$.

(iii) $dim\ V = dim\ V^*$ *for finite dimensional vector spaces.*

Proof: (1) We want to prove that $\{\Delta_i^j\}$ is a linearly independent set. Suppose not. Let $\sum a_{ij}\Delta_i^j = 0$ be the zero linear transformation. Then we have

$$\sum a_{ij}\Delta_i^j(v_s) = \sum_i a_{is}w_i = 0$$

Since $\{w_j\}$ is a basis, then we must have,

$$a_{is} = 0, \qquad \forall\ i, s$$

We claim that $\{\Delta_i^j\}$ is a generating set. Let $T \in Hom(V, W)$. Let

$$T(v_s) = \sum_i b_{is}w_i$$

Define $T' = T - \sum b_{ij}\Delta_i^j$. Then we have for any s, the following

$$\begin{aligned} T'(v_s) &= T(v_s) - \sum b_{ij}\Delta_i^j(v_s) \\ &= \sum b_{is}w_i - \sum b_{is}w_i \\ &= 0 \end{aligned}$$

Therefore, we have $T = \sum b_{ij}\Delta_i^j$. ∎

(2) It is easy and left to the reader.

Remark: The above theorem fails in the infinite dimensional case. Let **V**=**W** and *dim* **V**=∞. Then $\{\Delta_i^j\}$ is not a generating set. For instance, the identity linear transformation I can not be written as a finite sum $\sum_{finite} b_{ij}\Delta_i^j$. ∎

Definition 4.7. *Let T be a linear transformation from* **V** *to* **W**. *The* ***transpose*** *of T, T^t, is a linear transformation from* $\mathbf{W}^*$ *to* $\mathbf{V}^*$ *defined as*

$$T^t(f)(v) = f(T(v)), \qquad \forall\, f \in \mathbf{W}^*, v \in \mathbf{V}$$

∎

We have the following theorem,

Theorem 4.6. *Let T be a linear transformation from finite dimensional vector spaces* **V** *to* **W** *with $rank(T) = r$. Let $\{v_1, v_2, \cdots, v_n\}$ be a basis for* **V**, *$w_i = T(v_i)$ for $i = 1, 2, \cdots, r$, and $\{w_1, w_2, \cdots, w_m\}$ be a basis for* **W**. *Furthermore, we have the following,*

$$T(v_i) = \sum_{j=1}^{r} a_{ij} T(v_j)$$

Then we have

$$T^t(w_k^*) = 0, \qquad \forall\, k > r$$

$$T^t(w_k^*) = \sum_{i=1}^{r} a_{ik} v_i^*, \qquad \forall\, k \le r$$

Especially, we have $\{v_1^*, v_2^*, \cdots, v_r^*\}$ as basis for $T^t(\mathbf{W}^*)$, therefore, $rank(T) = rank(T^t)$.

Proof: Clearly, we have

$$T^t(w_k^*)(v_i) = \sum_{j=1}^{r} a_{ij} w_k^*(w_j) = 0, \qquad \forall\, k > r$$

$$T^t(w_k^*)(v_i) = \sum_{i=1}^{r} a_{ij} \delta_j^k = a_{ik}, \qquad \forall\, k \le r$$

$$(T^t(w_k^*) - \sum_{i=1}^{r} a_{ik} v_i^*))(v_s)$$
$$= a_{sk} - a_{sk} = 0, \qquad \forall\, s$$

Naturally, $\{v_1^*, v_2^*, \cdots, v_r^*\}$ is a linearly independent set. Moreover it generates $T^t(\mathbf{W}^*)$ as just proved, therefore, $rank(T) = rank(T^t)$. ∎

Example 11. *(1) Let* $\mathbf{V}=\mathbf{K}^n$ *and* $\mathbf{W}=\mathbf{K}^m$*. Let* $T \in FL(n, m\mathbf{K})$*, and* $e_i^{(n)}, e_j^{(m)}$ *be the standard bases of* $\mathbf{K}^n$*,* $\mathbf{K}^m$*. Let the action of* T *be the usual matrix multiplications as follows,*

$$T(v) = (a_{ij})(b_j) = \begin{pmatrix} a_{11} & a_{12} & \cdot & \cdot & a_{1n} \\ a_{21} & a_{22} & \cdot & \cdot & a_{2n} \\ \cdot & \cdot & \cdot & \cdot & \cdot \\ a_{m1} & a_{m2} & \cdot & \cdot & a_{mn} \end{pmatrix} \begin{pmatrix} b_1 \\ b_2 \\ \cdot \\ \cdot \\ b_n \end{pmatrix} = \begin{pmatrix} c_1 \\ c_2 \\ \cdot \\ c_m \end{pmatrix}$$

Let $e_i^{(n)*}, e_j^{(m)*}$ *be the dual bases. Then the elements in* $\mathbf{V}^*$ *and* $\mathbf{W}^*$ *may be expressed as the rows with the linear functional actions expressed as the matrix multiplications as follows,*

$$v^*(v) = (\textstyle\sum_i b_i' e_i^{(n)*})(\sum_j b_j e_j^{(n)}) = \begin{pmatrix} b_1' & b_2' & \cdot & \cdot & b_n' \end{pmatrix} \begin{pmatrix} b_1 \\ b_2 \\ \cdot \\ \cdot \\ b_n \end{pmatrix} \quad \textstyle\sum_i b_i' b_i$$

$$w^*(w) = (\textstyle\sum_i c_i' e_i^{(m)*})(\sum_j c_j e_j^{(m)}) = \begin{pmatrix} c_1' & c_2' & \cdot & c_m' \end{pmatrix} \begin{pmatrix} c_1 \\ c_2 \\ \cdot \\ c_m \end{pmatrix} \quad \textstyle\sum_i c_i' c_i$$

Then the transpose T^t *of* T *can be expressed as*

$$T^t(w^*) = (c_i')(a_{ij}) = \begin{pmatrix} c_1' & c_2' & \cdot & c_m' \end{pmatrix} \begin{pmatrix} a_{11} & a_{12} & \cdot & \cdot & a_{1n} \\ a_{21} & a_{22} & \cdot & \cdot & a_{2n} \\ \cdot & \cdot & \cdot & \cdot & \cdot \\ a_{m1} & a_{m2} & \cdot & \cdot & a_{mn} \end{pmatrix}$$

$$= \begin{pmatrix} b_1' & b_2' & \cdot & \cdot & b_n' \end{pmatrix}$$

(2) Let us continue our discussion. Let C_i *be the column vectors and* R_j *be the row vectors of the matrix* T *as follows,*

$$C_i = \begin{pmatrix} a_{1i} \\ a_{2i} \\ \cdot \\ a_{mi} \end{pmatrix}, \qquad R_j = \begin{pmatrix} a_{j1} & a_{j2} & \cdot & \cdot & a_{jn} \end{pmatrix}$$

Then it is easy to see that range(T*) is generated by all column vectors, and range(*T^t*) is generated by all row vectors. Usually we define dim range(*T*) =rank(*T*)=column rank* *of T, and dim range(T^t)=row rank of T. We have column rank= row rank for finite dimensional spaces.* ∎

In the tradition of geometry, we fix a point in the n-dimensional ***affine space*** $\mathbb{A}^n$ as the origin, then the affine space becomes an n-dimensional vector space $\mathbf{K}^n$. In the tradition of ***analytic geometry***, we may further fix a ***coordinate system*** in $\mathbf{K}^n$. Then the vector space becomes computable. This is the meaning of the following definition,

Definition 4.8. *Let T be a linear transformation from $\mathbf{V}$ to $\mathbf{W}$. If T is a bijective map, then we say T is an **isomorphism**. If a linear transformation ρ is an isomorphism from $\mathbf{V}$ to $\mathbf{W}=\prod_{i\in I}\mathbf{K}$ for some index set I, then we say that ρ defines a **coordinate system** for $\mathbf{V}$.* ∎

Discussion

(1) If a linear transformation T from $\mathbf{V}$ to $\mathbf{W}$ is an isomorphism, then the inverse map T^{-1} exists and is an isomorphic linear transformation.

(2) Let a linear transformation T from $\mathbf{V}$ to $\mathbf{W}$ be an isomorphism, and $\{v_i\}$ be a basis of $\mathbf{V}$. Then $\{T(v_i)\}$ is a basis of $\mathbf{W}$. Therefore, we have

$$dim\ V = dim\ W$$

(3) Let $\{v_i\}_{i\in I}$ be a basis of $\mathbf{V}$. Given any $u \in \mathbf{V}$ with

$$u = \sum_{finite} a_i v_i, \qquad a_i \in \mathbf{K}$$

we define

$$\rho_v(u) = (a_i)_{i\in I} \in \prod_{i\in I}\mathbf{K}$$

Then ρ defines a coordinate system for $\mathbf{V}$. We said the coordinate system is spanned by the basis $\{v_i\}$. We shall think the coordinate system as defined by taking the vector v_i as the unit vector on the i-th axis.

(4) We are especially interested in the finite dimensional case. Let $\{v_1, v_2, \cdots, v_n\}$ be a basis of $\mathbf{V}$. Given any $u \in \mathbf{V}$ with

$$u = a_1v_1 + a_2v_2 + \cdots + a_nv_n, \qquad a_i \in \mathbf{K}$$

we define

$$\rho_v(u) = \begin{pmatrix} a_1 \\ a_2 \\ \cdot \\ \cdot \\ a_n \end{pmatrix} \in \mathbf{K}^n$$

The above is the coordinate of u with respect to the coordinate system which takes v_i as the unit vector on the i-th axis. ∎

Now we will introduce analytic method to $Hom(V, W)$ for two finite dimensional vector spaces **V** and **W**. Let us consider the numerical vector space $\mathbf{K}^n$. We shall use the standard basis $\{e_i^{(n)}\}$ defined as follows,

$$e_i^{(n)} = (0, \cdots, 0, 1, 0 \cdots, 0), \qquad \text{where the only 1 appears at the } i\text{-th place.}$$

Let E_i^j be the linear transformation from $\mathbf{K}^n$ to $\mathbf{K}^m$ defined by

$$E_i^j(e_s^{(n)}) = \delta_s^j e_i^{(m)}$$

We have the following Theorem,

Theorem 4.7. *(1) Let us use the above notations. There is a canonical isomorphic linear transformation χ from $Hom(\mathbf{K}^n, \mathbf{K}^m)$ to $FL(n, m, \mathbf{K})$ defined as follows,*

$$\chi(E_i^j) = \begin{pmatrix} 0, & 0, & \cdots & 0, & \cdots & 0 \\ 0, & 0, & \cdots & 0, & \cdots & 0 \\ \cdot & \cdot & \cdots & \cdot & \cdots & \cdot \\ 0, & 0, & \cdots & 1, & \cdots & 0 \\ \cdot & \cdot & \cdots & \cdot & \cdots & \cdot \\ 0, & 0, & \cdots & 0, & \cdots & 0 \end{pmatrix}$$

where the right hand side is the elementary matrix with the only 1 appears at the i-th column and j-th row. From now on, we will canonically identify $Hom(\mathbf{K}^n, \mathbf{K}^m)$ and $FL(n, m, \mathbf{K})$ through χ.

*(2) Let **V** and **W** be finite dimensional. Let $\{v_1, v_2, \cdots, v_n\}$, $\{w_1, w_2, \cdots, w_m\}$ be bases of **V**, **W** respectively. Let us use the notations of Theorem 4.5.. Let ρ_w^v be the linear transformation from $Hom(V, W)$ to $FL(n, m, \mathbf{K})$ defined by,*

$$\rho_w^v(\Delta_i^j) = E_i^j = \begin{pmatrix} 0, & 0, & \cdots & 0, & \cdots & 0 \\ 0, & 0, & \cdots & 0, & \cdots & 0 \\ \cdot & \cdot & \cdots & \cdot & \cdots & \cdot \\ 0, & 0, & \cdots & 1, & \cdots & 0 \\ \cdot & \cdot & \cdots & \cdot & \cdots & \cdot \\ 0, & 0, & \cdots & 0, & \cdots & 0 \end{pmatrix}$$

Then we always have,

$$T(u) = \rho_w^{-1}(\rho_w^v(T)\rho_v(u)), \qquad \forall\, T \in Hom(V, W)$$

Namely, the following diagram is commutative,

$$\begin{array}{ccc} V & \xrightarrow{T} & W \\ \rho_v \downarrow & & \uparrow \rho_w^{-1} \\ \mathbf{K}^n & \xrightarrow{\rho_w^v(T)} & \mathbf{K}^m \end{array}$$

(3) If $\mathbf{V}=\mathbf{W}$, *then* ρ_v^v *saves the multiplication, namely,*

$$\rho_v^v(T_1T_2) = \rho_v^v(T_1)\rho_v^v(T_2).$$

Let $FL(n,\mathbf{K}) = FL(n,n,\mathbf{K})$. *Then* ρ_v^v *is a ring isomorphism from* $Hom(V,V)$ *to* $FL(n,\mathbf{K})$.

(4) Furthermore, if $\mathbf{V}=\mathbf{W}$, *and let* $\{v_1, v_2, \cdots, v_n\}$, $\{w_1, w_2, \cdots, w_n\}$ *be two bases of* $\mathbf{V}$. *Let* A *be the coefficient matrix, namely,*

$$w_i = \sum_j a_{ij}v_j$$

$$A = (a_{ij})^t$$

Then we have

$$\rho_v\rho_w^{-1} = A$$

Proof: The proof is routine and left as an exercise to the reader. ∎

The coordinate of a non-zero vector u can be arbitrary (cf Exercise 1). However, depending on the two bases $\{v_i\}, \{w_j\}$ for $\mathbf{V}$ and $\mathbf{W}$ respectively, the matrix forms of a fixed linear transformation from $\mathbf{V}$ to $\mathbf{W}$ may vary within certain limits. Depending on the bases, sometimes one form of T is better than other one. Let us consider the following Example.

Example 12. (Fibonacci[4] Sequence). *Let us consider the following sequence* $\{a_i\} = \{1, 1, 2, 3, 5, 8, \cdots\}$ *of integers,*

$$a_1 = 1, a_2 = 1, a_n = a_{n-1} + a_{n-2}, \quad \forall\, n \geq 3$$

We wish to find a_n *directly without finding all preceding* a_i. *We may transform the above construction into one in* $\mathbf{R}^2$. *Let us group* $\{a_i\}$ *by pairs as follows,*

$$\begin{pmatrix} a_1 \\ a_2 \end{pmatrix} = \begin{pmatrix} 1 \\ 1 \end{pmatrix}, \qquad \begin{pmatrix} a_{2m+1} \\ a_{2m+2} \end{pmatrix} = \begin{pmatrix} a_{2m-1} + a_{2m} \\ a_{2m-1} + 2a_{2m} \end{pmatrix} = \begin{pmatrix} 1 & 1 \\ 1 & 2 \end{pmatrix} \begin{pmatrix} a_{2m-1} \\ a_{2m} \end{pmatrix}$$

[4] Leonardo of Pisa, also known as Fibonacci, Italian Mathematician 1180-1250.

Then we have the following,

$$A = \begin{pmatrix} 1 & 1 \\ 1 & 2 \end{pmatrix}, \qquad \begin{pmatrix} a_{2m+1} \\ a_{2m+2} \end{pmatrix} = A^m \begin{pmatrix} 1 \\ 1 \end{pmatrix}$$

It is obviously hard to find A^m directly. Let us view the above as a linear transformation T represented by the matrix A. Let

$$u_0 = \begin{pmatrix} 1 \\ 1 \end{pmatrix}, \qquad u_m = T^m(u_0)$$

Let θ_1 (the golden mean) and θ_2 be defined as,

$$\theta_1 = \frac{1+\sqrt{5}}{2}, \qquad \theta_2 = \frac{1-\sqrt{5}}{2}$$

Let a new basis $\{w_1, w_2\}$ of $\mathbf{R}^2$ be given as follows,

$$w_1 = \begin{pmatrix} 1 \\ \theta_1 \end{pmatrix} = e_1^{(2)} + \theta_1 e_2^{(2)}, \qquad w_2 = \begin{pmatrix} 1 \\ \theta_2 \end{pmatrix} = e_1^{(2)} + \theta_2 e_2^{(2)}.$$

Then we have the following,

$$e_1^{(2)} = \begin{pmatrix} 1 \\ 0 \end{pmatrix} = -\tfrac{\theta_2}{\sqrt{5}} w_1 + \tfrac{\theta_1}{\sqrt{5}} w_2, \quad e_2^{(2)} = \begin{pmatrix} 0 \\ 1 \end{pmatrix} = \tfrac{1}{\sqrt{5}} w_1 - \tfrac{1}{\sqrt{5}} w_2.$$

$$\rho_w = \begin{pmatrix} -\frac{\theta_2}{\sqrt{5}} & \frac{1}{\sqrt{5}} \\ \frac{\theta_1}{\sqrt{5}} & -\frac{1}{\sqrt{5}} \end{pmatrix}, \qquad \rho_w^{-1} = \begin{pmatrix} 1 & 1 \\ \theta_1 & \theta_2 \end{pmatrix}$$

Then we have the following direct computations,

$$\begin{aligned}
& T(w_1) = (1+\theta_1) w_1, \qquad T(w_2) = (1+\theta_2) w_2 \\
& u_0 = \frac{\theta_1}{\sqrt{5}} w_1 - \frac{\theta_2}{\sqrt{5}} w_2, \qquad \rho_w(u_0) = \begin{pmatrix} \frac{\theta_1}{\sqrt{5}} \\ -\frac{\theta_2}{\sqrt{5}} \end{pmatrix} \\
& T^m(u_0) = \rho_w^{-1}(\rho_w^w(T^m)\rho_w(u_0)) \\
& \quad = \begin{pmatrix} 1 & 1 \\ \theta_1 & \theta_2 \end{pmatrix} \begin{pmatrix} 1+\theta_1 & 0 \\ 0 & 1+\theta_2 \end{pmatrix}^m \begin{pmatrix} \frac{\theta_1}{\sqrt{5}} \\ -\frac{\theta_2}{\sqrt{5}} \end{pmatrix} \\
& \quad = \begin{pmatrix} 1 & 1 \\ \theta_1 & \theta_2 \end{pmatrix} \begin{pmatrix} (1+\theta_1)^m & 0 \\ 0 & (1+\theta_2)^m \end{pmatrix} \begin{pmatrix} \frac{\theta_1}{\sqrt{5}} \\ -\frac{\theta_2}{\sqrt{5}} \end{pmatrix} \\
& \quad = \begin{pmatrix} \frac{\theta_1(1+\theta_1)^m - \theta_2(1+\theta_2)^m}{\sqrt{5}} \\ \frac{\theta_1^2(1+\theta_1)^m - \theta_2^2(1+\theta_2)^m}{\sqrt{5}} \end{pmatrix}
\end{aligned}$$

Therefore, we conclude,

$$a_{2m+1} = \frac{\theta_1(1+\theta_1)^m - \theta_2(1+\theta_2)^m}{\sqrt{5}}$$

$$a_{2m+2} = \frac{\theta_1^2(1+\theta_1)^m - \theta_2^2(1+\theta_2)^m}{\sqrt{5}}$$

We have found a way to compute the general elements of the Fibonacci sequence in one step. ∎

Let us find all matrix forms of a fixed linear transformation T with respect to different bases. We have the following lemma,

Lemma. *Let* **V** *be a finite dimensional vector space over* **K** *with two bases* $\{v_1, v_2, \cdots, v_n\}$, $\{v'_1, v'_2, \cdots, v'_n\}$. *Then we have,*

(1) *The map* $\rho_v \rho_{v'}^{-1} : \mathbf{K}^n \to \mathbf{K}^n$ *is an isomorphism. Hence* $\rho_v \rho_{v'}^{-1} = A \in FL(n, \mathbf{K})$. *Furthermore,* $A^{-1} = \rho_{v'} \rho_v^{-1}$.

(2) *For any given* $C \in FL(n, \mathbf{K})$ *such that* $C : \mathbf{K}^n \to \mathbf{K}^n$ *is an isomorphism, then there exits a basis* $\{u_1, u_2, \cdots, u_n\}$ *of* **V** *with* $\rho_v \rho_u^{-1} = C$, *and* $C^{-1} = \rho_u \rho_v^{-1}$.

Proof: The conclusion (1) is routine. Let us prove (2). Given $\{v_1, v_2, \cdots, v_n\}$ and C. Let us define ρ^{-1} and u_i by

$$\rho^{-1} = \rho_v^{-1} C$$

$$u_i = \rho^{-1}(e_i^{(n)})$$

Then we have

$$\rho_u(u_i) = e_i^{(n)}$$

$$\rho_u \rho^{-1} = \text{identity}$$

$$C = \rho_v \rho_u^{-1}$$

∎

Theorem 4.8. *Let* **V**, **W** *be finite dimensional vector spaces over* **K**. *We have*

(1) *Let* $\{v_1, \cdots, v_n\}$, $\{v'_1, \cdots, v'_n\}$ *be two bases of* **V**, *and* $\{w_1, \cdots, w_m\}$, $\{w'_1, \cdots, w'_m\}$ *be two bases of* **W**. *Let* $T \in Hom(V, W)$. *We have the following diagram,*

$$\begin{array}{ccc}
\mathbf{K}^n & \xrightarrow{\rho_{w'}^{v'}(T)} & \mathbf{K}^m \\
\rho_{v'} \uparrow & & \downarrow \rho_{w'}^{-1} \\
V & \xrightarrow{T} & W \\
\rho_v \downarrow & & \uparrow \rho_w^{-1} \\
\mathbf{K}^n & \xrightarrow{\rho_w^v(T)} & \mathbf{K}^m
\end{array}$$

Then we have

$$T = \rho_w^{-1}\rho_w^v(T)\rho_v = \rho_{w'}^{-1}\rho_{w'}^{v'}(T)\rho_{v'}$$
$$\rho_w^v(T) = (\rho_w\rho_{w'}^{-1})\rho_{w'}^{v'}(T)(\rho_{v'}\rho_v^{-1})$$
$$\rho_w^v(T) = A\rho_{w'}^{v'}(T)B^{-1}$$

where $A = \rho_w\rho_{w'}^{-1} : \mathbf{K}^m \to \mathbf{K}^m$ is an isomorphism, $B^{-1} = \rho_{v'}\rho_v^{-1} : \mathbf{K}^n \to \mathbf{K}^n$ is an isomorphism.

(2) Conversely, given any two isomorphisms $C : \mathbf{K}^m \to \mathbf{K}^m$, $D^{-1} : \mathbf{K}^n \to \mathbf{K}^n$, then there are bases $\{v_1^*, \cdots, v_n^*\}$ of **V** and $\{w_1^*, \cdots, w_m^*\}$ of **W** such that $C = \rho_w\rho_{w^*}^{-1}$ and $D^{-1} = \rho_{v^*}\rho_v^{-1}$ with

$$\rho_w^v(T) = C\rho_{w^*}^{v^*}(T)D^{-1}$$

Proof: Our theorem follows from the preceding Lemma. ∎

Corollary. *Suppose that* $\mathbf{V}=\mathbf{W}$, $\{v_i\} = \{w_i\}$ *and* $\{v'_i\} = \{w'_i\}$. *Then the two matrix forms of a linear transformation* $T \in Hom(V, V)$ *satisfy the following equation,*

$$\rho_v^v(T) = A\rho_{v'}^{v'}(T)A^{-1}, \qquad A = \rho_v\rho_{v'}^{-1}.$$

Conversely, if there is a matrix M *satisfying,*

$$\rho_v^v(T) = AMA^{-1}$$

Then there is a basis $\{v'_i\}$ *of* **V** *such that*

$$M = \rho_{v'}^{v'}(T)$$

Proof: It is left as an exercise to the reader. ∎

Definition 4.9. *Given two matrices* $M, N \in FL(n, \mathbf{K})$, *if there is an isomorphism* A *of* $\mathbf{K}^n$ *such that,*

$$N = AMA^{-1}$$

then M, N *are said to be similar to each other, in symbol,* $M \sim N$. ∎

Discussion

(1) The similarity defined above is an equivalence relation: we have

(i) $M = IMI^{-1}$, therefore $M \sim M$.

(ii) $M \sim N \Leftrightarrow N = AMA^{-1} \Leftrightarrow M = A^{-1}N(A^{-1})^{-1} \Leftrightarrow N \sim M$.

(iii) $M \sim N, N \sim R \Leftrightarrow N = AMA^{-1}, R = BNB^{-1} \Rightarrow R = (BA)M(BA)^{-1} \Rightarrow M \sim R$.

(2) As indicated by Example 12., among the possible matrix forms of a linear transformation T, some form is better than the others. This leads to the selection of a simplest representation, the *canonical form*, of an equivalent class determined by a linear transformation T. The problem of canonical form will be treated in the next two sections. ∎

Example 13. *(1) Let us consider a homogeneous differential equation with constant coefficients as follows,*

$$L(D)f_1(x) = D^n f_1(x) + a_{n-1}D^{n-1}f_1(x) + \cdots + a_1 Df_1(x) + a_0 f_1(x) = 0$$

Let us introduce functions $\{f_2(x), \cdots, f_n(x)\}$ *as follows,*

$$\begin{aligned} Df_1(x) &= f_2(x) \\ Df_2(x) &= f_3(x) \\ &\cdots \\ Df_{n-1}(x) &= f_n(x) \\ Df_n(x) &= D^n f_1(x) = -a_{n-1}f_n(x) - \cdots - a_1 f_2(x) - a_0 f_1(x) \end{aligned}$$

Then the original differential equation may be rewritten as

$$D\begin{pmatrix} f_1 \\ f_2 \\ \cdot \\ \cdot \\ f_{n-1} \\ f_n \end{pmatrix} = \begin{pmatrix} 0 & 1 & 0 & \cdots & 0 \\ 0 & 0 & 1 & \cdots & 0 \\ \cdot & \cdot & \cdot & \cdots & \cdot \\ \cdot & \cdot & \cdot & \cdots & \cdot \\ 0 & 0 & 0 & \cdots & 1 \\ -a_0 & -a_1 & -a_3 & \cdots & -a_{n-1} \end{pmatrix} \begin{pmatrix} f_1 \\ f_2 \\ \cdot \\ \cdot \\ f_{n-1} \\ f_n \end{pmatrix} = M \begin{pmatrix} f_1 \\ f_2 \\ \cdot \\ \cdot \\ f_{n-1} \\ f_n \end{pmatrix}$$

In other words, a homogeneous differential equation with constant coefficients $L(D) = 0$ *can be understood as a system of linear homogeneous differential equations of order 1 as follows*

$$D\begin{pmatrix} f_1 \\ f_2 \\ \cdot \\ \cdot \\ f_{n-1} \\ f_n \end{pmatrix} = M \begin{pmatrix} f_1 \\ f_2 \\ \cdot \\ \cdot \\ f_{n-1} \\ f_n \end{pmatrix}$$

(2) Let us consider the following system of linear homogeneous differential equations of order 1,

$$D\begin{pmatrix} f_1 \\ f_2 \end{pmatrix} = \begin{pmatrix} 1 & 2 \\ 2 & 1 \end{pmatrix} \begin{pmatrix} f_1 \\ f_2 \end{pmatrix} = M \begin{pmatrix} f_1 \\ f_2 \end{pmatrix}$$

We have the following computation,

$$M = \begin{pmatrix} 1 & 2 \\ 2 & 1 \end{pmatrix} = \begin{pmatrix} \frac{1}{\sqrt{2}} & \frac{-1}{\sqrt{2}} \\ \frac{1}{\sqrt{2}} & \frac{1}{\sqrt{2}} \end{pmatrix} \begin{pmatrix} 3 & 0 \\ 0 & -1 \end{pmatrix} \begin{pmatrix} \frac{1}{\sqrt{2}} & \frac{1}{\sqrt{2}} \\ \frac{-1}{\sqrt{2}} & \frac{1}{\sqrt{2}} \end{pmatrix}$$
$$= ANA^{-1}$$

Let

$$A^{-1}\begin{pmatrix} f_1 \\ f_2 \end{pmatrix} = \begin{pmatrix} g_1 \\ g_2 \end{pmatrix}, \quad \begin{pmatrix} f_1 \\ f_2 \end{pmatrix} = A\begin{pmatrix} g_1 \\ g_2 \end{pmatrix}.$$

Then we have

$$A^{-1}D\begin{pmatrix} f_1 \\ f_2 \end{pmatrix} = D(A^{-1}\begin{pmatrix} f_1 \\ f_2 \end{pmatrix}) = D\begin{pmatrix} g_1 \\ g_2 \end{pmatrix} = N\begin{pmatrix} g_1 \\ g_2 \end{pmatrix} = \begin{pmatrix} 3g_1 \\ -g_2 \end{pmatrix}$$

Namely,

$$Dg_1 = 3g_1, \qquad Dg_2 = -g_2$$

We have (cf Example 6.),

$$g_1 = k_1e^{3x}, \qquad g_2 = k_2e^{-x}$$

Therefore, we have

$$f_1 = \frac{1}{\sqrt{2}}k_1e^{3x} - \frac{1}{\sqrt{2}}k_2e^{-x}, \quad f_2 = \frac{1}{\sqrt{2}}k_1e^{3x} + \frac{1}{\sqrt{2}}k_2e^{-x}.$$

∎

Exercises

(1) Let **V** be a n-dimensional vector space. Let $u \neq 0$ be a vector. For any coordinate $(a_1, a_2, \cdots, a_n) \neq (0, 0, \cdots, 0)$, there is a basis $\{v_1, v_2, \cdots, v_n\}$ such that $\rho_v(u) = (a_1, a_2, \cdots, a_n)$.

(2) Let **V**, **W** be vector spaces, and T a linear transformation from **V** to **W**. Show that the canonical map from **V** to $\mathbf{V}/ker(T)$ as commutative groups is a linear map.

(3) Give an example of an infinite dimensional vector space **V** over the real field **R** and two linear transformations T_1, T_2 from **V** to **V** such that
  (i) T_1 is surjective but not injective.
  (ii) T_2 is injective but not surjective.

(4) Let **V** be a finite dimensional vector space, and T a linear transformation from **V** to **W**. Show that nullity(T)+rank(T) $=dim$ **V**.

(5) Let $\mathbf{V}=\mathbf{R}^2$. Give an example of a linear transformation T from $\mathbf{V}$ to $\mathbf{V}$ such that the null space of T is the range of T.

(6) Let $\mathbf{V}$ be a finite dimensional vector space, and T a linear transformation from $\mathbf{V}$ to $\mathbf{V}$. Suppose that rank(T)= rank(T^2). Show that $\mathbf{V}$= (null space of T)$\oplus$(range of T).

(7) Let $\mathbf{V}=\mathbf{R}^3$, and linear functionals f_1, f_2, f_3 be defined as

$$f_1((x_1,x_2,x_3)) = x_1 + 2x_2 + x_3$$
$$f_2((x_1,x_2,x_3)) = x_2 - x_3$$
$$f_3((x_1,x_2,x_3)) = x_1 - x_2$$

Show that $\{f_1, f_2, f_3\}$ is a basis of $\mathbf{V}^*$.

(8) Let T be a linear transformation from a finite dimensional vector space $\mathbf{V}$ to $\mathbf{V}$. Show that for any positive integer k, we always have

$$rank(T^{k+1}) + rank(T^{k-1}) \geq 2\ rank(T^k)$$

(9) Give an example of a linear transformation $T : \mathbf{C}^2 \to \mathbf{C}^2$ with only finitely many invariant subspaces.

(10) Let T be a linear transformation from a vector space $\mathbf{V}$ to itself. Prove that T is a scalar times the identity map I if and only if every subspace is an invariant subspace.

(11) Given an $n \times n$ matrix A in the following block form,

$$A = \begin{pmatrix} B & C \\ 0 & D \end{pmatrix}$$

where B, D are square matrices. Show that $det\ A = (det\ B)(det\ D)$.

(12) Show that $\rho_v^v(T) = I$ the identity matrix $\Leftrightarrow T$ is the identity linear transformation.

(13) Let $\{v_1, \cdots, v_n\}, \{w_1, \cdots, w_n\}$ be two bases of $\mathbf{V}$. Let the relation between them be

$$v_i = \sum_j a_{ij} w_j, \qquad w_i = \sum_j b_{ij} v_j.$$

Let A, B be the matrices $(a_{ij}), (b_{ij})$. Then we have $AB = BA = I$ the identity matrix.

(14) Let us use the notations of the preceding exercise. Let T be a linear transformation with,

$$T(v_i) = \sum_j m_{ij} v_j, \qquad T(w_i) = \sum_j n_{ij} w_j.$$

Let M, N be the matrices $(m_{ij}), (n_{ij})$. Then we have

$$\rho_v^v(T) = M, \quad \rho_w^w(T) = N, \quad M = (B^t)N(B^t)^{-1}$$

(15) Let the matrix A be given as follows,

$$A = \begin{pmatrix} \frac{9}{2} & -\frac{7}{2} \\ -\frac{7}{2} & \frac{9}{2} \end{pmatrix}$$

Find B such that $B^3 = A$.

(16) Let T be a linear transformation T from a finite dimensional vector space **V** to **V**. Show that if $T = AB - BA$, then $det\ T = 0$, where A, B are linear transformations from **V** to **V**.

(17) A linear transformation T from a vector space **V** to **V** is said to be ***nilpotent*** if there is a positive integer m such that $T^m = 0$. Show that if $T = AB - BA$, then $I - T$ is not nilpotent, where A, B are linear transformations from **V** to **V**.

(18) Let **V** be a finite dimensional vector space over **R**. Let $p_1, p_2, \cdots, p_k$ be projections of **V**, namely they are linear transformations from **V** to **V** and satisfying $p_i^2 = p_i$. Show that $p_i p_j = 0$ for all $i \neq j$.

---

§4 Module and Module over P.I.D.

In this section, we will study an important concept, ***module***, in algebra. The modules are related to commutative groups, rings, and vector spaces. The theory of modules is inspired by the theories of commutative groups, rings, and vector spaces. Many times the theorems of these algebraic structures are unified in the theory of modules. For instance, two important theorems, the ***fundamental theorem of finitely generated abelian groups*** (cf Theorem 4.15) and the ***theorem of Jordan canonical forms*** (cf Theorem 4.25) for matrices, can be uniformly treated by the theory of ***modules***. For simplicity, we restrict our attentions to commutative rings in the following definition.

Definition 4.10. *Let* **R** *be a commutative ring. A non-empty set M is said to be a* ***module*** *over* **R**, *or an* **R**-*module, if the following conditions are satisfied,*

(1) *There is a binary operation + in M. With respect to + , M is a commutative group. Let the zero element be* 0.

(2) *There is a scalar multiplication $\cdot$, namely, for $r \in$* **R** *and $m \in M$, we have $r \cdot m \in M$. Furthermore, the scalar multiplication is* ***unimodulo***, *namely*

$$1 \cdot m = m, \qquad \forall\, m \in M$$

where 1 *is the unit* **R**.

(3) *The associative law holds for the four operations, the addition and multiplication of* **R**, *the addition and the scalar multiplication of M,*

$$(r_1 r_2)m = r_1(r_2 m), \quad (r_1 + r_2)m = r_1 m + r_2 m$$
$$r(m_1 + m_2) = rm_1 + rm_2$$

∎

Discussion

(1) The definition of module is similar to the definition of vector space, except the scalar set is a commutative ring for modules, instead of fields for vector spaces. Therefore, every vector space over a field **R** is a **R**-module.

(2) Every commutative ring **R** is a module over itself.

(3) The scalar multiplication $r \cdot m$ is sometimes simply written as rm. The ring **R** is sometimes referred as the *coefficient ring*. ∎

Example 14. *(1) Let* **G** *be any commutative group. Then* **G** *is a* **Z***-module is the following way,*

$$n \cdot g = \overbrace{g + \cdots + g}^{n \text{ times}}, \qquad \text{for } n \geq 0.$$
$$n \cdot g = -((-n) \cdot g), \qquad \text{for } n < 0.$$

Any theorem about modules over P.I.D. will apply to a commutative group **G**.

(2) Let **G** *be a commutative group, or a* **Z***-module. An interesting idea is the order of an element g, $ord(g)$, which was defined to be the least positive integer n, if there is any, such that $ng = 0$. In the case of a* **R***-module M, we may define the **annihilator** of $m \in M$, $ann(m) = \{r : r \in \mathbf{R}, rm = 0\}$. Then it is easy to see that $ann(m)$ is an ideal of* **R***, and in the case of commutative group* **G***, $ann(g) = (ord(g)) \subset \mathbf{Z}$. In general, if* **R** *is a P.I.D., then $ann(m)$ is always a principal ideal, thus $ann(m)$ fully generalizes the concept of order from commutative group to* **R***-module M.* ∎

Definition 4.11. *Let* **S**⊂**M**. *If we have*

$$m = \sum_{finite} r_i s_i, \qquad \forall\, m \in M, \text{ where } r_i \in \mathbf{R}$$

Then **S** *is a **generating set** of M and we write [**S**]=M. If there is a finite set* **S** *such that [**S**]=M, then M is said to be a **finitely generated module**. If we have*

$$\sum_{finite} r_i s_i = 0, \quad r_i \in \mathbf{R} \Rightarrow r_i = 0\ \forall\, i$$

then we say that **S** *is a **linearly independent set**. If* **S** *is a linearly independent generating set of M, then* **S** *is said to be a **basis** of M.* ∎

Discussion

(1) Here we see a main difference between modules and vector spaces. As pointed out in §3, the cardinalities of all minimal generating sets of a vector space **V** are the same,

which is defined to be the dimension of the vector space **V**. For modules, the cardinalities of all minimal generating sets may be different as in the following example.

Let us consider **Z** as a module over **Z**. Then both $\{1\}$ and $\{2, 3\}$ are minimal generating sets. They have different cardinalities.

(2) Let us consider the commutative group $\mathbf{G}=\mathbf{Z}/n\mathbf{Z}$ where n is an integer ≥ 2. Clearly we have

$$ng = 0, \qquad \forall\, g \in \mathbf{G}$$

Therefore, the only linearly independent set is the empty set $\emptyset$. Henceforth, there is no basis for **G**.

∎

Example 15. *(1) Let $\mathbf{K}^n$ be a n-dimensional vector space over a field $\mathbf{K}$. Let $M = \mathbf{K}^n$ as a commutative group, and $\mathbf{R}=\mathbf{K}\,[x]$ the polynomial ring of one variable over the field $\mathbf{K}$. Given any $A \in FL(n, \mathbf{K})$, we will use A to define a $\mathbf{R}$-module structure on M as follows,*

$$f(x)m = f(A)m, \qquad \forall\, f(x) \in \mathbf{K}[x], m \in M.$$

Note that for different A, the underlying module (as a commutative group) M and the coefficient ring $\mathbf{K}[x]$ are identical, the differences are in the scalar multiplication $xm = Am$. The module structure is determined by the matrix A. We expect to find some important properties of A by examining the module structure. Note that $\mathbf{K}[x]$ is a P.I.D.. Any theorem about module over P.I.D. can be applied to this case.

(2) Let us consider three simple cases. Let $n = 2$, $\mathbf{V}=\mathbf{K}^2$ and

$$\text{Case (i): } A_1 = I = \begin{pmatrix} 1 & 0 \\ 0 & 1 \end{pmatrix}$$

$$\text{Case (ii): } A_2 = J = \begin{pmatrix} 0 & 1 \\ 1 & 0 \end{pmatrix}$$

$$\text{Case (iii): } A_3 = \begin{pmatrix} 1 & 1 \\ 0 & 1 \end{pmatrix}$$

Case (i): We have

$$f(x)m = (\sum a_i x^i)m = (\sum a_i I^i)m = (\sum a_i)m = f(1)m$$
$$(x-1)m = 0, \qquad \forall\, m \in M.$$

It is easy to see that any generating set of M requires at least two elements, and the only linearly independent set is the empty set $\emptyset$.

Case (ii): We have

$$x \cdot \mathbf{e}_1^{(2)} = x\begin{pmatrix} 1 \\ 0 \end{pmatrix} = \begin{pmatrix} 0 & 1 \\ 1 & 0 \end{pmatrix}\begin{pmatrix} 1 \\ 0 \end{pmatrix} = \begin{pmatrix} 0 \\ 1 \end{pmatrix} = \mathbf{e}_2^{(2)}$$
$$x \cdot \mathbf{e}_2^{(2)} = x\begin{pmatrix} 0 \\ 1 \end{pmatrix} = \begin{pmatrix} 0 & 1 \\ 1 & 0 \end{pmatrix}\begin{pmatrix} 0 \\ 1 \end{pmatrix} = \begin{pmatrix} 1 \\ 0 \end{pmatrix} = \mathbf{e}_1^{(2)}$$
$$(x^2 - 1)m = 0, \qquad \forall\, m \in M.$$

Therefore, M is generated by $\{e_1^{(2)}\}$, and there is no linearly independent set except the empty set $\emptyset$.

Case (iii): We have

$$\begin{aligned}
x \cdot e_1^{(2)} &= x\begin{pmatrix}1\\0\end{pmatrix} = \begin{pmatrix}1 & 1\\0 & 1\end{pmatrix}\begin{pmatrix}1\\0\end{pmatrix} = \begin{pmatrix}1\\0\end{pmatrix} = e_1^{(2)}\\
x \cdot e_2^{(2)} &= x\begin{pmatrix}0\\1\end{pmatrix} = \begin{pmatrix}1 & 1\\0 & 1\end{pmatrix}\begin{pmatrix}0\\1\end{pmatrix} = \begin{pmatrix}1\\1\end{pmatrix} = e_1^{(2)} + e_2^{(2)}\\
(x-1)^2 m &= 0, \qquad \forall\, m \in M.
\end{aligned}$$

Therefore, M is generated by $\{e_2^{(2)}\}$, and there is no linearly independent set except the empty set $\emptyset$. Note that in the three cases, the module structures are essentially different. ∎

The following routine definitions about *submodules*, *quotient modules*, *homomorphisms*, *direct products*, and *direct sums* are motivated by the corresponding ones in group theory.

Definition 4.12. *Let M be a $\mathbf{R}$-module. If a non-empty subset $N \subset M$ is a $\mathbf{R}$-module with respect to the same $+$ and $\cdot$, then N is said to be a submodule of M.* ∎

Discussion

(1) Let a commutative group (vector space, commutative ring respectively) M be considered as a module. Then any subgroup (subspace, ideal respectively) is a submodule. ∎

Definition 4.13. *Let M, M' be $\mathbf{R}$-modules, and $\rho : M \to M'$ be a map. We have the following,*

(1) *If ρ is a group homomorphism from the commutative group M to the commutative group M', and linear with respect to the scalar multiplication, namely*

$$\rho(rm) = r\rho(m), \qquad \forall\, r \in \mathbf{R}, m \in M$$

then ρ is said to be (module) homomorphism.

(2) *A bijective module homomorphism ρ is called an isomorphism. If there is an isomorphism ρ from M to M', then M is said to be isomorphic to M'.*

(3) *Let ρ be a module homomorphism from M to M'. Then the kernel of ρ, $ker(\rho)$, and the image of ρ, $im(\rho)$, are defined as*

$$\begin{aligned}
ker(\rho) &= \{m : m \in M, \rho(m) = 0\}\\
im(\rho) &= \{m' : m' \in M', \text{ there is } m \in M \text{ with } \rho(m) = m'\}
\end{aligned}$$ ∎

It is left to the reader to verify that the following definition is *well-defined.*

Definition 4.14. *Let N be a submodule of M. On the quotient group M/N, let us define scalar multiplication, $\cdot$, as follows,*

$$r \cdot [m] = [rm]$$

Then the **R***-module formed this way is called the* ***quotient module****, M/N, of M over N. The map $\chi : M \to M/N$ defined by*

$$\chi(m) = [m]$$

is called the ***canonical homomorphism*** *from M to M/N.* ∎

The following theorem imitates the *first isomorphism theorem of groups* (cf Corollary of Theorem 2.9.).

Theorem 4.9. *Let $\rho : M \to M'$ be a module homomorphism. Let $\chi : M \to M/(ker(\rho))$ be the canonical homomorphism. Then the map $\overline{\rho} : M/(ker(\rho)) \to M'$ defined by*

$$\overline{\rho}([m]) = \rho(m)$$

is an isomorphism from $M/(ker(\rho))$ to $im(\rho)$ satisfying.

$$\overline{\rho}\chi = \rho$$

Proof: It is left to the reader as an exercise. ∎

Definition 4.15. *Let M_i for $i \in I$ be a set of* **R***-modules. We have*

(1) *Let $\oplus_i M_i$ be the direct sum of M_i as groups. Define scalar multiplications as*

$$r(m_i) = (rm_i), \qquad \forall\, (m_i) \in \oplus_i M_i$$

Then $\oplus_i M_i$ is a **R***-module. It is called the* ***direct sum*** *of M_i.*

(2) *Let $\prod_i M_i$ be the direct product of M_i as groups. Define scalar multiplications as*

$$r(m_i) = (rm_i), \qquad \forall\, (m_i) \in \prod_i M_i$$

Then $\prod_i M_i$ is a **R***-module. It is called the* ***direct product*** *of M_i.*

∎

Theorem 4.10. *Let M_1, M_2 be submodules of M. Let χ_i be the canonical homomorphism from M to M/M_i for $i = 1, 2$. Let χ be the homomorphism from M to $(M/M_1) \oplus (M/M_2)$ defined by*

$$\chi(m) = (\chi_1(m), \chi_2(m)) \in (M/M_1) \oplus (M/M_2)$$

Then we have (i) $M_1 \cup M_2$ generates M and (ii) $M_1 \cap M_2 = \{0\}$ if and only if χ is an isomorphism. In this case, we have

$$\chi_1 : M_2 \approx M/M_1, \qquad \text{and} \qquad \chi_2 : M_1 \approx M/M_2$$

We say M is the (interior) direct sum of M_1 and M_2.

Proof: ($\Rightarrow$) (i) If $\chi(m) = ([m]_1, [m]_2) = (0, 0)$, then $m \in M_1 \cap M_2 = \{0\}$. Therefore,$m = 0$ and χ is injective. (ii) Given any $([m]_1, [n]_2) \in (M/M_1) \oplus (M/M_2)$, since $M_1 \cup M_2$ generates M, we have $m = m_1 + m_2, n = n_1 + n_2$ where $m_1, n_1 \in M_1, m_2, n_2 \in M_2$. Therefore, $[m]_1 = [m_2]_1, [n]_2 = [n_1]_2$. Let $m = m_2 + n_1$. Then we have $\chi(m) = ([m]_1, [n]_2)$.
($\Leftarrow$) Left to the reader. ∎

Corollary. *Let a module M be the direct sum of two submodules M_1, M_2. Let π_i be the projection from M to M_i, namely,*

$$m = m_1 + m_2, \quad m_i \in M_i, \quad \Rightarrow \pi_i(m) = m_i$$

For any submodule N, let $N_i = M_i \cap N$, then $N = N_1 \oplus N_2$ if and only if $\pi_1(N) = M_1 \cap N$. Furthermore, suppose that $N = N_1 \oplus N_2$. Then we have the isomorphism $M/N \approx (M_1/N_1) \oplus (M_2/N_2)$.

Proof: The *only if* part is trivial. Let us prove the *if* part. We clearly have $N_1 \cap N_2 = \emptyset$. It suffices to show $N_1 \cup N_2$ generates N. For any $n \in N$, let

$$n = m_1 + m_2, \quad m_i \in M_i$$

We have

$$m_1 \in \pi(N_1) = M_1 \cap N = N_1 \subset N$$
$$m_2 = n - m_1 \in N \cap M_2 = N_2$$

Therefore, $N_1 \cup N_2$ generates N.

Let us prove the last part of the Corollary. Let the canonical maps $\chi_i : M_i \to M_i/N_i$ be denoted by $\chi_i(m_i) = [m_i]_i$. Let us define a map $\rho : M \to (M_1/N_1) \oplus (M_2/N_2)$ as follows,

$$\rho(m) = [m_1]_1 + [m_2]_2, \qquad \text{if } m = m_1 + m_2, \quad m_1 \in M_1, m_2 \in M_2$$

It suffices to show that ρ is surjective and $ker(\rho) = N$. These are trivial. ∎

Imitating the theory of vector spaces, we have the following definition and two theorems.

Definition 4.16. *If there is a basis $\{m_i\}_{i\in I}$ for a* **R**-*module M, then M is said to be a free module.* ∎

Theorem 4.11. *A* **R**-*module M is a free module with a basis $\{m_i\}_{i\in I}$ if and only if M is isomorphic to the direct sum $\oplus_{i\in I}\mathbf{R}$.*

Proof: (⇒) Let us define a map ρ as follows,

$$\rho(m) = (r_i) \in \oplus_{i\in I}\mathbf{R}, \text{ if } m = \sum_i r_i m_i$$

Note that it follows from the linearly independent property of the basis $\{m_i\}$ that the expression of m in term of the basis is unique. Therefore, the map ρ is well-defined. It is routine to prove that ρ is an isomorphism.

(⇐) Let $e_j^I \in \oplus_{i\in I}\mathbf{R}$ be defined as

$$e_j^I = (\delta_j^i), \qquad \text{where } \delta_j^i = \begin{cases} 1, & if\ i = j \\ 0, & if\ i \neq j \end{cases}$$

Let the isomorphism be ρ. Let $m_j = \rho^{-1}(e_j^I)$. We claim that $\{m_j\}$ is a basis for M. The proof is routine and left to the reader. ∎

Theorem 4.12. *Let M be a free module. Then any two bases of M have the same cardinality. The common cardinality is called the rank of M.*

Proof: Let I be a maximal ideal of **R**. Since **R** is commutative, then $\mathbf{R}/I$ is a field **F**. Let the canonical map from **R** to $\mathbf{R}/I$ be π. Let $N = IM = \{\sum_{finite} i_j n_j : i_j \in I,\ n_j \in M.\}$. It is easy to see that N is a submodule of M. Let us consider the module M/N and the canonical map $\chi : M \to M/N$. Let us write $\chi(m) = [m]$.

(1) Note that M/M is a commutative group. We claim M/N has a natural **K**-vector space structure defined as follows,

$$k[m] = [\pi^{-1}(k)m], \qquad \forall\, k \in \mathbf{K},\ m \in M$$

It is routine to show that the above definition is well-defined.

(2) Let $\{m_s\}$ be any basis of M. We claim that $\{[m_s]\}$ is a basis of the **K**-vector space M/N. It is easy to see that $\{[m_s]\}$ is a generating set for M/N. Let us prove that it is a linearly independent set. Suppose not. Then there exists a relation of the following form,

$$\sum_{finite} k_s[m_s] = 0, \qquad k_s \in \mathbf{K}.$$

Let $\pi(r_s) = k_s$. Then we have

$$\sum_s r_s m_s = \sum i_j m_j \in N$$
$$\sum_s r_s m_s - \sum i_j m_j = 0$$

Since $\{m_s\}$ is a linearly independent set, the above expression for 0 must be unique and $r_s \in I$ for all s. In other words, $k_s = \pi(r_s) = 0$ for all s, and the original relation is trivial.

(3) We have shown that the cardinality of any basis $\{m_s\}$ of M is the dimension of the $\mathbf{K}$-vector space M/N. Therefore, all bases of M must have the same cardinality. ∎

Remark. *The above theorem is false for modules over non-commutative rings.* ∎

The above materials are formal generalizations of group theory, ring theory and vector space theory. In the remaining part of this section, we will concentrate on *finitely generated modules* over P.I.D.. Let L be a finitely generated module over a P.I.D. $\mathbf{R}$. Let $\{\ell_1, \ell_2, \cdots, \ell_n\}$ be a generating set of L. Let us consider the map ρ from $\mathbf{R}^n$ to L defined as

$$\rho(e_i^{(n)}) = \ell_i$$
$$\rho((r_1, \cdots, r_n)) = \sum_i r_i m_i$$

It is easy to see that ρ is surjective. Therefore, we have

$$L \approx \mathbf{R}^n/(ker(\rho))$$

The study of a general $\mathbf{R}$-module L becomes a study of submodules $ker(\rho)$ of the free module $\mathbf{R}^n$. We have the following theorem about the submodules of a free module over a P.I.D.,

Theorem 4.13. *Let $\mathbf{R}$ be a P.I.D., M be a finitely generated free module over $\mathbf{R}$, and N a non-zero submodule of M. Then there exist a basis, $\{m_1, m_2, \cdots, m_n\}$, of M and $c_1, c_2, \cdots, c_\ell \in \mathbf{R}$ such that*

(1) $c_1 \mid c_2 \mid \cdots \mid c_\ell, \quad \ell \leq n.$
(2) *N is generated by $c_1 m_1, c_2 m_2, \cdots, c_\ell m_\ell$.*

Proof: Let us use mathematical induction on the rank n of the free module M. Let us consider the case $n = 1$. Let $M = \mathbf{R}m_1$, and $I = \{r : r \in \mathbf{R}, rm_1 \in N\}$. It follows at once that I is an ideal of $\mathbf{R}$. Since $\mathbf{R}$ is a P.I.D., then $I = (c_1)$ and $N = \mathbf{R}\, c_1 m_1$. Our theorem is proved in this case by taking $\ell = 1$.

Let us assume that $n > 1$ and the theorem is true for free modules of smaller rank.

(1) Let us consider all possible free bases $\{m^*_1, m^*_2, \cdots, m^*_n\}$ of M. Let $\chi_i^{m^*}$ be the projection of M to the i-th axis of this coordinate system, namely

$$\chi_i^{m^*}(m) = c_i, \quad \text{if } m = \sum_j c_j m^*_j$$

Then it is clear that $\chi_i^{m^*}(N)$ is an ideal of $\mathbf{R}$ for any i and basis $\{m^*_1, m^*_2, \cdots, m^*_n\}$. Since $\mathbf{R}$ is a P.I.D., then it is a Noetherian ring. Therefore, among the non-empty collection $\{\chi_i^{m^*}(N)\}$ of ideals, there must be a maximal one, I^*. It is obvious that $I^* \neq \{0\}$. Let us assume that $I^* = \chi_1^{m^*}(N) = (c_1)$.

(2) We want to improve the basis $\{m^*_1, m^*_2, \cdots, m^*_n\}$. It follows from the definition of $\chi_i^{m^*}$ that there is $m \in N$ with

$$(1) \qquad m = c_1 m^*_1 + c^*_2 m^*_2 + \cdots + c^*_n m^*_n$$

We *claim* that

$$(2) \qquad c_1 \mid c^*_i, \forall\, 1 < i \leq n$$

Note then that $c^*_i = c_1 d^*_i, \forall\, 1 < i \leq n$. Let

$$\overline{m}_1 = m^*_1 + \sum_2^n d^*_i m^*_i, \quad m_i = m^*_i, \quad \text{for } i = 2, \cdots, n$$

Then it is easy to see that (i) $\{\overline{m}_1, \overline{m}_2, \cdots, \overline{m}_n\}$ is a basis of M, and (ii) the expression of $m \in N$ becomes

$$(1') \qquad m = c_1 \overline{m}_1$$

The advantage of this new basis $\{\overline{m}_1, \cdots, \overline{m}_n\}$ is the following: let M_1 be the submodule generated by $\{\overline{m}_1\}$ and M_2 be the submodule generated by $\{\overline{m}_2, \cdots, \overline{m}_n\}$. Then we have $\chi_1^{\overline{m}}(N) = N \cap M_1$, and the condition of Corollary of Theorem 4.10. is satisfied. Its conclusion, $N = (N \cap M_1) \oplus (N \cap M_2)$, will provide a base for induction.

Let us prove our claim. We shall prove $c_1 \mid c^*_2$. The same prove will show that $c_1 \mid c^*_i$. Let $d = (c_1, c^*_2)$. We will show that d is associated with c_1. We have

$$(3) \qquad \begin{aligned} d &= \alpha c_1 + \beta c^*_2, & \alpha, \beta \in \mathbf{R} \\ c_1 &= \delta d, \;\; c^*_2 = \epsilon d, & \delta, \epsilon \in \mathbf{R} \end{aligned}$$

From Eq (3), we get

$$(4) \qquad 1 = \alpha\delta + \beta\epsilon$$

Let $\{\overline{\overline{m}}_1, \overline{\overline{m}}_2, \cdots, \overline{\overline{m}}_n\}$ be defined as

(5) $$\overline{\overline{m}}_1 = \delta m^*{}_1 + \epsilon m^*{}_2, \overline{\overline{m}}_2 = -\beta m^*{}_1 + \alpha m^*{}_2, \overline{\overline{m}}_i = m^*{}_i, \quad \forall\, i > 2.$$

With the help of Eq (4), we may solve the above system of equations and get the inverse expressions as follows,

(6) $$m^*{}_1 = \alpha \overline{\overline{m}}_1 - \epsilon \overline{\overline{m}}_2, m^*{}_2 = \beta \overline{\overline{m}}_1 + \delta \overline{\overline{m}}_2, m^*{}_i = \overline{\overline{m}}_i, \quad \forall\, i > 2.$$

A direct computation shows that $\{\overline{\overline{m}}_1, \overline{\overline{m}}_2, \cdots, \overline{\overline{m}}_n\}$ is a linearly independent generating set of M, therefore, a basis of M. Let us consider the expression of $m \in N$ with respect to this new basis,

$$\begin{aligned} m &= c_1 m^*{}_1 + c^*{}_2 m^*{}_2 + \cdots + c^*{}_n m^*{}_n \\ &= d\overline{\overline{m}}_1 + (-\epsilon c_1 + \delta c^*{}_2)\overline{\overline{m}}_2 + \cdots + c^*{}_n \overline{\overline{m}}_n \end{aligned}$$

Therefore, we have $\chi_1^{\overline{\overline{m}}}(N) \ni d$ and $I^* = (c_1) \subset (d) \subset \chi_1^{\overline{\overline{m}}}(N)$. Since I^* is assumed to a maximal ideal, then we must have $I^* = (c_1) = (d) = \chi_1^{\overline{\overline{m}}}(N)$. Therefore, we have c_1 is associated with d and the following,

$$c_1 \mid d \mid c^*{}_2$$

(3) Let us use the notation of the beginning of (2). Let $N_1 = N \cap M_1 = \mathbf{R} c_1 \overline{\overline{m}}_1$ and $N_2 = N \cap M_2$. Then we have $N = N_1 \oplus N_2$. If $N_2 = \{0\}$, then our proof is finished. Let us assume that $N_2 \neq \{0\}$. Note that M_2 is a free module of rank $n-1$. It follows from the induction hypothesis that there is a basis $\{m_2, m_3, \cdots, m_n\}$ and elements $c_2, c_3, \cdots, c_\ell \in \mathbf{R}$ such that

(7) $$\begin{aligned} &c_2 \mid c_3 \mid \cdots \mid c_\ell \\ &N_2 \text{ is generated by } c_2 m_2, c_2 m_3, \cdots, c_\ell m_\ell \end{aligned}$$

Let $m_1 = \overline{\overline{m}}_1$. Then it suffices to prove

$$c_1 \mid c_2$$

It follows from Eq (7) that $c_2 m_2 \in N_2 \subset N$. We have,

$$m' = m + c_2 m_2 = c_1 m_1 + c_2 m_2 \in N_1 \oplus N_2 = N$$

The proof of Eq (1) implies that $c_1 \mid c_2$. ∎

Theorem 4.14. (The Fundamental Theorem of Finitely Generated Modules over P.I.D.). *Let L be a finitely generated module over a P.I.D. $\mathbf{R}$. Then there exist non-invertible elements $c_1, c_2, \cdots, c_\ell \in \mathbf{R}$, such that*

(1) $c_1 \mid c_2 \mid \cdots \mid c_\ell$

(2) *L is isomorphic to* $\mathbf{R}/(c_1) \oplus \mathbf{R}/(c_2) \oplus \cdots\ \mathbf{R}/(c_\ell) \oplus \overbrace{\mathbf{R} \oplus \cdots \oplus \mathbf{R}}^{m \text{ times}}$.

Proof: Let $\{\ell_1, \ell_2, \cdots, \ell_n\}$ be a generating set of L. Let $\{e_1^{(n)}, e_2^{(n)}, \cdots, e_n^{(n)}\}$ be the standard basis of $M = \mathbf{R}^n = \prod_1^n \mathbf{R}$, namely $e_i^{(n)} = (0, \cdots, 0, 1, 0, \cdots, 0)$ with the only 1 appearing at the i-th place. Let us define the homomorphism $\rho : M \to L$ by $\rho(e_i^{(n)}) = \ell_i$. It is easy to see that ρ is surjective. Let the kernel of ρ, $ker(\rho)$, be N. Then we have the following isomorphism,

$$L \approx M/N$$

It follows from the preceding theorem that there is a basis, $\{m_1, m_2, \cdots, m_n\}$, of M and $c_1, c_2, \cdots, c_\ell \in \mathbf{R}$ such that

(1) $c_1 \mid c_2 \mid \cdots \mid c_\ell, \quad \ell \leq n$.

(2) N is generated by $c_1 m_1, c_2 m_2, \cdots, c_\ell m_\ell$.

As in the proof of the preceding theorem, let $M_1 = \mathbf{R}m_1$, M_2 be the submodule generated by $\{m_2, m_3, \cdots, m_n\}$, $N_1 = N \cap M_1$, and $N_2 = N \cap M_2$. Then we have $M = M_1 \oplus M_2, N = N_1 \oplus N_2$. It follows from the Corollary of Theorem 4.10. that

$$M/N \approx (M_1/N_1) \oplus (M_2/N_2) \approx (\mathbf{R}/(c_1)) \oplus (M_2/N_2)$$

Repeating the above procedure, we conclude that

$$M/N \approx \mathbf{R}/(c_1) \oplus \mathbf{R}/(c_2) \oplus \cdots \mathbf{R}/(c_\ell) \oplus \overbrace{\mathbf{R} + \cdots + \mathbf{R}}^{m \text{ times}}.$$

In the above expression, if c_i is a unit, then $\mathbf{R}/(c_i) = \{0\}$ and can be discarded. Our theorem is established. ∎

Theorem 4.15. (The Fundamental Theorem of Finitely Generated Commutative Groups). *Let G be a finitely generated commutative group. Then there exist non-invertible integers $c_1, c_2, \cdots, c_\ell \in \mathbf{Z}$, such that*

(1) $c_1 \mid c_2 \mid \cdots \mid c_\ell$

(2) *G is isomorphic to* $\mathbf{Z}/(c_1) \oplus \mathbf{Z}/(c_2) \oplus \cdots \mathbf{Z}/(c_\ell) \oplus \overbrace{\mathbf{Z} \oplus \cdots \oplus \mathbf{Z}}^{m \text{ times}}$.

Proof: This is a special case of the preceding Theorem. ∎

Definition 4.17. *The elements $c_1, c_2, \cdots, c_\ell \in \mathbf{R}$ in the Theorem 4.14. (respectively, Theorem 4.15.) are called the **torsion divisors** of the module L (respectively, the commutative group $\mathbf{G}$). In the case of commutative groups, the torsion divisors are sometimes called **torsion numbers**. The isomorphism of Theorem 4.14. (respectively, Theorem 4.15.) establishes a decomposition of M (respectively, $\mathbf{G}$), the **torsion decomposition** of M (respectively, $\mathbf{G}$).* ∎

Example 16. *(1) Let us consider a commutative group $\mathbf{G}$ of order 100. It follows from Theorem 4.15. that*

$$\mathbf{G} \approx \oplus_i \mathbf{Z}/(c_i).$$

It is obvious that $ord(\mathbf{G}) = 100 = \prod_i |c_i|$. The conclusion (1) of Theorem 4.15. implies that the are only four selections of $\{c_i\}$: (i) $c_1 = 2, c_2 = 50$, (ii) $c_1 = 5, c_2 = 20$, (iii) $c_1 = 10, c_2 = 10$, (iv) $c_1 = 100$. Thus we have found all commutative groups of order 100 up to isomorphisms.

(2) Let us continue our study of Example 15. (2). Let us assume that the characteristic of the field $\mathbf{K}$ is not 2. Recall that we have three cases:

$$\text{Case (i): } A_1 = I = \begin{pmatrix} 1 & 0 \\ 0 & 1 \end{pmatrix}$$

$$\text{Case (ii): } A_2 = J = \begin{pmatrix} 0 & 1 \\ 1 & 0 \end{pmatrix}$$

$$\text{Case (iii): } A_3 = \begin{pmatrix} 1 & 1 \\ 0 & 1 \end{pmatrix}$$

Case (i): It is easy to see that the $\mathbf{K}[x]$-module $\mathbf{K}^2$ is generated by $\{e_1^{(2)}, e_2^{(2)}\}$. We have

$$\mathbf{K}^2 \approx (\mathbf{K}[x]/(x-1)) \oplus (\mathbf{K}[x]/(x-1))$$

The torsion divisors are $(x-1)$ and $(x-1)$.

Case (ii): It is easy to see that the $\mathbf{K}[x]$-module $\mathbf{K}^2$ is generated by $\{e_1^{(2)}\}$. We have

$$\mathbf{K}^2 \approx (\mathbf{K}[x]/(x^2-1))$$

The torsion divisor is (x^2-1).

Case (iii): It is easy to see that the $\mathbf{K}[x]$-module $\mathbf{K}^2$ is generated by $\{e_2^{(2)}\}$. We have

$$\mathbf{K}^2 \approx (\mathbf{K}[x]/((x-1)^2))$$

The torsion divisor is $(x-1)^2$. ∎

Furthermore, we will prove that the torsion divisors $c_1, c_2, \cdots, c_\ell$ are uniquely determined by the isomorphic class of the module M or the commutative group $\mathbf{G}$. Then all cases of Example 4.16. will be non-isomorphic, since their torsion divisors are distinct. Let us introduce the following definition,

Definition 4.18. *Let $\mathbf{R}$ be a commutative ring and M a module over $\mathbf{R}$. Let $\mathbf{S} \subset M$. Then the* ***annihilator*** *of $\mathbf{S}$, $ann(S) = \{r : r \in R, rs = 0, \forall\ s \in S\}$. If $\mathbf{S}=\{m\}$, then $ann(m) = ann(\{m\})$ is sometimes call the* ***order ideal*** *of m. If $ann(m) \neq (0)$, then m is called a* ***torsion element****.* ∎

Discussion

(1) Let a commutative group $\mathbf{G}$ be considered as a module over $\mathbf{Z}$. Then an element g is a torsion element if and only if the order of g is finite.

(2) Suppose that $\mathbf{R}$ is an integral domain. then the set of all torsion elements is a submodule, the ***torsion submodule***: let m_1, m_2 be torsion elements, $0 \neq r_1 \in ann(m_1)$, $0 \neq r_2 \in ann(m_2)$, then we have,

$$r_1 r_2(m_1 + m_2) = 0$$
$$r_1(cm_1) = c(r_1 m_1) = 0$$

Similarly, all torsion elements of a commutative group form a subgroup, the ***torsion subgroup***.

(3) Let p be a prime of an integral domain $\mathbf{R}$. Then all elements in L which are annihilated by a power of p, namely $\{\ell : p^m\ell = 0$ for some positive integer $m.\}$, is a submodule of L. It is called *p-submodule* of L. If $L=$ a p-submodule, then L is called a p-module. The concept of p-modules is a generalization of p-groups in the group theory.

(4) The number m of the free copies of $\mathbf{R}$ (respectively, $\mathbf{Z}$) in the statement of Theorem 4.14. (respectively, Theorem 4.15.) is called the ***Betti number*** of L (respectively, $\mathbf{G}$).

(5) A module M without torsion elements, a ***torsion free module***, may not be free. For instance, $\mathbf{Q}$ is a torsion free module over $\mathbf{Z}$ but not free. Another example is the ideal $(x, y) \subset \mathbf{R}[x, y]$. It is a torsion free module over $\mathbf{R}[x, y]$, while not a free module. ∎

Theorem 4.16. *Let* $\mathbf{R}$ *be a P.I.D.,* L *a finitely generated module over* $\mathbf{R}$*, and* N *the torsion module of* L*. Then the quotient module* L/N *is a free module of rank = Betti number of* L*. Therefore, the Betti number is uniquely determined by the module* L*.*

Proof: Let us use the notations of Theorem 4.14.. It is easy to see that

(1) N is isomorphic to $\mathbf{R}/(c_1)\oplus\mathbf{R}/(c_2)\oplus\cdots\ \mathbf{R}/(c_\ell)$.

(2) L/N is isomorphic to $\overbrace{\mathbf{R}\oplus\cdots\oplus\mathbf{R}}^{m \text{ times}}$.

Therefore, our theorem is proved. ∎

We shall clarify the structure of the torsion submodule of L. First, we shall study the further decompositions of $\mathbf{R}/(c_i)$. It is nothing but a generalization of the ***Chinese Remainder Theorem*** (cf Theorem 1.9.).

Theorem 4.17. (Chinese Remainder Theorem). *Suppose that* $\mathbf{R}$ *is a P.I.D.. Let* $c \in \mathbf{R}$ *with* c *neither* 0 *nor an unit. Let the following be the decomposition of* c *as a product of prime elements,*

$$c = \delta \prod_{i=1}^{q} p_i^{s_i}, \qquad \delta \text{ a unit}, p_i, p_j \text{ non-associative, if } i \neq j.$$

Then we have the following isomorphism

$$\mathbf{R}/(c) \approx \oplus_{i=1}^{q} \mathbf{R}/(p_i^{s_i})$$

Proof: The proof is similar to the proof of Theorem 1.9.. Let

$$d = \delta \prod_{i=2}^{q} p_i^{s_i}, \qquad c = p_1^{s_1} \cdot d$$

We will prove that

$$\mathbf{R}/(c) \approx \mathbf{R}/(p_1^{s_1}) \oplus \mathbf{R}/(d) \tag{1}$$

Then we repeat the above process to decompose d. After finitely many repetitions, our theorem will be proved.

Because there is no prime common divisors of $p_1^{s_1}$ and d, we have

$$(p_1^{s_1}, d) = 1$$
$$\alpha p_1^{s_1} + \beta d = 1, \qquad \text{for some } \alpha, \beta \in \mathbf{R}$$

Let $\chi_1 : \mathbf{R} \to \mathbf{R}/(p_1^{s_1})$, $\chi_2 : \mathbf{R} \to \mathbf{R}/(d)$, be the canonical maps. Define a map χ by the following,

$$\chi(r) = (\chi_1(r), \chi_2(r)) \in \mathbf{R}/(p_1^{s_1}) \oplus \mathbf{R}/(d)$$

We claim (i) χ is surjective, (ii) $ker(\chi) = (c)$. Note that then our statement (1) follows.

(i) For any r_1, r_2, let $r = r_1 \beta d + r_2 \alpha p_1^{s_1}$. Then we have

$$\chi_1(r) = \chi_1(r_1 \beta d) = \chi_1(r_1(1 - \alpha p_1^{s_1})) = \chi_1(r_1)$$
$$\chi_2(r) = \chi_2(r_2 \alpha p_1^{s_1}) = \chi_2(r_2(1 - \beta d)) = \chi_2(r_2)$$

Therefore. χ is surjective.

(ii) Let $r \in ker(\chi)$. Then we have

$$\chi_1(r) = 0 \Rightarrow r \in (p_1^{s_1}) \Rightarrow p_1^{s_1} \mid r$$
$$\chi_2(r) = 0 \Rightarrow r \in (d) \Rightarrow d$$

Since $\mathbf{R}$ is a U.F.D., we have $c \mid r$, and $r \in (c)$. On the other hand, $\chi(r) = 0$ for all $r \in (c)$. Therefore, we conclude that $ker(\chi) = (c)$. ∎

Definition 4.19. *Suppose that the following is an isomorphism,*

$$L \approx \bigoplus_{finite} \mathbf{R}/p_i^{s_i} \oplus \overbrace{\mathbf{R} \oplus \cdots \oplus \mathbf{R}}^{m \text{ times}}.$$

where all p_i, which may not be distinct, are primes of **R**. *Then we call the above an elementary decomposition of L. The primaries $\{p_i^{s_i}\}$ are called the elementary divisors of L.*

∎

Corollary. *Let* **R** *be a P.I.D., and L a finitely generated* **R**-*module. Then there is an elementary decomposition of L.*

Proof: It follows from Theorem 4.14. and Theorem 4.16. directly. ∎

We shall prove the uniqueness of torsion decomposition and elementary decomposition of a finitely generated module L over a P.I.D. **R**. We have the following lemmas,

Lemma. *Let* **R** *be a P.I.D., and L a finitely generated* **R**-*module. If L is a p-module, then the elementary decomposition of L is of the following form,*

$$L \approx \oplus_{i=1}^{q} \mathbf{R}/p^{s_i}$$

Moreover, the collection of integers $\{s_i\}$ is uniquely determined by L up to a reordering.

Proof: Let the following be an elementary decomposition of L,

$$L \approx \bigoplus_{finite} \mathbf{R}/p_i^{s_i} \oplus \overbrace{\mathbf{R} \oplus \cdots \oplus \mathbf{R}}^{m \text{ times}}.$$

If $m \neq 0$, then there are some elements which are not torsion elements. Since L is a torsion module, then $m = 0$. If there are some p_i not associated with p, say p_1 not associated with p, let $\chi_1 : L \to \mathbf{R}/(p_1^{s_1})$ be the map induced by the canonical map: $L \to \mathbf{R}/(p_1^{s_1})$. Let $p^s\chi_1(1) = 0$. Since p_s and $p_1^{s_1}$ are not associated, then there are α, β such that

$$\alpha p^s + \beta p_1^{s_1} = 1$$

Therefore, we have,

$$\chi_1(1) = 1 \cdot \chi_1(1) = \chi_1(\alpha p^s + \beta p_1^{s_1}) = p^s \chi_1(1) = 0$$

A contradiction. We conclude that all p_i must be associated with p. Then we may replace all p_i by p and get the expression we want.

Let us prove the uniqueness. We claim

(i) $(p)L \approx \oplus_{i=1}^{q}(p)\mathbf{R}/p^{s_i} \approx \oplus_{i=1}^{q}\mathbf{R}/p^{(s_i-1)}$.

(ii) $L/(p)L \approx \overbrace{\mathbf{R}/(p) \oplus \cdots \oplus \mathbf{R}/(p)}^{q \text{ times}}$.

(iii) $\mathbf{R}/(\mathrm{p})$ is a field, and $L/(p)L$ is a q-dimensional vector space over $\mathbf{R}/(\mathrm{p})$.

Note that the uniqueness follows from the above: by a induction on *sum* s_i, it follows from (i) that the elementary decomposition of $(p)L$ is unique, and the collection $\{s_i : s_i > 1\}$ is unique. It follows from (iii) that the total number q is unique. Therefore, the number of $\{s_i : s_i = 1\}$ is determined by L. Henceforth, the collection $\{s_i\}$ is uniquely determined by L.

(i) The first isomorphism is trivial. We shall prove the second isomorphism. Let the elements of $\mathbf{R}/(p^{s_i})$ be $[r]_1$ and the elements of $\mathbf{R}/(p^{(s_1-1)})$ be $[r]_2$. Let us define

$$\rho : \mathbf{R}/(p^{(s_i-1)}) \to (p)\mathbf{R}/(p^{s_i})$$
$$\rho([r]_2) = [pr]_1$$

It is a simple exercise to show that ρ is an isomorphism.

(ii) Let us use the notations of (i). It follows from Corollary of Theorem 4.10. that

$$L/(p)L \approx \oplus_{i=1}^{q} \frac{\mathbf{R}/(p^{s_i})}{(p)\mathbf{R}/(p^{s_i})}$$

To prove (ii), it suffices to show

$$\frac{\mathbf{R}/(p^{s_i})}{(p)\mathbf{R}/(p^{s_i})} \approx \mathbf{R}/(p)$$

Let us define a map $\rho_i : \mathbf{R}/(p^{s_i}) \to \mathbf{R}/(p)$ as follows,

$$\rho_i([r]_1) = [r] \in \mathbf{R}/(p)$$

It is easy to see that ρ_i is a surjective homomorphism. We have

$$[r] = 0 \Leftrightarrow r \in (p) \Leftrightarrow r = pr' \Leftrightarrow [r]_1 = p[r']_1$$

Therefore, $ker(\rho_i) = (p)/(p^{s_i})$.

(iii) It is easy to see that (p) is maximal ideal. Therefore, $\mathbf{R}/(p)$ is a field. ∎

Theorem 4.18. *Let* $\mathbf{R}$ *be a P.I.D., and* L *a finitely generated* $\mathbf{R}$*-module. Then in any two elementary decompositions of* L,

$$L \approx \oplus_{finite}\mathbf{R}/p_i^{s_i} \oplus \overbrace{\mathbf{R} \oplus \cdots \oplus \mathbf{R}}^{m \text{ times}} \approx \oplus_{finite}\mathbf{R}/p'^{s'_i}_i \oplus \overbrace{\mathbf{R} \oplus \cdots \oplus \mathbf{R}}^{m' \text{ times}}.$$

we have the same Betti number $m = m'$*, and up to an reordering,* p_i *associated with* p'_i $(p_i \sim p'_i)$.

Proof. The uniqueness of the Betti number has been proved before. Let N be the torsion submodule of L. Let $\{q_j\}$ be a set of non-associated primes such that any p_i is associated to one of q_j. Let N_j be the q_j-submodule of L. Then we have

$$N_j \approx \oplus_{p_i \sim q_j}\mathbf{R}/p_i^{s_i}$$
$$N = \oplus_j N_j$$

It follows from the Lemma that the first decomposition is unique. It is easy to see the second decomposition is also unique. ∎

Corollary. *A finite commutative group G is cyclic if and only if the set $\{p_i\}$ deduced from the set of the elementary divisors $\{p_i^{s_i}\}$ of G consists of non-associated elements p_i.*

Proof: It follows from the Chinese Remaider Theorem (Theorem 4.16.). ∎

Remark: (1) The above theorem can be considered as a generalization of the integers.

(2) The Corollary can be used in the *field theory* (cf Chapter V) to prove that any finite multiplicative group of a field is commutative. ∎

Now we shall prove the uniqueness of the torsion decomposition of a finitely generated module over a P.I.D..

Theorem 4.19. *Let* **R** *be a P.I.D., and L a finitely generated* **R***-module. Then in any two torsion decompositions of L,*

$$L \approx \oplus_{i=1}^{\ell} \mathbf{R}/(c_i) \oplus \overbrace{\mathbf{R} \oplus \cdots \oplus \mathbf{R}}^{m \text{ times}} \approx \oplus_{i=1}^{\ell'} \mathbf{R}/(c_i') \oplus \overbrace{\mathbf{R} \oplus \cdots \oplus \mathbf{R}}^{m' \text{ times}}.$$
$$\text{where } c_1 \mid c_2 \mid \cdots \mid c_i \mid \cdots \mid c_\ell, \text{ and } c_1' \mid c_2' \mid \cdots \mid c_j' \mid \cdots \mid c_{\ell'}'.$$

we have the same Betti number $m = m'$, and c_i associated with c_i' $(c_i \sim c_i')$.

Proof: In the preceding Theorem, we have proved that the elementary divisors $\{p_i^{s_i}\}$ of L are uniquely determined. Note that we may replace p_i by any element of **R** which is associated to p_i. Therefore, we may assume that all p_i are either identical or non-associated. Let $\{p_1, p_2, \cdots, p_t\}$ be the set all distinct $p_i's$. We may arrange all elementary divisors of L in the following way in a matrix: on the j-th row, we put all elementary divisors of the form p_j^s in a decreasing order of the exponents,

$$\begin{array}{l} p_1 : p_1^{s_{1,1}}, p_1^{s_{1,2}}, \cdots, p_1^{s_{1,i}}, \cdots \\ p_2 : p_2^{s_{2,1}}, p_1^{s_{2,2}}, \cdots, p_1^{s_{2,i}}, \cdots \\ \cdots\cdots\cdots\cdots \\ \cdots\cdots\cdots\cdots \\ p_t : p_t^{s_{t,1}}, p_t^{s_{t,2}}, \cdots, p_t^{s_{t,i}}, \cdots \end{array}$$

It follows from the condition $c_1 \mid c_2 \mid \cdots \mid c_\ell$ that

$$c_\ell \sim \prod_{j=1}^{t} p_j^{s_{j,1}} \sim c_{\ell'}'$$

Similarly, we have

$$c_{(\ell-k)} \sim \prod_{j=1}^{t} p_j^{s_{j,(k+1)}} \sim c'_{(\ell'-k)}$$

∎

Example 17. *Let us count the number of all non-isomorphic commutative group of order n. Let the prime decomposition of n be*

$$n = \prod p_i^{s_i}$$

Let the partition function of s be P(s), namely P(s) is the number of ways of writing s into sums of positive integers. For instance,

$$1 = 1 \Rightarrow P(1) = 1,$$
$$2 = 2 = 1 + 1 \Rightarrow P(2) = 2,$$
$$3 = 3 = 2 + 1 = 1 + 1 + 1 \Rightarrow P(3) = 3,$$
$$4 = 4 = 3 + 1 = 2 + 2 = 2 + 1 + 1 = 1 + 1 + 1 + 1 \Rightarrow P(4) = 5,$$

It is easy to see the number of all non-isomorphic commutative group of order n is $\prod_i P(s_i)$. *For instance,* $n = 100 = 2^2 5^2$, *then we have the number of non-isomorphic commutative groups of order 100 is* $P(2)P(2) = 4$. ∎

Exercises

(1) Let M be a **R**-module. Let $0_R, 0_M$ be the zero elements of **R**, M respectively. Show that
  (i) $r0_M = 0_M, \forall\, r \in \mathbf{R}$.
  (ii) $0_R m = 0_M, \forall\, m \in M$.
  (iii) $(-r)m = r(-m) = -rm, \forall\, m \in M,\ r \in \mathbf{R}$.

(2) Let I be an ideal of a commutative ring **R**, and M be a **R**-module. Show that I is **R**-module, and IM is a submodule of M.

(3) Let M be a **R**-module, and N_1, N_2 be submodules of M. Show that $[N_1 : N_2] = \{r : r \in \mathbf{R}, rN_2 \subset N_1\}$ is an ideal of **R**.

(4) Show that a submodule N of a **R**-module is generated by a subset $\{m_i\} \Leftrightarrow N$ is the smallest submodule which contains $\{m_i\}$.

(5) Let **R** be a commutative ring. If every submodule of any free **R**-module is free, then **R** is a P.I.D..

(6) Show that **Q** is a torsion free **Z**-module while not a free **Z**-module.

(7) Let $\{m_1, m_2, \cdots, m_n\}$ be a basis of a free **R**-module M. Let another set of elements $\{s_1, s_2, \cdots, s_n\}$ of M be given with the following relation

$$s_i = \sum_j a_{ij} m_j, \qquad \text{where } a_{ij} \in \mathbf{R}.$$

Show that $\{s_1, s_2, \cdots, s_n\}$ is a basis of M if and only if the determinant of the coefficient matrix (a_{ij}) is a unit in **R**.

(8) Find the number of non-isomorphic commutative groups of order 936.

(9) Find the number of non-isomorphic commutative groups of order 72.

(10) Show that every $n \times n$ matrix A is similar to its transpose A^t.

(11) Express the commutative group $\mathbf{Z}^3/(f_1, f_2, f_3)$ where $f_1 = (2,4,6)$, $f_2 = (4,6,8)$, $f_3 = (3,4,5)$ as a direct sum of cyclic groups.

(12) Let $\mathbf{R}=\mathbb{R}[x]$, and f_1, f_2, f_3 be the following elements in $\mathbf{R}^3$,

$$f_1 = (x,1,0), \quad f_2 = (1,x,0), \quad f_3 = (0,0,x-1)$$

Express $\mathbf{R}^3/(f_1, f_2, f_3)\mathbf{R}^3$ in term of the Fundamental Theorem.

§5 Jordan Canonical Form

Let **K** be a field, and **V** a finite dimensional vector space over **K**. Let $T \in Hom(V,V)$ be a linear transformation from **V** to **V**.As we pointed out in Theorem 4.8., for any basis $\{v_1, v_2, \cdots, v_n\}$ of **V**, there are isomorphisms ρ_v, ρ_v^v such that the following diagram is commutative,

$$\begin{array}{ccc} V & \xrightarrow{\ T\ } & V \\ {\scriptstyle \rho_v}\downarrow & & \uparrow{\scriptstyle \rho_v^{-1}} \\ \mathbf{K}^n & \xrightarrow{\ \rho_v^v(T)\ } & \mathbf{K}^n \end{array}$$

The matrix form of T, $\rho_v^v(T)$, is numerical and computable. However, it depends on the selection of the basis $\{v_1, v_2, \cdots, v_n\}$. In this section, we will apply Theorem 4.14. to find a canonical selection of the basis $\{v_1, v_2, \cdots, v_n\}$ which depends on the selection of the linear transformation T. Then the matrix form of T, the *Jordan Canonical form*, will be simple and canonical, which means that every similar class of linear transformations will have a distinct Jordan canonical form. Thus two matrices are similar to each other if and only if their Jordan canonical forms are identical.

For simplicity of notations, we will assume that $\mathbf{V}=\mathbf{K}^n$. Then we have $T \in FL(n, \mathbf{K})$. A polynomial $f(x)$ is said to be *monic* if the coefficient of the highest degree term is 1.

Note then that every polynomial is associated to a unique monic polynomial. We shall rewrite Theorems 4.14. 4.17. 4.18. as follows,

Theorem 4.20. *Let us consider the vector space* $\mathbf{K}^n$ *over a field* $\mathbf{K}$. *Let the commutative ring be* $\mathbf{R}=\mathbf{K}[x]$, *given any* $A \in FL(n, \mathbf{K})$, *we will use* A *to define a* $\mathbf{R}$*-module structure on* $\mathbf{K}^n$ *as follows,*

$$f(x)m = f(A)m, \qquad \forall\, f(x) \in \mathbf{K}[x],\ m \in \mathbf{K}^n$$

There exist unique non-invertible monic polynomials $c_1(x), c_2(x), \cdots, c_\ell(x) \in \mathbf{K}[x]$, *the* **invariant factors** *of* A, *such that*

(1) $c_1(x) \mid c_2(x) \mid \cdots \mid c_\ell(x)$
(2) *The* $\mathbf{K}[x]$*-module* $\mathbf{K}^n$ *is isomorphic to* $\oplus_{i=1}^{\ell} \mathbf{K}[x]/(c_i(x))$.

Proof: It follows from Theorem 4.14. that we have a collection $\{c_1(x), \cdots, c_n(x)\}$ such that

(1) $c_1(x) \mid c_2(x) \mid \cdots \mid c_\ell(x)$
(2) The $\mathbf{K}[x]$-module $\mathbf{K}^n$ is isomorphic to $\oplus_{i+1}^{\ell} \mathbf{K}[x]/(c_i(x)) \oplus \overbrace{\mathbf{K}[x] \oplus \cdots \oplus \mathbf{K}[x]}^{m \text{ times}}$.

Since $\mathbf{K}^n$ is a finite dimensional vector space, and $\mathbf{K}[x]$ is an infinite dimensional vector space, then the copies of $\mathbf{K}[x]$ can not exist. Our theorem follows. ∎

Theorem 4.21. *Let us consider the vector space* $\mathbf{K}^n$ *over a field* $\mathbf{K}$. *Let the commutative ring be* $\mathbf{R}=\mathbf{K}[x]$, *given any* $A \in FL(n, \mathbf{K})$, *we will use* A *to define a* $\mathbf{R}$*-module structure on* $\mathbf{K}^n$ *as follows,*

$$f(x)m = f(A)m, \qquad \forall\, f(x) \in \mathbf{K}[x],\ m \in \mathbf{K}^n$$

There exist unique non-invertible irreducible monic polynomials $p_1(x), p_2(x), \cdots, p_q(x) \in \mathbf{K}[x]$, *some of which may be equal, and positive integers* $s_1, s_2, \cdots, s_q$ *such that we have the following* $\mathbf{K}[x]$*-module isomorphism,*

$$\mathbf{K}^n \approx \oplus_{i=1}^{q} \mathbf{K}[x]/(p_i^{s_i}(x))$$

The polynomials $\{p_i^{s_i}\}$ *are the* **elementary divisors** *of* A.

Proof: It is left to the reader as an exercise. ∎

Let us consider the above decompositions. Let $\{g_i(x)\}$ be either $\{c_i(x)\}$ or $\{p_i^{s_i}(x)\}$. Consider the following direct sum,

$$\oplus_i \mathbf{K}[x]/(g_i(x))$$

with the action of linear transformation A as the multiplication by x. Let

$$\mathbf{V}_i = \mathbf{K}[x]/(g_i(x))$$

Then we have

$$\mathbf{K}^n \approx \oplus_i \mathbf{V}_i$$

And each $\mathbf{V}_i$ is an *invariant subspace*, namely $A\mathbf{V}_i = x\mathbf{K}[x]/(g_i(x)) \subset \mathbf{K}[x]/(g_i(x)) = \mathbf{V}_i$. It is easy to see that after selecting bases from each $\mathbf{V}_i$, let the matrix forms of the restriction of A to $\mathbf{V}_i$ be B_i, then the matrix form of A with respect to the union of bases will be of the following block form,

$$\begin{pmatrix} B_1 & 0 & \cdot & \cdot & 0 \\ 0 & B_2 & 0 & \cdot & 0 \\ 0 & \cdot & \cdot & \cdot & 0 \\ 0 & \cdot & \cdot & \cdot & 0 \\ 0 & 0 & \cdot & 0 & B_s \end{pmatrix}$$

The question becomes finding the *block matrix* B_i. We have the following theorem for the companion block matrix,

Theorem 4.22. *Let a monic polynomial $g(x)$ be given as follows,*

$$g(x) = x^m - a_{m-1}x^{m-1} - \cdots - a_1 x - a_0$$

Then $\mathbf{K}[x]/(g(x))$ is an m-dimensional vector space over $\mathbf{K}$. Let $\chi : \mathbf{K}[x] \to \mathbf{K}[x]/(g(x))$ be the canonical map, and $\chi(h(x)) = [h(x)]$. Then $\{[1], [x], \cdots, [x^{m-1}]\}$ is a basis for the vector space, and $\rho_x^x(A)$ is of the following form,

$$\begin{pmatrix} 0 & 1 & 0 & \cdot & 0 \\ 0 & 0 & 1 & \cdot & 0 \\ 0 & 0 & 0 & \cdot & 0 \\ \cdot & \cdot & \cdot & \cdot & \cdot \\ a_0 & a_1 & a_2 & \cdot & a_{m-1} \end{pmatrix}$$

The polynomial $g(x)$ is said to be the corresponding polynomial of the above matrix, and the matrix is said to be the companion matrix of $g(x)$.

Proof. (i) We claim that $\{[1], [x], \cdots, [x^{m-1}]\}$ is a generating set of $\mathbf{K}[x]/(g(x))$. For any $[h(x)] \in \mathbf{K}[x]/(g(x))$, it follows from the Euclidean Algorithm that there exist $d(x)$ and $r(x)$ such that

$$h(x) = d(x)g(x) + r(x), \qquad deg\ r(x) < deg\ g(x) = m$$

Therefore, we have

$$[h(x)] = [d(x)g(x) + r(x)] = [r(x)] = \sum_{i=1}^{m-1} b_i[x^i]$$

(ii) We claim that $\{[1],[x],\cdots,[x^{m-1}]\}$ is a linearly independent set. Note the following simple computation,

$$\sum_{i=1}^{m-1} b_i[x^i] = [0] \Leftrightarrow [\sum_{i=1}^{m-1} b_i x^i] = [0] \Leftrightarrow \sum_{i=1}^{m-1} b_i x^i \in (g(x))$$

$$\Leftrightarrow g(x) \mid \sum_{i=1}^{m-1} b_i x^i \Leftrightarrow a_i = 0 \; \forall \, i$$

(iii) Let us write down the action of x on this vector space as follows,

$$\begin{aligned} x\cdot[1] &= 0\cdot[1] + 1\cdot[x] + 0\cdot[x^2] + \cdot + 0\cdot[x^{m-1}] \\ x\cdot[x] &= 0\cdot[1] + 0\cdot[x] + 1\cdot[x^2] + \cdot + 0\cdot[x^{m-1}] \\ x\cdot[x^2] &= 0\cdot[1] + 0\cdot[x] + 0\cdot[x^2] + \cdot + 0\cdot[x^{m-1}] \\ &\cdots\cdots\cdots\cdots\cdots\cdots \\ x\cdot[x^{m-1}] &= a_0\cdot[1] + a_1\cdot[x] + a_2\cdot[x^2] + \cdot + a_{m-1}\cdot[x^{m-1}] \end{aligned}$$

∎

Theorem 4.23. *Let* $\mathbf{K}$ *be a field, and* $A \in FL(n,\mathbf{K})$*. After selecting a suitable basis for* $\mathbf{K}^n$*, the matrix form of* A*, the rational form, will be as follows,*

$$\begin{pmatrix} B_1 & 0 & \cdot & \cdot & 0 \\ 0 & B_2 & 0 & \cdot & 0 \\ 0 & \cdot & \cdot & \cdot & 0 \\ 0 & \cdot & \cdot & \cdot & 0 \\ 0 & 0 & \cdot & 0 & B_s \end{pmatrix}$$

where the block matrices, B_i*, on the diagonal are of the form in Theorem 4.22., and the corresponding polynomials,* $c_i(x)$*, satisfy*

$$c_1(x) \mid c_2(x) \mid \cdots \mid c_\ell(x)$$

Two matrices are similar to each other if and only if they have the same rational form.

Proof. We have to prove the *only if* part of the last statement. Suppose that $A \sim B$ with the following equation,

$$A = CBC^{-1}$$

Let the multiplications of the two $\mathbf{K}[x]$-modules of $\mathbf{K}^n$ through the actions of A and B be defined respectively as follows,

$$\begin{aligned} x\cdot v &= A(v), \qquad \forall\, v \in \mathbf{K}^n \\ x * v &= B(v), \qquad \forall\, v \in \mathbf{K}^n \end{aligned}$$

Let us define a map ρ as

$$\rho(v) = C^{-1}(v), \qquad \forall\, v \in \mathbf{K}^n$$

Then we have

$$\rho(x \cdot v) = C^{-1}A(v) = BC^{-1}(v) = x * (\rho(v)), \qquad \forall\, v \in \mathbf{K}^n$$

It then follows easily that ρ is a $\mathbf{K}[x]$-isomorphism of those two $\mathbf{K}[x]$-modules. It follows from Theorem 4.20. that A, B have the same rational form. ∎

Among all fields the complex field C has an extraordinary position. Algebraically, the most important property of C is that it is *algebraically closed*: the only irreducible polynomials in C$[x]$ are linear polynomials. This is the *Fundamental Theorem of Algebra*. We will list it in below for reference. A short and elementary proof of it will be provided in Chapter V as Theorem 5.2.,

Theorem 4.24. (The Fundamental Theorem of Algebra). *A polynomial $f(x)$ is irreducible in* C$[x]$ *if and only of $f(x)$ is linear.* ∎

Similar to Theorem 4.22., we have the following theorem for the torsion decompositions,

Theorem 4.25. *Let a monic polynomial $g(x)$ be given as follows,*

$$g(x) = (x-c)^m$$

Then $\mathbf{K}[x]/(g(x))$ is an m-dimensional vector space over $\mathbf{K}$. Let $\chi : \mathbf{K}[x] \to \mathbf{K}[x]/(g(x))$ be the canonical map, and $\chi(h(x)) = [h(x)]$. Then $\{[1], [x-c], \cdots, [(x-c)^{m-1}]\}$ is a basis for the vector space, and $\rho_x^x(A)$ is of the following form,

$$\begin{pmatrix} c & 1 & 0 & \cdot & \cdot & 0 \\ 0 & c & 1 & \cdot & \cdot & 0 \\ 0 & 0 & c & \cdot & \cdot & 0 \\ \cdot & \cdot & \cdot & \cdot & \cdot & \cdot \\ \cdot & \cdot & \cdot & \cdot & \cdot & 1 \\ 0 & 0 & 0 & \cdot & 0 & c \end{pmatrix}$$

*The above matrix is called a **block matrix of the Jordan canonical form**. The polynomial $(x-c)^m$ is said to be the **corresponding polynomial** of the above matrix.*

Proof: The proof of $\{[1], [x-c], \cdots, [(x-c)^{m-1}]\}$ being a basis is the same as in the proof of Theorem 4.22..

Let us write down the action of x on this vector space as follows,

$$\begin{aligned} x \cdot [1] &= c \cdot [1] + 1 \cdot [x-c] + 0 \cdot [(x-c)^2] + \cdot + \cdot + 0 \cdot [(x-c)^{m-1}] \\ x \cdot [x-c] &= 0 \cdot [1] + c \cdot [x-c] + 1 \cdot [(x-c)^2] + \cdot + \cdot + 0 \cdot [(x-c)^{m-1}] \\ x \cdot [(x-c)^2] &= 0 \cdot [1] + 0 \cdot [(x-c)] + c \cdot [(x-c)^2] + \cdot + \cdot + 0 \cdot [(x-c)^{m-1}] \\ &\cdots\cdots\cdots\cdots\cdots\cdots \\ &\cdots\cdots\cdots\cdots\cdots\cdots \\ x \cdot [(x-c)^{m-1}] &= 0 \cdot [1] + 0 \cdot [x-c] + 0 \cdot [(x-c)^2] + \cdot + \cdot + c \cdot [(x-c)^{m-1}] \end{aligned}$$

The coefficient matrix is what we want. ∎

Similar to Theorem 4.23. we have the following theorem,

Theorem 4.26. (Jordan Canonical form). *Let* $\mathbf{K}$ *be a field, and* $A \in FL(n, \mathbf{K})$. *Suppose that the elementary divisors of* A *are powers of linear polynomials (This is certainly true if* $\mathbf{K}$=$\mathbf{C}$*). Then after selecting a suitable basis for* $\mathbf{K}^n$*, the matrix form of* A*, the Jordan canonical form, will be as follows,*

$$\begin{pmatrix} J_1 & 0 & \cdot & \cdot & 0 \\ 0 & J_2 & 0 & \cdot & 0 \\ 0 & \cdot & \cdot & \cdot & 0 \\ 0 & \cdot & \cdot & \cdot & 0 \\ 0 & 0 & \cdot & 0 & J_s \end{pmatrix}$$

where the block matrices, J_i*, on the diagonal are the block matrices of Jordan canonical form in Theorem 4.25.. Certainly, the order of* J_i'*s may be changed by selecting different bases for* $\mathbf{K}^n$*. Two matrices are similar to each other if and only if their Jordan canonical forms are identical up to a reordering of the block matrices of Jordan canonical form.*

Proof: Similar to the proof of Theorem 4.23.. ∎

Example 18. *(1) Let* $\mathbf{K}$=$\mathbf{C}$*, and* $\mathbf{V}$=$\mathbf{C}^2$*. Let* A *be any linear transformation of* $\mathbf{C}^2$*. Then the Jordan canonical form of* A *must be one of the following two,*

$$\begin{pmatrix} c_1 & 0 \\ 0 & c_2 \end{pmatrix}, \quad \begin{pmatrix} c_1 & 1 \\ 0 & c_1 \end{pmatrix}$$

(2) Let us consider the following system of linear homogeneous differential equations of order 1 over complex numbers $\mathbf{C}$ *(cf Example 13. (2)),*

$$Df_1(x) = a_{11}f_1(x) + a_{12}f_2(x)$$
$$Df_2(x) = a_{21}f_1(x) + a_{22}f_2(x)$$

We may rewrite the above system in the following way,

$$D\begin{pmatrix} f_1 \\ f_2 \end{pmatrix} = \begin{pmatrix} a_{11} & a_{12} \\ a_{21} & a_{22} \end{pmatrix} \begin{pmatrix} f_1 \\ f_2 \end{pmatrix} = A\begin{pmatrix} f_1 \\ f_2 \end{pmatrix}$$

There are two possibilities:

$$(i) \qquad A = C\begin{pmatrix} c_1 & 0 \\ 0 & c_2 \end{pmatrix} C^{-1},$$

$$(ii) \qquad A = C\begin{pmatrix} c_1 & 1 \\ 0 & c_1 \end{pmatrix} C^{-1}$$

In either case, let

$$C^{-1}\begin{pmatrix} f_1 \\ f_2 \end{pmatrix} = \begin{pmatrix} g_1 \\ g_2 \end{pmatrix},$$

Then we have

$$(i) \qquad D\begin{pmatrix} g_1 \\ g_2 \end{pmatrix} = \begin{pmatrix} c_1 & 0 \\ 0 & c_2 \end{pmatrix} \begin{pmatrix} g_1 \\ g_2 \end{pmatrix}$$

$$(ii) \qquad D\begin{pmatrix} g_1 \\ g_2 \end{pmatrix} = \begin{pmatrix} c_1 & 1 \\ 0 & c_1 \end{pmatrix} \begin{pmatrix} g_1 \\ g_2 \end{pmatrix}$$

The case (i) is reduced to the following system,

$$Dg_1(x) = c_1 g_1(x)$$
$$Dg_2(x) = c_2 g_2(x)$$

The method of Example 6. can be applied to yield,

$$g_1(x) = k_1 e^{c_1 x}$$
$$g_2(x) = k_2 e^{c_2 x}$$

Let us consider the case (ii). We have the following system,

$$Dg_1(x) = c_1 g_1(x) + g_2(x)$$
$$Dg_2(x) = c_1 g_2(x)$$

The method of Example 6. yields

$$g_2(x) = k_1 e^{c_1 x}$$
$$(D - c_1)^2 g_1(x) = 0$$
$$g_1(x) = k_2 e^{c_1 x} + k_3 x e^{c_1 x}$$
$$Dg_1(x) = c_1(k_2 e^{c_1 x} + k_3 x e^{c_1 x}) + k_3 e^{c_1 x}$$
$$k_3 = k_1$$

∎

The field of real numbers **R** is important in applications. One of the important algebraic theorems of **R** is the following theorem which can be deduced from the Fundamental Theorem of Algebra, its proof is referred to Chapter V,

Theorem 4.27. *The set of monic irreducible polynomials in* $\mathbf{R}[x]$ *is the union of the following two sets*

$$\{x + r : r \in \mathbf{R}\}, \quad \{(x - r)^2 + s^2 : r, s \in \mathbf{R}, s \neq 0.\}$$

∎

We have the following two theorems of *Jordan Canonical Form of Real Matrices*, their proofs are left as exercises to the reader.

Theorem 4.28. *Let i be the imaginary number $\sqrt{-1}$. Given an irreducible polynomial $p(x) = (x-r)^2 + s^2$ $(s \neq 0)$, for any positive integer n, let us define $f_n(x) \in \mathbf{C}[x], g_n(x), h_n(x) \in \mathbf{R}[x]$ as follows,*

$$f_n(x) = [(x-r)+s]^n = g_n(x) + i\; h_n(x)$$

Let m be a fixed positive integer. Let $\chi : \mathbf{R}[x] \to \mathbf{R}[x]/(p(x)^m)$ be the canonical map, and $\chi(q(x)) = [q(x)]$. Furthermore, for $n = 1, 2, \cdots, m$, let the following be defined,

$$v_{2m-2n+2} = [(g_n(x) + h_n(x))p(x)^{m-n}]$$
$$v_{2m-2n+1} = [(g_n(x) - h_n(x))p(x)^{m-n}]$$

Then $\{v_1, v_2, \cdots, v_{2m}\}$ is a basis for the vector space $\mathbf{R}[x]/(p(x)^m)$, and the matrix form of the linear transformation of multiplication by x is as follows,

$$\begin{pmatrix} r & -s & 1 & 0 & 0 & 0 & \cdot & \cdot & 0 & 0 \\ s & r & 0 & 1 & 0 & 0 & \cdot & \cdot & 0 & 0 \\ 0 & 0 & r & -s & 1 & 0 & \cdot & \cdot & 0 & 0 \\ 0 & 0 & s & r & 0 & 1 & \cdot & \cdot & 0 & 0 \\ \cdot & \cdot & \cdot & \cdot & \cdot & \cdot & \cdot & \cdot & \cdot & \cdot \\ \cdot & \cdot & \cdot & \cdot & \cdot & \cdot & \cdot & \cdot & \cdot & \cdot \\ 0 & 0 & 0 & 0 & 0 & 0 & \cdot & \cdot & 1 & 0 \\ 0 & 0 & 0 & 0 & 0 & 0 & \cdot & \cdot & 0 & 1 \\ 0 & 0 & 0 & 0 & 0 & 0 & \cdot & \cdot & r & -s \\ 0 & 0 & 0 & 0 & 0 & 0 & \cdot & \cdot & s & r \end{pmatrix}$$

*The above matrix is called a **block matrix of the Jordan canonical form of real matrix**. The polynomial $[(x-r)^2 + s^2]^m$ is said to be the **corresponding polynomial** of the above matrix.* ∎

Theorem 4.29. (Jordan Canonical form of real matrices). *Let $A \in FL(n, \mathbf{R})$. Then after selecting a suitable basis for $\mathbf{K}^n$, the matrix form of A, the Jordan canonical form of real matrices, will be as follows,*

$$\begin{pmatrix} J_1 & 0 & \cdot & \cdot & 0 \\ 0 & J_2 & 0 & \cdot & 0 \\ 0 & \cdot & \cdot & \cdot & 0 \\ 0 & \cdot & \cdot & \cdot & 0 \\ 0 & 0 & \cdot & 0 & J_s \end{pmatrix}$$

where the block matrices, J_i, on the diagonal are either the block matrices of Jordan canonical form in Theorem 4.25., or the block matrices of Jordan canonical form of real matrices in Theorem 4.28.. Certainly, the order of J_i's may be changed by selecting different bases for $\mathbf{K}^n$. Two matrices are similar to each other if and only if their Jordan canonical forms of real matrices are identical up to a reordering of the block matrices of Jordan canonical form of real matrices. ∎

Example 19. *(1) Let* $\mathbf{K}=\mathbf{R}$, *and* $\mathbf{V}=\mathbf{R}^3$. *Let A be any linear transformation of* $\mathbf{R}^3$. *Then the Jordan canonical form of real matrix A must be one of the following four,*

$$\begin{pmatrix} r_1 & 0 & 0 \\ 0 & r_2 & 0 \\ 0 & 0 & r_3 \end{pmatrix}, \begin{pmatrix} r_1 & 1 & 0 \\ 0 & r_1 & 0 \\ 0 & 0 & r_2 \end{pmatrix} \begin{pmatrix} r_1 & 1 & 0 \\ 0 & r_1 & 1 \\ 0 & 0 & r_1 \end{pmatrix} \begin{pmatrix} r & -s & 0 \\ s & r & 0 \\ 0 & 0 & r_2 \end{pmatrix}$$

(2) Let $\mathbf{K}=\mathbf{R}$, *and* $\mathbf{V}=\mathbf{R}^4$. *Let A be any linear transformation of* $\mathbf{R}^4$. *Then the Jordan canonical form of real matrix A must be one of the following nine,*

$$\begin{pmatrix} r_1 & 0 & 0 & 0 \\ 0 & r_2 & 0 & 0 \\ 0 & 0 & r_3 & 0 \\ 0 & 0 & 0 & r_4 \end{pmatrix}, \begin{pmatrix} r_1 & 1 & 0 & 0 \\ 0 & r_1 & 0 & 0 \\ 0 & 0 & r_2 & 0 \\ 0 & 0 & 0 & r_3 \end{pmatrix}, \begin{pmatrix} r_1 & 1 & 0 & 0 \\ 0 & r_1 & 0 & 0 \\ 0 & 0 & r_2 & 1 \\ 0 & 0 & 0 & r_2 \end{pmatrix},$$

$$\begin{pmatrix} r_1 & 1 & 0 & 0 \\ 0 & r_1 & 1 & 0 \\ 0 & 0 & r_1 & 0 \\ 0 & 0 & 0 & r_2 \end{pmatrix}, \begin{pmatrix} r_1 & 1 & 0 & 0 \\ 0 & r_1 & 1 & 0 \\ 0 & 0 & r_1 & 1 \\ 0 & 0 & 0 & r_1 \end{pmatrix}, \begin{pmatrix} r & -s & 0 & 0 \\ s & r & 0 & 0 \\ 0 & 0 & r_1 & 0 \\ 0 & 0 & 0 & r_2 \end{pmatrix},$$

$$\begin{pmatrix} r & -s & 0 & 0 \\ s & r & 0 & 0 \\ 0 & 0 & r_1 & 1 \\ 0 & 0 & 0 & r_1 \end{pmatrix}, \begin{pmatrix} r_1 & -s_1 & 0 & 0 \\ s_1 & r_1 & 0 & 0 \\ 0 & 0 & r_2 & -s_2 \\ 0 & 0 & s_2 & r_2 \end{pmatrix}, \begin{pmatrix} r & -s & 1 & 0 \\ s & r & 0 & 1 \\ 0 & 0 & r & -s \\ 0 & 0 & s & r \end{pmatrix},$$

(3) Let us solve the following system of differential equations of real functions,

$$D\begin{pmatrix} g_1 \\ g_2 \\ g_3 \\ g_4 \end{pmatrix} = \begin{pmatrix} r & -s & 1 & 0 \\ s & r & 0 & 1 \\ 0 & 0 & r & -s \\ 0 & 0 & s & r \end{pmatrix} \begin{pmatrix} g_1 \\ g_2 \\ g_3 \\ g_4 \end{pmatrix}$$

An argument similar to Example 6. will produce the following solutions for the equations,

$$g_3(x) = k_3e^{rx}\cos sx - k_4e^{rx}\sin sx$$
$$g_4(x) = k_4e^{rx}\cos sx + k_3e^{rx}\sin sx$$
$$g_1(x) = (k_3x + k_1)e^{rx}\cos sx - (k_4x + k_2)e^{rx}\sin sx$$
$$g_2(x) = (k_4x + k_2)e^{rx}\cos sx + (k_3x + k_1)e^{rx}\sin sx$$

It is not hard to generalize the above method to solve any system of homogeneous differential equations with constant coefficients of order 1 of real functions. ∎

Exercises

(1) Find the rational forms of the following two matrices over $\mathbf{Q}$,

$$\begin{pmatrix} 0 & 1 & 0 \\ 4 & 0 & 0 \\ 0 & 0 & -3 \end{pmatrix}, \quad \begin{pmatrix} 0 & 1 & 0 \\ 0 & 0 & 1 \\ 12 & 4 & -3 \end{pmatrix}$$

and show that they are similar.

(2) Let $A \in FL(8, \mathbf{C})$. Suppose that $rank(A+I) = 6$, $rank((A+I)^2) = 5$, $rank((A+I)^k) = 4$ for all $k \geq 3$, $rank(A-2I) = 7$, $rank((A-2I)^k) = 6$ for all $k \geq 2$, and $rank((A-3I)^k) = 6$, for all $k \geq 1$. Find the Jordan canonical form of A.

$$A = \begin{pmatrix} A_1 & 0 & \cdot & \cdot & 0 \\ 0 & A_2 & 0 & \cdot & 0 \\ 0 & \cdot & \cdot & \cdot & 0 \\ 0 & \cdot & \cdot & \cdot & 0 \\ 0 & 0 & \cdot & 0 & A_s \end{pmatrix}, \quad B = \begin{pmatrix} B_1 & 0 & \cdot & \cdot & 0 \\ 0 & B_2 & 0 & \cdot & 0 \\ 0 & \cdot & \cdot & \cdot & 0 \\ 0 & \cdot & \cdot & \cdot & 0 \\ 0 & 0 & \cdot & 0 & B_s \end{pmatrix}$$

where A_i, B_i are $n_i \times n_i$ square matrices. Prove that A is similar to B if and only if A_i is similar to B_i for all i.

(3) Find the Jordan canonical form of the following matrix

$$\begin{pmatrix} 3 & -1 & 0 & 0 \\ 1 & 0 & 0 & 0 \\ 0 & 1 & 0 & 0 \\ 0 & 0 & 0 & 4 \end{pmatrix}$$

(4) Find the Jordan canonical form of the following matrix

$$\begin{pmatrix} -1 & 1 & 1 \\ -3 & 2 & 2 \\ -1 & 1 & 1 \end{pmatrix}$$

§6 Characteristic Polynomial

In this section, we will assume that the reader is familiar with the definition of *determinant*[5] and its basic properties as follows,

(i) Let $A = (a_{ij}) \in FL(n, \mathbf{K})$. Then its determinant, $det\ A$, is given by

$$det\ A = \sum Sign(i_1, i_2, \cdots, i_n) a_{1i_1} a_{2i_2} \cdots a_{ni_n}$$

[5]The concept of determinant was due to Japanese Mathematician Seki Kōwa 1683, and German Mathematician Leibniz 1693.

where

$$Sign(i_1, i_2, \cdots, i_n) = \begin{cases} 0, & \text{if some } i_j = i_k \text{ for } j \neq k \\ 1, & \text{if } (i_1, i_2, \cdots, i_n) \text{ is an even permutation} \\ -1, & \text{if } (i_1, i_2, \cdots, i_n) \text{ is an odd permutation} \end{cases}$$

(ii) $det\ (AB) = (det\ A)(det\ B)$.

(iii) $det\ A = 0 \Leftrightarrow A$ is not an isomorphism$\Leftrightarrow A$ is not injective$\Leftrightarrow A$ is not surjective.

We assume the reader knows $AB = I$ the identity matrix $\Leftrightarrow BA = I$.

Definition 4.20. *Let $A \in FL(n, \mathbf{K})$ where $\mathbf{K}$ is a field. The characteristic polynomial*[6]*, $\chi_A(x)$, of A is defined to be det $(xI - A)$, where as usual, I is the unit matrix. The roots of $\chi_A(x)$ are called the characteristic values (eigenvalues or proper values) of A.* ∎

We give another definition of characteristic values in the following theorem,

Theorem 4.30. *An element $\lambda \in \mathbf{K}$ is a characteristic value of A if and only if there is a non-zero vector $v \in \mathbf{K}^n$ such that*

$$Av = \lambda v$$

The vector v which satisfies the above equation will be called a characteristic vector (eigenvector or proper vector) of A associated with λ.

Proof: ($\Rightarrow$) Suppose that λ is a characteristic value of A. Then we have

$$det\ (\lambda I - A) = 0$$

Therefore, $\lambda I - A$ is not an injective linear transformation. Let $0 \neq v \in ker(\lambda I - A)$. Then we have

$$(\lambda I - A)v = 0, \qquad Av = \lambda I v = \lambda v.$$

($\Leftarrow$) Suppose that there is a $0 \neq v \in \mathbf{K}^n$ with

$$Av = \lambda v$$

Therefore, we have

$$(\lambda I - A)v = 0$$

It follows that $(\lambda I - A)$ is not injective and

$$det\ (\lambda I - A) = 0$$

Therefore, λ is a characteristic value of A. ∎

[6]Due to Cayley.

heorem 4.31. *Let A, B be similar to each other. Then we have their characteristic ɔlynomials, $\chi_A(x), \chi_B(x)$, equal. Therefore, all matrix forms of a linear transformation T ave the same characteristic polynomial which will be called the characteristic polynomial, $_T(x)$, of T.*

roof: Let $A = CBC^{-1}$. Then we have

$$\begin{aligned} det\ (xI - CAC^{-1}) &= det\ (C(xI - A)C^{-1}) \\ &= det\ (C) det\ (xI - A) det\ (C)^{-1} \\ &= det\ (xI - A) \end{aligned}$$

∎

iscussion

(1) Let $A = (a_{ij})$, and $\chi_A(x) = x^n - a_1x^{n-1} + a_2x^{n-2} + \cdots + (-1)^n a_n$. Then it is easy to e that $a_1 = \sum_i a_{ii}$, the *trace* of A, in symbol $tr(A)$, and $a_n = det\ A$, the determinant of . Therefore, all similar matrices have the same trace and determinant. Note that trace a linear functional on the vector space $FL(n, \mathbf{K})$. ∎

heorem 4.32. *Let the invariant factors of A be $c_1(x), c_2(x), \cdots, c_s(x)$. Then the char- cteristic polynomial of A is*

$$det\ (xI - A) = c_1(x)c_2(x) \cdots c_s(x)$$

roof: It follows from Theorem 4.31. that we may assume that A is of the rational form,

$$\begin{pmatrix} B_1 & 0 & \cdot & \cdot & 0 \\ 0 & B_2 & 0 & \cdot & 0 \\ 0 & \cdot & \cdot & \cdot & 0 \\ 0 & \cdot & \cdot & \cdot & 0 \\ 0 & 0 & \cdot & 0 & B_s \end{pmatrix}$$

We have,

$$det\ (xI - A) = det\ \left[\begin{pmatrix} xI_1 & 0 & \cdot & \cdot & 0 \\ 0 & xI_2 & 0 & \cdot & 0 \\ 0 & \cdot & \cdot & \cdot & 0 \\ 0 & \cdot & \cdot & \cdot & 0 \\ 0 & 0 & \cdot & 0 & xI_s \end{pmatrix} - \begin{pmatrix} B_1 & 0 & \cdot & \cdot & 0 \\ 0 & B_2 & 0 & \cdot & 0 \\ 0 & \cdot & \cdot & \cdot & 0 \\ 0 & \cdot & \cdot & \cdot & 0 \\ 0 & 0 & \cdot & 0 & B_s \end{pmatrix} \right]$$

Therefore, we have,

$$det\ (xI - A) = \prod_{i=1}^{s} det\ (xI_i - B_i)$$

It suffices to prove $det\ (xI_i - B_i) = c_i(x)$. For simplicity, let us take $B = B_i, I = I_i, c(x) = c_i(x) = x^m - a_{m-1}x^{m-1} - \cdots - a_1x - a_0$. We have

$$det\ (xI - B) = \begin{pmatrix} x & -1 & 0 & \cdot & \cdot & 0 \\ 0 & x & -1 & \cdot & \cdot & 0 \\ 0 & 0 & x & \cdot & \cdot & 0 \\ \cdot & \cdot & \cdot & \cdot & \cdot & \cdot \\ 0 & \cdot & \cdot & \cdot & x & -1 \\ -a_0 & -a_1 & -a_2 & \cdot & -a_{m-2} & x - a_{m-1} \end{pmatrix}$$

Multiplying the 2nd column by x, the 3rd column by x^2, $\cdots$, the m-th column by x^{m-1}, and then adding all them to the first column, we get

$$det\ (xI - B) = \begin{pmatrix} 0 & -1 & 0 & \cdot & \cdot & 0 \\ 0 & x & -1 & \cdot & \cdot & 0 \\ 0 & 0 & x & \cdot & \cdot & 0 \\ \cdot & \cdot & \cdot & \cdot & \cdot & \cdot \\ 0 & \cdot & \cdot & \cdot & x & -1 \\ c(x) & -a_1 & -a_2 & \cdot & -a_{m-2} & x - a_{m-1} \end{pmatrix}$$

We expand the above determinant by the 1st row, then the 2nd row, $\cdots$, the last row, and get

$$det\ (xI - B) = c(x)$$

This is what we want. ∎

Example 20. *Let* **K** *be a field of characteristic not 2. Find the characteristic polynomial, characteristic values, characteristic vectors, and the Jordan canonical form B of the following matrix A,*

$$A = \begin{pmatrix} 1 & 1 \\ -1 & 3 \end{pmatrix}$$

Let us find its characteristic polynomial,

$$det\ (xI - A) = (x - 1)(x - 3) + 1 = x^2 - 4x + 4 = (x - 2)^2$$

We know that the Jordan canonical form of A must be one of the following two,

$$\begin{pmatrix} 2 & 0 \\ 0 & 2 \end{pmatrix}, \quad \begin{pmatrix} 2 & 1 \\ 0 & 2 \end{pmatrix}$$

Let us find a characteristic vector w_1 of A,

$$(2I - A)w_1 = \begin{pmatrix} 2-1 & -1 \\ 1 & 2-3 \end{pmatrix} \begin{pmatrix} y_1 \\ y_2 \end{pmatrix} = \begin{pmatrix} y_1 - y_2 \\ y_1 - y_2 \end{pmatrix} = \begin{pmatrix} 0 \\ 0 \end{pmatrix}$$

$$w_1 = \begin{pmatrix} 1 \\ 1 \end{pmatrix}$$

Since all characteristic vectors form a 1-dimensional subspace, the Jordan canonical form of A must be the 2nd one. Furthermore, let us find the matrix C such that $A = CBC^{-1}$. Let w_2 be a vector satisfying

$$(2I - A)w_2 = \begin{pmatrix} 2-1 & -1 \\ 1 & 2-3 \end{pmatrix} \begin{pmatrix} z_1 \\ z_2 \end{pmatrix} = \begin{pmatrix} z_1 - z_2 \\ z_1 - z_2 \end{pmatrix} = -w_1 = \begin{pmatrix} -1 \\ -1 \end{pmatrix}$$

$$w_2 = \begin{pmatrix} 1 \\ 2 \end{pmatrix}$$

Recall Theorem 4.8. (2). Let v be the standard basis for $\mathbf{K}^2$, w be the basis $\{w_1, w_2\}$. Then we have

$$e_1^{(2)} = 2w_1 - w_2$$
$$e_2^{(2)} = -w_1 + w_2$$

Therefore, we have

$$C^{-1} = \rho_v = \begin{pmatrix} 2 & -1 \\ -1 & 1 \end{pmatrix}$$
$$C = \rho_v^{-1} = \begin{pmatrix} 1 & 1 \\ 1 & 2 \end{pmatrix}$$

We have

$$\begin{pmatrix} 1 & 1 \\ -1 & 3 \end{pmatrix} = \begin{pmatrix} 1 & 1 \\ 1 & 2 \end{pmatrix} \begin{pmatrix} 2 & 1 \\ 0 & 2 \end{pmatrix} \begin{pmatrix} 2 & -1 \\ -1 & 1 \end{pmatrix}$$

∎

Definition 4.21. *Let $A \in FL(n, \mathbf{K})$, $\mathbf{V} = \mathbf{K}^n$, and λ a characteristic value of A. We have*

(i) *The subspace $\mathbf{V}_\lambda = \{v : Av = \lambda v\}$ is called the characteristic subspace of λ.*
(ii) *The geometric multiplicity of $\lambda = \dim V_\lambda$.*
(iii) *The algebraic multiplicity of λ is the multiplicity of $(x - \lambda)$ in the characteristic polynomial, $\chi_A(x)$.*

∎

Lemma. *Let $\{\lambda_j\}$ be the set of characteristic values of A. Then we have*

$$V_{\lambda_j} \cap (\sum_{i \neq j} V_{\lambda_i}) = \{0\}$$

Proof. Suppose that the above equation is not true for some j. Then there is a non-trivial relation of the following form involving the least number of $v_i \in V_{\lambda_i}$,

(1) $$\sum v_i = 0$$

Say $v_1 \neq 0$, applying A to the above Equation (1), we get

$$\lambda_1 v_1 + \sum_{i\neq 1} \lambda_i v_i = 0 \tag{2}$$

Multiplying Equation (1) by λ_1 and subtracting from Equation (2), we get

$$\sum_{i\neq 1}(\lambda_i - \lambda_1)v_i = \sum_{i\neq 1} v_i' = 0$$

We get a new equation involving less elements. Contradiction. ∎

Theorem 4.33. *Let $A \in FL(n, \mathbf{K})$, $\mathbf{V} = \mathbf{K}^n$, and λ a characteristic value of A. We have*

(i) *The geometric multiplicity of $\lambda \leq$ the algebraic multiplicity of λ.*

(ii) *The matrix A can be **diagonalized**, namely A is similar to a diagonal matrix, $\Leftrightarrow$ the last invariant factor c_ℓ can be factored into product of distinct linear monic polynomials.*

(iii) *The matrix A can be diagonalized $\Leftrightarrow \sum($ geometric multiplicity$) = n$.*

Proof: (i) Selecting a suitable basis of $\mathbf{V}$, we may assume that A is in its rational form. Suppose $(x - \lambda) \mid c_i(x)$ and $c_i(x) = (x - \lambda)^q s(x), s(\lambda) \neq 0$. It is a consequence of the Chinese Remainder Theorem that

$$\mathbf{K}[x]/(c_i(x)) \approx \mathbf{K}[x]/((x - \lambda)^q) \oplus \mathbf{K}[x]/(s(x)).$$

The matrix form of the action of multiplying by x, namely the action of A, to the first component in the above direct sum is as follows,

$$\begin{pmatrix} \lambda & 1 & 0 & \cdot & \cdot & 0 \\ 0 & \lambda & 1 & \cdot & \cdot & 0 \\ 0 & 0 & \lambda & \cdot & \cdot & 0 \\ \cdot & \cdot & \cdot & \cdot & \cdot & \cdot \\ 0 & \cdot & \cdot & \cdot & \lambda & 1 \\ 0 & 0 & 0 & \cdot & 0 & \lambda \end{pmatrix}$$

It follows from the above form that the characteristic space of λ in this component is 1-dimensional. Therefore, we may conclude that

$$dim\ V_\lambda = Card(\{i : (x - \lambda) \mid c_i(x)\})$$

Henceforth, (i) is established.

(ii) ($\Rightarrow$) Let us continue our discussion. From (i), it is easy to see that $c_i(x)$ must be a product of distinct linear monic polynomials for all i.

($\Leftarrow$) We have $c_i(x) \mid c_\ell(x)$ for all i. Therefore, $c_i(x)$ must be a product of distinct linear monic polynomials for all i. Therefore, the rational form of A is diagonal.

(iii) ($\Rightarrow$) It follows from (ii) that all invariant factors, $c_i(x)$, of A will be products of distinct linear monic polynomials. A direct computation will produce $\sum$ (geometric multiplicity) $= n$.

($\Leftarrow$) Let $\{\lambda_j\}$ be the set of characteristic values of A. Let $\mathbf{U}=\sum_j V_{\lambda_j}$. It follows from the preceding Lemma that $\mathbf{V}\supset \mathbf{U}=\oplus_j V_{\lambda_j}$. Since $dim\ \mathbf{V}= dim\ \mathbf{U}=n$, then we have $\mathbf{V}=\oplus_j V_{\lambda_j}$. Therefore, the union of bases of V_{λ_j} will be a basis of $\mathbf{V}$. It is easy to see that the matrix form of A is diagonal. ∎

There is another important polynomial for a matrix $A \in FL(n, \mathbf{K})$. Let us consider the map $\rho : \mathbf{K}[x] \to \mathbf{K}[A]$, with $\rho(x) = A$. Then the kernel of ρ, $ker(\rho) = \{f(x) : f(A) = 0\}$, is an ideal of $\mathbf{K}[x]$. Since $\mathbf{K}[x]$ is a P.I.D., then $ker(\rho) = (g(x))$ for some monic polynomial $g(x)$. We define

Definition 4.22. *The minimal polynomial of A is the monic polynomial $g(x)$ such that $ker(\rho) = (g(x))$ where ρ is defined above.* ∎

Discussion

(1) If $f(A) = 0$, then $f(x) \in ker(\rho)$, therefore, $g(x) \mid f(x)$. ∎

Theorem 4.34. *Let $A \in FL(n, \mathbf{K})$. Then the minimal polynomial of A is the last invariant factor, $c_\ell(x)$, of A.*

Proof: After selecting a suitable basis of $\mathbf{K}^n$, we may assume the action of A is the multiplication of x on the following vector space,

$$\oplus_{i=1}^{\ell} \mathbf{K}[x]/(c_i(x))$$

Clearly, on the i-th component of the direct sum above, we have $f(x) = 0$ if and only if $f(x) \in (c_i(x))$, namely $c_i(x) \mid f(x)$. Furthermore, we have $c_1(x) \mid c_2(x) \mid \cdots \mid c_\ell(x)$. Therefore, $f(x) = 0$ on the whole vector space only if $c_\ell(x) \mid f(x)$. The converse is obvious. ∎

Corollary 1. *The characteristic polynomial, $\chi_A(x)$, and the minimal polynomial, $c_\ell(x)$, of A have the same set of roots.* ∎

Corollary 2. *Let $\mathbf{K}$=C and $A^m = I$ for some m. Then A can be diagonalized.*

Proof: We have $(x^m - 1) \in ker(\rho)$. Note that the derivative of $(x^m - 1)$ is mx^{m-1} which has no common root with $(x^m - 1)$. Therefore, $(x^m - 1)$ has no multiple root, and $c_\ell(x) \mid (x^m - 1)$ has no multiple root. Furthermore, C is algebraically closed, $c_\ell(x)$ can be factored into product of irreducible polynomials, every irreducible polynomial in C$[x]$ is linear. Henceforth, $c_\ell(x)$ can be factored into product of distinct linear monic polynomials. Our Corollary follows. ∎

Example 21. *(1) Let us consider the 3-dimensional real space* $\mathbf{R}^3$. *Let* $A \in FL(3,\mathbf{R})$. *Then the characteristic polynomial,* $\chi_A(x)$, *of A is a polynomial of degree 3. We know every polynomial of odd degree must have a real root. Therefore, A has at least one real characteristic value. This observation checks with the Jordan canonical forms of real matrices.*

(2) Let us solve the following system of homogeneous linear differential equations with constant coefficients of real functions,

$$D\begin{pmatrix} f_1 \\ f_2 \\ f_3 \end{pmatrix} = \begin{pmatrix} a_{11} & a_{12} & a_{13} \\ a_{21} & a_{22} & a_{23} \\ a_{31} & a_{32} & a_{33} \end{pmatrix}\begin{pmatrix} f_1 \\ f_2 \\ f_3 \end{pmatrix}$$

It follows from (1) that there must be a solution of the form $ke^{\lambda x}$.

(3) Let use the assumption of (1). A linear transformation A preserves the orientation if $\det A > 0$. *A linear transformation A reverses the orientation if* $\det A < 0$. *Let A be a linear transformation which preserves the orientation. We claim that it must have a positive characteristic value. It follows from (1) that there must be a real characteristic value* λ_1. *If* $\lambda_1 > 0$, *then we are done. Otherwise, we have,*

$$\chi_A(x) = (x - \lambda_1)(x^2 + ax + b)$$
$$\det A = \lambda b > 0$$

Therefore, we have $b < 0$. *The discriminant of* $x^2 + ax + b$ *is* $a^2 - 4b > 0$. *Therefore,* $x^2 + ax + b$ *has two real roots, and*

$$\chi_A(x) = (x - \lambda_1)(x - \lambda_2)(x - \lambda_3)$$
$$\det A = \lambda_1\lambda_2\lambda_3 > 0$$

Therefore, one of λ_i *must be positive.*

(4) Let us apply uniform pressure to an elastic material. We may assume it is produced by a linear transformation T of $\mathbf{R}^3$. *Furthermore, we may assume the linear transformation preserves the orientation of* $\mathbf{R}^3$. *It follows from (3) that there is at least a direction, as determined by a characteristic space, which is preserved by the linear transformation.* ∎

Exercises

(1) Let $A \in FL(n, K)$ be a linear transformation. A is said to be ***nilpotent*** if there is a positive integer m such that $A^m = 0$. Show that if A is nilpotent and diagonalizable, then $A = 0$.

(2) Use an example to show that not all $A \in FL(n, R)$ with $A^m = I$ can be diagonalizable over R.

(3) A set of matrices $A_i \in FL(n, K)$ can be ***simultaneously diagonalizable*** if there is a basis $\{v_1, v_2, \cdots, v_n\}$ of K^n such that the matrix forms of all A_i are diagonal. Prove that any finite subgroup of $FL(n, C)$ can be simultaneously diagonalizable.

(4) Let $A \in FL(n, R)$ with $A^m = I$ for some m. Show that $A^2 = I$.

(5) Let A, B be two linear transformations of a finite dimensional vector space V. Suppose that $AB = BA$. Prove that A, B have the same set of characteristic values.

(6) Let A be a linear transformation of a finite dimensional vector space V. Suppose that A is invertible. Let λ be a characteristic value of A. Prove that λ^{-1} is a characteristic value of A^{-1}.

(7) Extend Example 21. to R^n where n is an odd integer.

(8) Classify all $A \in FL(3, C)$ up to similarity with $A^3 = 1$.

(9) Determine all possible Jordan canonical forms for a linear transformation $T \in FL(n, C)$ if
  (i) the characteristic polynomial of T is $(\lambda - 2)^2(\lambda - 3)^3$.
  (ii) the minimal polynomial of T is $(\lambda - 2)^2(\lambda - 3)$.

(10) Find the characteristic polynomial, the minimal polynomial, the rational form, the Jordan canonical form of the following matrix $A \in FL(3, C)$,

$$A = \begin{pmatrix} 1 & 2 & 1 \\ 0 & 1 & 0 \\ 0 & 0 & 1 \end{pmatrix}$$

(11) Determine the possible Jordan canonical forms of $A \in FL(5, C)$ whose minimal polynomial is $(\lambda - 1)^2$.

(12) An $n \times n$ matrix $A = (a_{ij})$ is called ***upper triangular*** if $a_{ij} = 0$ for all $i > j$, it is called ***lower triangular*** if $a_{ij} = 0$ for all $i < j$, and it is called ***triangular*** if it is either upper triangular or lower triangular. The matrix A is called ***triangularable*** if it is similar to a triangular matrix. Prove that A is triangular if and only if the minimal polynomial of A is a product of linear polynomials.

(13) Find the characteristic values of the following matrix A,

$$A = \begin{pmatrix} 6 & 0 & 5 \\ 0 & 7 & 0 \\ 5 & 0 & 6 \end{pmatrix}$$

(14) Let $A = (a_{ij}) \in FL(n, K)$ where K is a field. Suppose that for all i, we have $\sum_j a_{ij} = 1$. Show that 1 is a characteristic value of A.

(15) Let $A \in FL(n, \mathbf{K})$. Show that A is nilpotent if and only if the characteristi polynomial of A is $x^n = 0$.
(16) Let $\mathbf{R}$ be a commutative ring with identity. Let A, B be $n \times n$ matrices with entrie in $\mathbf{R}$. Suppose that $AB = I$ the identity matrix. Show that $BA = I$.
(17) Prove the Hamilton-Cayley theorem: any square matrix A satisfies its characteristi polynomial.

---

§7 Inner Product and Bilinear form

In this section and the next section, we will assume that the field $\mathbf{K}$ is either the rea field, $\mathbb{R}$, or the complex field, $\mathbb{C}$.

Definition 4.23. *Let* $\mathbf{V}$ *be a finite dimensional vector space over* $\mathbf{K}$. *A map* $\kappa : V \oplus V \to \mathbf{K}$ *is said to be a* ***bilinear form*** *if*

$$\kappa((r_1v_1 + r_2v_2, u)) = r_1\kappa((v_1, u)) + r_2\kappa((v_2, u))$$
$$\kappa((u, r_1v_1 + r_2v_2)) = r_1\kappa((u, v_1)) + r_2\kappa((u, v_2))$$

The map κ *is said to be a* ***symmetric form*** *if*

$$\kappa((v, u)) = \kappa((u, v))$$

The map κ *is said to be a* ***skew symmetric form*** *if*

$$\kappa((v, u)) = -\kappa((u, v))$$

The map κ *is said to be a* ***hermitian form***[7] *if*

$$\kappa((r_1v_1 + r_2v_2, u)) = r_1\kappa((v_1, u)) + r_2\kappa((v_2, u))$$
$$\kappa((v, u)) = \overline{\kappa((u, v))}$$

where is $\overline{(\ )}$ *is the usual conjugation on the complex numbers, namely* $\overline{a + bi} = a - bi$. *Note that the complex conjugation induces the identity map on the real field* $\mathbb{R}$. *The map* κ *is said to be* ***non-degenerate*** *if*

$$\kappa((v, u)) = 0, \quad \forall\, u \in V \Rightarrow v = 0$$
$$\kappa((v, u)) = 0, \quad \forall\, v \in V \Rightarrow u = 0$$

The map is said to be ***positive definite*** *if*

$$\kappa((v, v)) \in \mathbb{R}, \quad \kappa((v, v)) \geq 0 \text{ and } \kappa((v, v)) = 0 \Leftrightarrow v = 0$$ ∎

---

[7] After Hermite:French Mathematician 1822-1905.

Discussion

(1) If the coefficient field is the real field, **R**, then a form, κ, is hermitian if and only if it is symmetric bilinear.

(2) If the form κ is clear in our discussion, we sometimes write $\kappa((v,u))$ as $[v,u]$.

(3) Let κ be hermitian. Then we have,

$$[v, au] = \overline{[au, v]} = \overline{a[u, v]} = \bar{a}\overline{[u, v]} = \bar{a}[v, u]$$

In some book, a hermitian form is said to be ***anti-linear*** to the second variable u.

(4) Let κ be hermitian. Then we always have $[u,u] = \overline{[u,u]}$. Therefore, $[u,u] \in \mathbf{R}$ a real number.

(5) It is easy to see that

$$[v, u \pm w] = [v, u] \pm [v, w]$$

∎

Definition 4.24. *(1) A hermitian non-degenerate positive definite form on* **V** *is called an* ***inner product***[8] *of* **V**. *A vector space with an inner product is called an* ***inner product space***. *Note that in the case of real field* **R**, *an inner product is a symmetric non-degenerate positive definite bilinear form. (2) A hermitian non-degenerate form on* **V** *is called a* ***quasi-inner product***. ∎

Discussion

(1) Let κ be a quasi-inner product and $\mathbf{S}\subset\mathbf{V}$ be any subset. Let $\mathbf{S}^{\perp} = \{v : [v,s] = 0\ \forall\, s \in \mathbf{S}\}$. Then it is easy to see that $\mathbf{S}^{\perp}$ is a subspace of **V**.

(2) If **U** is a subspace of **V** and κ is a quasi-inner product, then we always have $\mathbf{U}\subset\mathbf{U}^{\perp\perp}$. ∎

Example 22. [9] *(1) Let* $\mathbf{V}=\mathbf{R}^3$. *We will define the* ***usual inner product*** *as follows, let*

$$v = \begin{pmatrix} a_1 \\ a_2 \\ a_3 \end{pmatrix}, \qquad u = \begin{pmatrix} b_1 \\ b_2 \\ b_3 \end{pmatrix}$$

then $[v,u]$ *is as follows*

$$[v,u] = v^t u = \begin{pmatrix} a_1 \\ a_2 \\ a_3 \end{pmatrix}^t \begin{pmatrix} b_1 \\ b_2 \\ b_3 \end{pmatrix} = \begin{pmatrix} a_1 & a_2 & a_3 \end{pmatrix} \begin{pmatrix} b_1 \\ b_2 \\ b_3 \end{pmatrix} = a_1b_1 + a_2b_2 + a_3b_3$$

---

[8] Pioneered by German Mathematician Grassmann 1809-1877.

[9] Pioneered by American Scientist Gibbs 1839-1903.

(2) *We may use the above defined inner product* [,] *to systematize many familiar concepts.*

(i) *The length of a vector,* $|v|$, *is* $\sqrt{a_1^2 + a_2^2 + a_3^2} = \sqrt{[v, v]}$.

(ii) *The distance between the end points of two vectors,* v, u, *is* $\sqrt{[v - u, v - u]}$.

(iii) *The angle,* θ, *spanned by two vectors,* v, u, *is*

$$\theta = cos^{-1} \frac{[v, u]}{\sqrt{[v, v][u, u]}}$$

(iv) *It follows from (iii) the condition for two vectors,* v, u, *to be perpendicular to each other is* $[v, u] = 0$.

(v) *The sphere of radius* r *is given by the following vector equation,*

$$[v, v] = r^2$$

∎

Example 23. (1) *Let* V=$\mathbf{R}^n$. *We will define the* ***usual inner product*** *as follows, let*

$$v = \begin{pmatrix} a_1 \\ \cdot \\ \cdot \\ a_n \end{pmatrix}, \quad u = \begin{pmatrix} b_1 \\ \cdot \\ \cdot \\ b_n \end{pmatrix}$$

then $[v, u]$ *is as follows*

$$[v, u] = v^t u = \begin{pmatrix} a_1 \\ \cdot \\ \cdot \\ a_n \end{pmatrix}^t \begin{pmatrix} b_1 \\ \cdot \\ \cdot \\ b_n \end{pmatrix} = (a_1 \quad \cdot \quad \cdot \quad a_n) \begin{pmatrix} b_1 \\ \cdot \\ \cdot \\ b_n \end{pmatrix} = a_1 b_1 + \cdots + a_n b_n$$

(2) *We may use the above defined inner product* [,] *to define many familiar concepts.*

(i) *The length of a vector,* $|v|$, *is* $\sqrt{a_1^2 + \cdots + a_n^2} = \sqrt{[v, v]}$.

(ii) *The distance between the end points of two vectors,* v, u, *is* $\sqrt{[v - u, v - u]}$.

(iii) *The angle,* θ, *spanned by two vectors,* v, u, *is*

$$\theta = cos^{-1} \frac{[v, u]}{\sqrt{[v, v][u, u]}}$$

(iv) *It follows from (iii) that the condition for two vectors,* v, u, *to be perpendicular to each other is* $[v, u] = 0$.

(v) The sphere of radius r is given by the following vector equation,

$$[v, v] = r^2$$

∎

Example 24. *In many interesting applications, we have only a quasi-inner product, the positive definite condition is not satisfied. In Physics, we have a $(3+1)$-dimensional vector space, $\mathbf{R}^3 \oplus \mathbf{R}$, the Minkowski space. The first 3-dimensional component is the space and the second component $\mathbf{R}$ is the time. Let*

$$v = \begin{pmatrix} a_1 \\ a_2 \\ a_3 \\ a_4 \end{pmatrix}, \qquad u = \begin{pmatrix} b_1 \\ b_2 \\ b_3 \\ b_4 \end{pmatrix}$$

Then we have the Minkowski metric $[v, u]$ defined as follows,

$$[v, u] = a_1 b_1 + a_2 b_2 + a_3 b_3 - c^2 a_4 b_4$$

where c is the speed of light. The solution of $[v, v] = 0$ is a cone, the light cone in Physics.

∎

Definition 4.25. *Given a quasi-inner product κ, we shall define the its matrix form with respect to a basis of $\mathbf{V}$. Let $\{v_1, v_2, \cdots, v_n\}$ be a basis of $\mathbf{V}$. Let*

$$a_{ij} = [v_i, v_j], \qquad A = \begin{pmatrix} a_{11} & \cdot & \cdot & a_{1n} \\ \cdot & \cdot & \cdot & \cdot \\ \cdot & \cdot & \cdot & \cdot \\ a_{n1} & \cdot & \cdot & a_{nn} \end{pmatrix}$$

Then we define the matrix form, $A_{\kappa,v}$, of κ with respect to the basis $\{v_1, v_2, \cdots, v_n\}$ to be

$$A_{\kappa,v} = ([v_i, v_j]) = A = (a_{ij}) = \begin{pmatrix} a_{11} & \cdot & \cdot & a_{1n} \\ \cdot & \cdot & \cdot & \cdot \\ \cdot & \cdot & \cdot & \cdot \\ a_{n1} & \cdot & \cdot & a_{nn} \end{pmatrix}$$

∎

Definition 4.26. *(1) Given a matrix $A \in FL(n, \mathbf{K})$, the conjugate of the transpose of A, in symbol $A^* = \overline{A^t}$, is called the adjoint of A. (2) A is said to be hermitian (or self-adjoint) if A equals its adjoint A^*.* ∎

Discussion

(1) If **K**=**R**, then a matrix is hermitian (or self-adjoint) if and only if it is symmetric, namely $A = A^t$ the transpose of A. ∎

The following theorem establishes a relation between quasi-inner products and hermitian matrices.

Theorem 4.35. *Given a basis $\{v_1, v_2, \cdots, v_n\}$ of* **V***, then we have the following,*

(1) *Let $\kappa(u,u') = [u,u']$ be a quasi-inner product on* **V** *with its matrix form $A_{\kappa,v}$. Then we have*

$$\kappa(u,u') = [u,u'] = \begin{pmatrix} b_1 \\ \cdot \\ \cdot \\ b_n \end{pmatrix}^t \begin{pmatrix} a_{11} & \cdot & \cdot & a_{1n} \\ \cdot & \cdot & \cdot & \cdot \\ \cdot & \cdot & \cdot & \cdot \\ a_{n1} & \cdot & \cdot & a_{nn} \end{pmatrix} \begin{pmatrix} \overline{b'_1} \\ \cdot \\ \cdot \\ \overline{b'_n} \end{pmatrix} = \rho_v(u)^t A_{\kappa,v}\overline{\rho_v(u')}$$

$$\text{where } \rho_v(u) = \begin{pmatrix} b_1 \\ \cdot \\ \cdot \\ b_n \end{pmatrix}, \qquad \rho_v(u') = \begin{pmatrix} b'_1 \\ \cdot \\ \cdot \\ b'_n \end{pmatrix}$$

(2) *Let $\kappa(u_1,u_2) = [u_1,u_2]$ be a quasi-inner product on* **V***. Then the matrix form, $A_{\kappa,v}$, of κ with respect to the basis is hermitian.*

(3) *Let $\{w_1, w_2, \cdots, w_n\}$ be another basis. Then there exists an invertible matrix B such that*

$$A_{\kappa,w} = BA_{\kappa,v}B^*$$

(4) *Conversely, let B be an invertible matrix. Then there exists a new basis $\{w_1, w_2, \cdots, w_n\}$ such that the following relation holds,*

$$A_{\kappa,w} = BA_{\kappa,v}B^*$$

Proof: (1) It follows from the definition.

(2) It follows from the definition that

$$[v_i, v_j] = \overline{[v_j, v_i]}$$

Therefore, $A_{\kappa,v}$ is hermitian.

(3) We have

$$\begin{aligned} \kappa(u,u') = [u,u'] &= \rho_v(u)^t A_{\kappa,v}\overline{\rho_v(u')} \\ &= (\rho_v\rho_w^{-1}\rho_w(u))^t A_{\kappa,v}\overline{(\rho_v\rho_w^{-1}\rho_w(u'))} \\ &= \rho_w(u)^t(\rho_v\rho_w^{-1})^t A_{\kappa,v}\overline{(\rho_v\rho_w^{-1})}(\overline{\rho_w(u')}) \\ &= \rho_w(u)^t A_{\kappa,w}\overline{\rho_w(u')} \end{aligned}$$

By taking $\rho_w(u), \rho_w(u')$ through all elements in the standard basis $\{e_i^{(n)}\}$, it is easy to see that the last equation implies

$$(\rho_v \rho_w^{-1})^t A_{\kappa,v} \overline{(\rho_v \rho_w^{-1})} = A_{\kappa,w}$$

Let

$$B = (\rho_v \rho_w^{-1})^t$$

Then we have

$$B A_{\kappa,v} \overline{B^t} = B A_{\kappa,v} B^* = A_{\kappa,w}$$

(4) Let us define

$$\rho_w^{-1} = \rho_v^{-1} B^t$$

Then $\rho_w^{-1} : \mathbf{K}^n \to \mathbf{V}$ an isomorphism. Let

$$w_i = \rho_w^{-1}(e_i^{(n)})$$

It is easy to see that $\{w_1, w_2, \cdots, w_n\}$ is the new basis sought. ∎

Definition 4.27. *If the following conditions are satisfied, the the basis $\{v_1, v_2, \cdots, v_n\}$ is said to be an orthonormal basis of the quasi-inner product, κ, if*

$$[v_i, v_j] = \pm\delta_{ij},$$
$$\delta_{ij} = \begin{cases} 1 & if\ i = j \\ 0 & if\ i \neq j \end{cases}$$

It is easy to see that $\{v_1, v_2, \cdots, v_n\}$ is an orthonormal basis of κ if and only if the matrix form, $A_{\kappa,v}$, of κ is

$$A_{\kappa,v} = \begin{pmatrix} \pm1 & 0 & \cdot & 0 \\ 0 & \pm1 & \cdot & 0 \\ \cdot & \cdot & \cdot & \cdot \\ 0 & 0 & \cdot & \pm1 \end{pmatrix}$$

∎

We have the following lemma and theorem,

Lemma. *Let κ be a quasi-inner product on $\mathbf{V}$ and $\mathbf{U}$ a subspace of $\mathbf{V}$. If $\mathbf{V}=\mathbf{U}\oplus\mathbf{U}^\perp$. Then the restrictions of κ to $\mathbf{U}$, $\mathbf{U}^\perp$ are quasi-inner products of $\mathbf{U}$, $\mathbf{U}^\perp$ respectively.*

Proof: Clearly the restriction of κ to $\mathbf{U}$ is hermitian. If it is degenerate, then there exists $0 \neq u \in \mathbf{U}$ such that

$$[u, u'] = 0, \qquad \forall\, u' \in \mathbf{U}$$

Since any $v \in \mathbf{V}$ can be written as $v = v' + v'', v' \in \mathbf{U}, v'' \in \mathbf{U}^{\perp}$, then we have $[u, v] = [u, v'] + [u, v''] = 0$. Therefore, κ is degenerate on $\mathbf{V}$. A contradiction. Similarly, we prove that the restriction of κ to $\mathbf{U}^{\perp}$ is a quasi-inner product. ∎

Discussion

(1) Suppose that $\mathbf{V}=\mathbf{U}\oplus\mathbf{U}^{\perp}$. Let $v \in \mathbf{V}$ be any element. Then $v = u + u'$ uniquely with $u \in \mathbf{U}$ $u' \in \mathbf{U}^{\perp}$. We define u = the *projection* of v to $\mathbf{U}$. ∎

Theorem 4.36. *Let κ be a quasi-inner product of a finite dimensional vector space $\mathbf{V}$. Then there is an orthonormal basis with respect to κ.*

Proof: If $\mathbf{V}$ is 1-dimensional, let $0 \neq u \in \mathbf{V}$. Note that $[u, u] \in \mathbf{R}$. If it is zero, then we have $[bu, cu] = 0$ for all $bu, cu \in \mathbf{V}$, and κ is degenerated. This is impossible. Therefore, $[u, u] = \pm c\bar{c}$ for some $c \neq 0$. Let $v = c^{-1}u$. Then we have $\{v\}$ an orthonormal basis for $\mathbf{V}$.

Suppose that *dim* $\mathbf{V}= n \geq 2$. We will find a 1-dimensional subspace $\mathbf{U}$ such that $\mathbf{V}=\mathbf{U}\oplus\mathbf{U}^{\perp}$.

(1) We claim that there is a basis $\{w_1, w_2, \cdots, w_n\}$ such that $[w_1, w_1] \neq 0$. Picking up any basis $\{u_1, u_2, \cdots, u_n\}$. We have two cases: (i) $[u_j, u_j] = 0$ for all j, (ii) $[u_j, u_j] \neq 0$ for some j.

(i) Since κ is non-degenerate, there must be a j such that $[u_1, u_j] \neq 0$. We may assume that $[u_1, u_2] \neq 0$. If $[u_1, u_2]$ is not pure imaginary, then let

$$w_1 = u_1 + u_2, w_2 = u_1 - u_2, w_3 = u_3, \cdots, w_n = u_n$$

then we have

$$\begin{aligned}[w_1, w_1] &= [u_1, u_1] + [u_2, u_1] + [u_1, u_2] + [u_2, u_2] \\ &= \overline{[u_1, u_2]} + [u_1, u_2] \neq 0\end{aligned}$$

If $[u_1, u_2]$ is pure imaginary, then let

$$w_1 = u_1 + iu_2, w_2 = u_1 - iu_2, w_3 = u_3, \cdots, w_n = u_n$$

then we have

$$\begin{aligned}[w_1, w_1] &= [u_1, u_1] + [iu_2, u_1] + [u_1, iu_2] + [u_2, u_2] \\ &= \overline{[u_1, iu_2]} + [u_1, iu_2] = -2i[u_1, u_2] \neq 0\end{aligned}$$

(ii) We may assume that $j = 1$ and $w_i = u_i$.

(2) Let $\mathbf{U}$ be the subspace generated by $\{w_1\}$. We claim that $\mathbf{V}=\mathbf{U}\oplus\mathbf{U}^{\perp}$. We have to prove the following two statements: (i) $\mathbf{U}+\mathbf{U}^{\perp}=\mathbf{V}$, (ii) $\mathbf{U}\cap\mathbf{U}^{\perp} = \{0\}$.

Given any element $v \in \mathbf{V}$, let $[w_1, v] = b$, and

$$\frac{b}{[w_1, w_1]} = \bar{c},$$
$$u' = v - cw_1.$$

We have

$$[w_1, u'] = [w_1, v] - [w_1, cw_1]$$
$$= b - b = 0$$

Therefore, $u' \in \mathbf{U}^{\perp}$, and $v = cw_1 + u' \in \mathbf{U} + \mathbf{U}^{\perp}$. The statement (i) is proved. Let $u \in \mathbf{U} \cap \mathbf{U}^{\perp}$. Then $u = aw_1 = u_1 \subset \mathbf{U} \cap \mathbf{U}^{\perp}$. We claim $[u, v] = 0$ for any element $v \in \mathbf{V}$. Let $v = cw_1 + u'$ as before. It follows that $[u, v] = [u, cw_1] + [u, u'] = [u_1, cw_1] + [aw_1, u'] = 0 + 0 = 0$.

Now we have found a 1-dimensional subspace $\mathbf{U}$ such that $\mathbf{V} = \mathbf{U} \oplus \mathbf{U}^{\perp}$. Note that $\dim \mathbf{U}^{\perp} = n - 1$, and it follows from the preceding Lemma that the restriction of κ to $\mathbf{U}^{\perp}$ is a quasi-inner product. By induction on the dimension, we may assume that $\mathbf{U}^{\perp}$ has a basis $\{v_2, v_3, \cdots, v_n\}$ with

$$[v_i, v_j] = \pm\delta_{ij},$$
$$\delta_{ij} = \begin{cases} 1 & if\ i = j \\ 0 & if\ i \neq j \end{cases}$$

Note that $0 \neq [w_1, w_1] \in \mathbf{R}$. Let $q^2 = \pm[w_1, w_1]$

$$v_1 = \frac{1}{q} w_1$$

It is easy to see that $\{v_1, v_2, \cdots, v_n\}$ is an orthonormal basis for κ. ∎

Theorem 4.37. *Let κ be an inner product of* $\mathbf{V}$, *and* $\mathbf{U} \subset \mathbf{V}$ *a subspace. Then we have*

(1) *There exists an orthonormal basis $\{v_1, v_2, \cdots, v_n\}$ for* $\mathbf{V}$ *with*

$$[v_i, v_j] = \delta_{ij},$$
$$\delta_{ij} = \begin{cases} 1 & if\ i = j \\ 0 & if\ i \neq j \end{cases}$$

(2) *The restrictions of κ to* $\mathbf{U}$ *and* $\mathbf{U}^{\perp}$ *are inner products.*

(3) $\mathbf{V} = \mathbf{U} \oplus \mathbf{U}^{\perp}$.

(4) *Let $\{w_1, w_2, \cdots, w_m\}$ be an orthonormal basis for* $\mathbf{U}$. *Then $u = \sum_{j=1}^{m} [v, w_j] w_j$ is the projection of v to* $\mathbf{U}$.

Proof: (1) It follows from the preceding theorem that there exists an orthonormal basis $\{v_1, v_2, \cdots, v_n\}$ for κ. If $[v_i, v_i] = -1$ for some i, then κ is not positive definite.

(2) The restrictions of κ to $\mathbf{U}$ and $\mathbf{U}^{\perp}$ are certainly hermitian. If it is degenerated on one of them, say on $\mathbf{U}$, then we have for some $u \neq 0$, $[u^*, u] = 0\ \forall\ u^* \in \mathbf{U}$. Therefore, $[u, u] = 0$. Impossible. Henceforth, it is non-degenerated. The positive definite property is inherited.

(3) (4) Let $u' = v - u$ with u defined in (4) of the theorem. We have

$$[u', w_j] = [v - u, w_j] = [v, w_j] - [v, w_j] = 0, \quad \forall\ j = 1, 2, \cdots, m$$

It is easy to see that $[u', u] = 0 \ \forall\ u \in U$. Therefore $u' \in U^{\perp}$. We conclude $v = u + u' \in U + U^{\perp}$. As pointed out in the proof of the preceding theorem that it is easy to see $U \cap U^{\perp} = \{0\}$. We conclude $V = U \oplus U^{\perp}$, and u is the projection of v to U. ∎

Example 25. *Let $P_n = \{f(x) : f(x) \in \mathbf{R}[x], deg\ f(x) < n\}$. Then $dim\ P_n = n$. Let an inner product, κ, be defined as*

$$[f(x), g(x)] = \int_0^1 f(x)g(x)\ dx$$

We wish to find an orthonormal basis. Let $\{v_1, v_2, \cdots, v_n\}$ be an orthonormal basis, a simple observation is that the projection of P_n to P_n produces

$$f(x) = [f(x), v_1]v_1 + [f(x), v_2]v_2 + \cdots + [f(x), v_n]v_n \tag{1}$$

Let $\{u_1, u_2, \cdots, u_n\}$ be any basis, say $\{1, x, \cdots, x^{n-1}\}$. Then we may proceed to construct an orthonormal basis from it as follows.

Since we have

$$[1, 1] = \int_0^1 1\ dx = 1$$

we may take $v_1 = 1$. Let us pick v_2 from the subspace P_2 generated by $\{1, x\}$. Since we have

$$[x, 1] = \int_0^1 x\ dx = \frac{1}{2}$$
$$[x - \frac{1}{2}1, x - \frac{1}{2}] = \int_0^1 (x - \frac{1}{2})^2\ dx = \frac{1}{12}$$

we may take $v_2 = \sqrt{12}(x - 1/2)$. Similarly, we may take v_3 from P_3. Then we make the following computations,

$$[x^2, 1] = \frac{1}{3}$$
$$[x^2, \sqrt{12}(x - \frac{1}{2})] = (\frac{1}{12})^{3/2}$$
$$[x^2 - (\frac{1}{12})^2(x - \frac{1}{2}) - \frac{1}{3},\ x^2 - (\frac{1}{12})^2(x - \frac{1}{2}) - \frac{1}{3}] = \frac{17 \cdot 6421}{2^{10} \cdot 3^5 5} = \frac{1}{c^2}$$

Therefore, we may take $v_3 = c(x^2 - (\frac{1}{12})^2(x - \frac{1}{2}) - \frac{1}{3})$. Successively, we may find v_i. ∎

Example 26. *In $\mathbf{R}^3$, let us use the usual inner product. We wish to find an orthonormal basis for the subspace* U *generated by $u_1 = (1, 1, 0), u_2 = (0, 1, 1)$. We have the following computation,*

$$[u_1, u_1] = 2$$

We may take $v_1 = \frac{1}{\sqrt{2}} u_1$. *We have further computations,*

$$[u_2, v_1] = \frac{1}{\sqrt{2}}$$
$$[u_2 - \frac{1}{\sqrt{2}} v_1,\ u_2 - \frac{1}{\sqrt{2}} v_1] = \frac{3}{2}$$

We may take

$$v_2 = \sqrt{\frac{3}{2}}(u_2 - \frac{1}{\sqrt{2}} v_1) = (-\frac{1}{\sqrt{6}}, \frac{1}{\sqrt{6}}, \sqrt{\frac{2}{3}})$$

Then $\{v_1, v_2\}$ *is an orthonormal basis for* **U**. ∎

Definition 4.28. *A matrix* $A \in FL(n, \mathbf{K})$ *is said to be a* ***unitary matrix*** *if* $AA^* = I$. *If* **K**=**R**, *then a unitary matrix is called an* ***orthogonal matrix***. ∎

Discussion

(1) Let $A = (a_{ij})$, and the (column) vectors a_j be defined as

$$a_j = \begin{pmatrix} a_{1j} \\ a_{2j} \\ \cdot \\ \cdot \\ a_{nj} \end{pmatrix}$$

Then A may be written as $(a_1, a_2, \cdots, a_n)$. The unitary condition means, for the usual inner product [ ] in $\mathbf{K}^n$, the following

$$[a_i, a_j] = \delta_{ij}$$

In other words, let us define a coordinate system with a_i as the i-th coordinate vector, then those coordinate vectors have unit lengths and are perpendicular to each other, thus they form an *orthonormal* basis, and the matrix A is said to be *orthogonal.* ∎

Theorem 4.38. *Let* κ *be an inner product on a vector space* **V**, *and* $\{v_1, v_2, \cdots, v_n\}$ *an orthonormal basis. A basis* $\{w_1, w_1, \cdots, w_n\}$ *is an orthonormal basis if and only if the coefficient matrix* A *is unitary, where* A *is defined as*

$$w_i = \sum_j a_{ij} v_j$$
$$A = (a_{ij})$$

Proof. It is easy and left to the reader as an exercise. ∎

Exercises

(1) Show that for some quasi-inner product κ, $\mathbf{U}^{\perp\perp} \neq \mathbf{U}$. (Hint: Modify Example 24..).

(2) Find a quasi inner product κ such that for some subspace $\mathbf{U}$, we have $\mathbf{V} \neq \mathbf{U} \oplus \mathbf{U}^{\perp}$.

(3) Let Δabc be a triangle and p the middle point of the segment $\overline{bc}$. Show that

$$\overline{ab}^2 + \overline{ac}^2 = 2(\overline{ap} + \overline{bp})$$

(4) Let $\mathbf{V}=FL(n,\mathbf{K})$ and $\mathbf{U}=\{A : A \in V, tr(A) = 0\}$. Let κ be a form on $\mathbf{V}$ defined as

$$\kappa(A,B) = n \cdot tr(AB) - tr(A)tr(B)$$

Show that κ is degenerated on $\mathbf{V}$ and non-degenerated on $\mathbf{U}$.

(5) Let T be a linear transformation from an inner product space $\mathbf{V}$ to $\mathbf{V}$. Suppose that $T^2 = T$ and $[T(v), T(v)] \leq [v, v]$ for all $v \in \mathbf{V}$. Show that the null space and the range of T are perpendicular.

(6) Let κ be a skew symmetric form on $\mathbf{R}^3$. Prove that there are linear functionals f_1, f_2 such that

$$\kappa(u,v) = f_1(u)f_2(v) - f_1(u)f_2(v)$$

(7) Let $\mathbf{K}$ be $\mathbf{R}$. A matrix $A \in FL(n,\mathbf{R})$ is said to be ***positive definite*** if for any $u, w \in \mathbf{R}^n$, we have

$$u^tAw \geq 0$$
$$u^tAu = 0 \Leftrightarrow u = 0$$

For any quasi inner product κ defined on $\mathbf{R}^n$ with

$$[u,w] = u^tAw$$

Show that κ is an inner product $\Leftrightarrow$ A is positive definite.

(8) Let the ***signature*** of a quasi inner product κ with respect to an orthonormal basis be the difference of the number of $+1$ and the number of -1 on the diagonal of the matrix form of κ with respect to the basis. Show that the signature is independent of the basis.

(9) Find the signature of the following matrix $A \in FL(3,\mathbf{C})$,

$$A = \begin{pmatrix} 1 & -3 & 2 \\ -3 & 7 & -5 \\ 2 & -5 & 8 \end{pmatrix}$$

(10) Let $\mathbf{V}$ be a finite dimensional complex inner product space, and T a linear transformation from $\mathbf{V}$ to $\mathbf{V}$. Prove that there is an orthonormal basis of $\mathbf{V}$ such that the matrix form of T is upper triangular.

(11) Prove that for every complex $n \times n$ matrix A, there is a unitary matrix U such that $U^{-1}AU$ is upper triangular.

§8 Spectral Theory

In many continuous phenomena, there are some discrete values happening naturally. For instance the rotation axis of earth rotations, the spectrum[10] of light etc. Many times the set of discrete values is the set of characteristic values. We have the following definition,

Definition 4.29. *Let T be a linear transformation of a vector space* $\mathbf{V}$. *Then the set of all characteristic values of T is called the **spectrum set** of T.* ∎

As pointed out in §7, we shall assume that the field $\mathbf{K}$ is the real field $\mathbf{R}$ or the complex field $\mathbf{C}$. Furthermore, we shall fix an inner product, κ, on the n-dimensional vector space $\mathbf{V}$. We will discuss the spectral theorems of ***self-adjoint linear transformations*** and ***unitary linear transformations*** (cf Definition 4.30.) in this section.

Theorem 4.39. *The dual space* $\mathbf{V}^*$, $Hom(V,K)$, *equals* $\{f_v : v \in V \text{ and } f_v(u) = [u,v]\ \forall\, u \in V.\}$. *Furthermore,* $\dim Hom(V,K) = n$.

Proof: It is clear that $f_v \in \mathbf{V}^*$. It is easy to see $f_{v_1} + f_{v_2} = f_{v_1+v_2}$. We define $af_v = f_{\bar{a}v}$. Then it is easy to see that $\{f_v\}$ is a subspace of $Hom(V,K)$. Let $\{v_1, v_2, \cdots, v_n\}$ be any orthonormal basis of $\mathbf{V}$. We claim (i) $\{f_{v_1}, f_{v_2}, \cdots, f_{v_n}\}$ generates $Hom(V,K)$, (ii) $\{f_{v_1}, f_{v_2}, \cdots, f_{v_n}\}$ is a linearly independent set. Note that our theorem will be proved.

(i) Let $g \in Hom(V,K)$, and $b_i = g(v_i)$. Let $v = \sum_j \overline{b_j} v_j$, and $u = \sum_i c_i v_i$ be any element in $\mathbf{V}$. Then we have

$$\begin{aligned}(g - f_v)(u) &= (g - f_v)(\sum_i c_i v_i) \\ &= \sum_i c_i (g - f_v)(v_i) \\ &= \sum_i c_i (b_i - [v_i, v]) \\ &= \sum_i c_i (b_i - b_i) = 0\end{aligned}$$

Therefore, $g = f_v$.

(ii) Let $\sum_i c_i f_{v_i} = 0$ be a non-trivial relation. Let $u = \sum_i \overline{c_i} v_i \neq 0$. Then we have $\sum_i c_i f_{v_i} = f_u = 0$, and $f_u(u) = [u,u] \neq 0$. A contradiction. ∎

Definition 4.30. *(1) Let T be a linear transformation on* $\mathbf{V}$. *Note that f_{T^*w} defined by*

$$f_{T^*w}(u) = [T(u), w]$$

[10] W. Heisenberg originally expressed *quantum mechanics* in matrix form, the spectrum of light corresponding to characteristic values of some matrices.

is a linear functional, therefore, $f_{T^*w} = f_{w'}$ *for a unique* w'. *The* ***adjoint transformation***, T^*, *of* T *is the linear transformation defined by* $T^*w = w'$ *as above.*
(2) *If* $T^* = T$, *then* T *is a* ***self-adjoint*** *linear transformation.*
(3) *If* $TT^* = I$ *the identity map, then* T *is a* ***unitary transformation.*** ∎

Discussion
(1) We always have $[T(u), w] = [u, T^*(w)]$.
(2) We always have,

$$(T_1 + T_2)^* = T_1^* + T_2^* \qquad T^{**} = T$$
$$(aT)^* = \bar{a}T^* \qquad (T_1T_2)^* = T_2^*T_1^*$$

(3) A linear transformation T is self-adjoint if and only if $[T(u), w] = [u, T(w)]$.
(4) A linear transformation T is unitary if and only if $[T(u), T(w)] = [u, w]$. In the case that $\mathbf{K}=\mathbf{R}$, a unitary transformation is also called an orthogonal transformation.
(5) Let T be any rigid motion of the n-dimensional affine space $\mathbf{R}^n$ which fixes the origin. By an argument similar to Section 1, Chapter II, we conclude that T maps lines to lines and preserves the absolute values of angles. Therefore, it must be a unitary (orthogonal) transformation. ∎

Theorem 4.40. *Let* $\{v_1, v_2, \cdots, v_n\}$ *be an orthonormal basis. Then (1) a linear transformation* T *is self-adjoint if and only if the matrix form* $\rho_v^v(T)$ *is self-adjoint (hermitian), and (2) a linear transformation* T *is unitary if and only if the matrix form* $\rho_v^v(T)$ *is unitary.*

Proof: Since the basis is orthonormal, the matrix form of the inner product is the identity matrix I. We have

$$[u, w] = \rho_v(u)^t\overline{\rho_v(w)}$$

Let $\rho_v^v(T) = A$ (resp. $\rho_v^v(T^*) = A*$) be the matrix forms of T (resp. T^*) with respect to the basis.
(1) ($\Rightarrow$) Suppose that $A = A^*$. Then we have $A^t = \overline{A}$, and

$$\begin{aligned}[T(u), w] &= (A\rho_v(u))^t\overline{\rho_v(w)} = \rho_v(u)^tA^t\overline{\rho(w)} \\ &= \rho_v(u)\overline{A\rho_v(w)} = [u, T(w)]\end{aligned}$$

($\Leftarrow$) Same as above.
(2) ($\Rightarrow$) Suppose that $AA^* = I$. Then we have $A^*A = I$, $A^t\overline{A} = I$, and

$$\begin{aligned}[T(u), T(w)] &= (A\rho_v(u))^t\overline{A\rho_v(w)} = \rho_v(u)^tA^t\overline{A\rho(w)} \\ &= \rho_v(u)\overline{\rho_v(w)} = [u, w]\end{aligned}$$

($\Leftarrow$) Same as above. ∎

We will prove a series of lemmas for self-adjoint linear transformations.

Lemma 1. *Let T be a self-adjoint linear transformation. Then the characteristic values of T are all real.*

Proof: Let λ be a characteristic value of T and u a characteristic vector associated with λ. Note that λ may be zero, while u is not zero. Then we have

$$\begin{aligned}[T(u), u] &= [\lambda u, u] = \lambda[u, u] \\ &= [u, T(u)] = \overline{\lambda}[u, u]\end{aligned}$$

Since $[u, u] \neq 0$, it follows from the above equation that

$$\lambda = \overline{\lambda}$$

Therefore, λ is real. ∎

Lemma 2. *Let T be a self-adjoint linear transformation. Then the characteristic polynomial $\chi_T(x)$ is a product of linear real polynomials.*

Proof: If $\mathbf{K}$=$\mathbf{C}$, then the characteristic polynomial $\chi_T(x) = \prod_i (x - \alpha_i)$ is a product of linear polynomials. Since every root α_i is a characteristic root, then they are real.

Suppose that $\mathbf{K}$=$\mathbf{R}$. We will *complexify* the vector space $\mathbf{V}$ to $\mathbf{V_C}$ as follows; for the basis $\{v_1, v_2, \cdots, v_n\}$, let us extend the coefficients b_i in the general sum $\sum_i b_i v_i$ to all elements in $\mathbf{C}$. It is not hard to see that $\mathbf{V_C}$ is a complex vector space. Extend the inner product and the linear transformation T for b_j, c_k complex numbers as follows,

$$\begin{aligned}[\sum_j b_j v_j, \sum_k c_k v_k] &= \sum b_j \overline{c_k} \delta_{jk} \\ T(\sum_j b_j v_j) &= \sum_j b_j T(v_j)\end{aligned}$$

Then it is easy to see that the extended transformation is a self-adjoint linear transformation. Furthermore, the matrix form $\rho_v^v(T)$ is not changed. Therefore, the characteristic polynomial is not changed, and it is a product of real linear polynomials. ∎

Lemma 3. *Let T be a self-adjoint linear transformation, and $\lambda_1 \neq \lambda_2$ be two distint characteristic values of T. Then the two characteristic spaces $\mathbf{V}_{\lambda_1}$ and $\mathbf{V}_{\lambda_2}$ are perpendicular, namely for $u_1 \in \mathbf{V}_{\lambda_1}$ and $u_2 \in \mathbf{V}_{\lambda_2}$, we always have $[u_1, u_2] = 0$.*

Proof: We have

$$\begin{aligned}[T(u_1), u_2] &= \lambda_1[u_1, u_2] \\ &= [u_1, T(u_2)] = \lambda_2[u_1, u_2]\end{aligned}$$

Therefore, we have $[u_1, u_2] = 0$. ∎

Lemma 4. *Let T be a self-adjoint linear transformation. Then there exists a basis such that the matrix form of T is diagonalized.*

Proof: It does not matter if $\mathbf{K}$ is real $\mathbf{R}$ or complex $\mathbf{C}$. In any case, it follows from Lemma 2 that the characteristic polynomial is a product of real linear polynomials, then there exists a basis $\{w_1, w_2, \cdots, w_n\}$ such that the matrix form of T is the Jordan canonical form as follows,

$$\begin{pmatrix} J_1 & 0 & \cdot & \cdot & 0 \\ 0 & J_2 & 0 & \cdot & 0 \\ 0 & \cdot & \cdot & \cdot & 0 \\ 0 & \cdot & \cdot & \cdot & 0 \\ 0 & 0 & \cdot & 0 & J_s \end{pmatrix}$$

where the block matrices J_i must be as follows,

$$\begin{pmatrix} c_i & 1 & 0 & \cdot & \cdot & 0 \\ 0 & c_i & 1 & \cdot & \cdot & 0 \\ 0 & 0 & c_i & \cdot & \cdot & 0 \\ \cdot & \cdot & \cdot & \cdot & \cdot & \cdot \\ \cdot & \cdot & \cdot & \cdot & \cdot & 1 \\ 0 & 0 & 0 & \cdot & 0 & c_i \end{pmatrix}$$

where c_i is a real number. We claim that every J_i is a 1×1 matrix. Suppose not. Then there is a corresponding system of equations,

$$T(w_s) = c_i w_s$$
$$T(w_{s+1}) = c_i w_{s+1} + w_s$$

for some suitable s. We have

$$\begin{aligned} [T(w_s), w_{s+1}] &= c_i[w_s, w_{s+1}] \\ &= [w_s, T(w_{s+1})] \\ &= c_i[w_s, w_{s+1}] + [w_s, w_s] \end{aligned}$$

Therefore, we conclude $[w_s, w_s] = 0$ which is impossible. ∎

Theorem 4.41. (Spectral Theorem Of Hermitian Matrix). *Let T be a self-adjoint linear transformation. Then there exists an orthonormal basis $\{v_1, v_2, \cdots, v_n\}$, and real numbers $\lambda_1, \lambda_2, \cdots, \lambda_n$ such that*

$$T(v_i) = \lambda_i v_i, \quad \forall\, i$$

In other words, the matrix form of T with respect to that basis $\{v_i\}$ is a real diagonal matrix.

Proof: It follows from Lemma 4 that there exists a basis $\{w_i\}$ such that the matrix form of T is diagonal. The numbers λ_i on the diagonal are characteristic values of T, therefore, all real. Let us group the numbers λ_i together so that identical ones are consecutive. Let $V_{\lambda_{s_j}}$ be the characteristic spaces for distinct λ_{s_j}. It follows from Lemma 3 that they are perpendicular. For each of them, let us find an orthonormal basis. The union of those bases is the basis of **V** we are looking for. ∎

Example 27. *Let us apply the above theorem to the classification of quadratic surfaces. Let the surface be defined by the following equation,*

$$f(x_1, x_2, \cdots, x_n) = \sum_{i \geq j} a_{ij} x_i x_j + \sum_i b_i x_i + c$$

We define a symmetric matrix $D = (d_{ij})$ as follows,

$$d_{ii} = a_{ii}, \qquad d_{ij} = d_{ji} = \frac{1}{2} a_{ij}, \quad \forall\, i > j$$

Then we have

$$\begin{aligned} f(x_1, x_2, \cdots, x_n) &= (x_1, x_2, \cdots, x_n) \begin{pmatrix} d_{11} & d_{12} & \cdot & \cdot & d_{1n} \\ d_{21} & d_{22} & \cdot & \cdot & d_{2n} \\ \cdot & \cdot & \cdot & \cdot & \cdot \\ \cdot & \cdot & \cdot & \cdot & \cdot \\ d_{n1} & d_{n2} & \cdot & \cdot & d_{nn} \end{pmatrix} \begin{pmatrix} x_1 \\ x_2 \\ \cdot \\ \cdot \\ x_n \end{pmatrix} + \textstyle\sum_i b_i x_i + c \\ &= (x)^t D(x) + (b)^t (x) + c \end{aligned}$$

Since D is symmetric, then it is self-adjoint. Therefore, it follows from the preceding theorem that there is an orthogonal matrix C and diagonal matrix E such that

$$D = C^{-1} E C$$

Note that an orthogonal matrix satisfies $D^* D = I$, namely $D^{-1} = D^* = D^t$ in the real case. The above equation can be rewritten as $D = C^t E C$. Henceforth, our original equation can be rewritten as

$$\begin{aligned} f(x_1, x_2, \cdots, x_n) &= (x)^t C^t E C(x) + (b)^t C^t C(x) + c \\ &= (C(x))^t E(C(x)) + (C(b))^t (C(x)) + c \end{aligned}$$

Let

$$\begin{pmatrix} y_1 \\ y_2 \\ \cdot \\ \cdot \\ y_n \end{pmatrix} = C \begin{pmatrix} x_1 \\ x_2 \\ \cdot \\ \cdot \\ x_n \end{pmatrix}, \qquad \begin{pmatrix} b'_1 \\ b'_2 \\ \cdot \\ \cdot \\ b'_n \end{pmatrix} = C \begin{pmatrix} b_1 \\ b_2 \\ \cdot \\ \cdot \\ b_n \end{pmatrix}.$$

Then we have the following canonical equation

$$f = \sum_i \lambda_i y_i^2 + \sum_i b'_i y_i + c$$

The numbers λ_i play an important role in the classification of surfaces. For instance, let $n = 2$, then we have curves. We have

(i) *Ellipse if $\lambda_1\lambda_2 > 0$. Furthermore, if $\lambda_1 = \lambda_2 \neq 0$, then it is a circle.*
(ii) *Hyperbola if $\lambda_1\lambda_2 < 0$.*
(iii) *Parabola if $\lambda_1\lambda_2 = 0$ and one of λ_1, λ_2 is not zero.*
(iv) *Degenerated case if $\lambda_1 = \lambda_2 = 0$.*

Note that in the three non-degenerated cases, the y_1-axis and the y_2-axis are perpendicular.

∎

In the following we will assume that the field **K**= **C**. In fact, Theorem 4.42. below is false for the real case **K**= **R** (cf Example 28..). We will prove a series of lemmas for the unitary linear transformations.

Lemma 5. *Let T be a unitary transformation. Then all characteristic values of T are with absolute value 1.*

Proof: Let λ be a characteristic value of T, and u a characteristic vector associated with λ. Then we have

$$[T(u), T(u)] = [\lambda u, \lambda u] = \lambda\overline{\lambda}[u, u] = [u, u]$$
$$\lambda\overline{\lambda} = |\lambda|^2 = 1, \quad |\lambda| = 1$$

∎

Lemma 6. *Let T be a unitary linear transformation, and $\lambda_1 \neq \lambda_2$ be two distint characteristic values of T. Then the two characteristic spaces $\mathbf{V}_{\lambda_1}$ and $\mathbf{V}_{\lambda_2}$ are perpendicular, namely for $u_1 \in \mathbf{V}_{\lambda_1}$ and $u_2 \in \mathbf{V}_{\lambda_2}$, we always have $[u_1, u_2] = 0$.*

Proof: We always have

$$[u_1, u_2] = [T(u_1), T(u_2)] = [\lambda_1 u_1, \lambda_2 u_2] = \lambda_1\overline{\lambda_2}[u_1, u_2] = \frac{\lambda_1}{\lambda_2}[u_1, u_2]$$

Since

$$\frac{\lambda_1}{\lambda_2} \neq 1$$

then we must have

$$[u_1, u_2] = 0$$

∎

Lemma 7. *Let T be a unitary linear transformation. Then there exists a basis such that the matrix form of T is diagonalized.*

Proof: The proof is similar to the proof of Lemma 4. Since we assume the field is the complex numbers **C**. It follows that the characteristic polynomial is a product of linear polynomials. Then there exist a basis $\{w_1, w_2, \cdots, w_n\}$ such that the matrix form of T is the Jordan canonical form as follows,

$$\begin{pmatrix} J_1 & 0 & \cdot & \cdot & 0 \\ 0 & J_2 & 0 & \cdot & 0 \\ 0 & \cdot & \cdot & \cdot & 0 \\ 0 & \cdot & \cdot & \cdot & 0 \\ 0 & 0 & \cdot & 0 & J_s \end{pmatrix}$$

where the block matrices J_i must be as follows,

$$\begin{pmatrix} c_i & 1 & 0 & \cdot & \cdot & 0 \\ 0 & c_i & 1 & \cdot & \cdot & 0 \\ 0 & 0 & c_i & \cdot & \cdot & 0 \\ \cdot & \cdot & \cdot & \cdot & \cdot & \cdot \\ \cdot & \cdot & \cdot & \cdot & \cdot & 1 \\ 0 & 0 & 0 & \cdot & 0 & c_i \end{pmatrix}$$

where c_i is a complex number with $|c_i| = 1$. We claim that every J_i is a 1×1 matrix. Suppose not. Then there is a corresponding system of equations,

$$\begin{aligned} T(w_s) &= c_i w_s \\ T(w_{s+1}) &= c_i w_{s+1} + w_s \end{aligned}$$

for some suitable s. We have

$$\begin{aligned} [w_s, w_{s+1}] &= [T(w_s), T(w_{s+1})] \\ &= [c_i w_s, c_i w_{s+1} + w_s] \\ &= c_i \overline{c_i} [w_s, w_{s+1}] + c_i [w_s, w_s] \\ &= [w_s, w_{s+1}] + c_i [w_s, w_s] \end{aligned}$$

Therefore, we conclude $c_i[w_s, w_s] = 0$, $[w_s, w_s] = 0$ which is impossible. ∎

Theorem 4.42. (Spectral Theorem Of Unitary Matrix). *Let the field* **K** *be the complex field* **C**, *and let T be a unitary linear transformation. Then there exists an orthonormal basis $\{v_1, v_2, \cdots, v_n\}$, and complex numbers $\lambda_1, \lambda_2, \cdots, \lambda_n$ with absolute values 1 such that*

$$T(v_i) = \lambda_i v_i, \quad \forall\, i$$

In other words, the matrix form of T with respect to that basis $\{v_i\}$ is a complex diagonal matrix with numbers of absolute value 1 on the diagonal.

Proof: It follows from Lemma 7 that there exists a basis $\{w_i\}$ such that the matrix form of T is diagonal. The numbers λ_i on the diagonal are characteristic values of T, therefore, they have absolute values 1. Let us group the numbers λ_i together so that identical ones are consecutive. Let $\mathbf{V}_{\lambda_{s_j}}$ be the characteristic spaces for distinct λ_{s_j}. It follows from Lemma 6 that they are perpendicular. For each of them, let us find an orthonormal basis. The union of those bases is the basis of $\mathbf{V}$ we are looking for. ∎

Example 29. *The above theorem is false in the real case. Let T be a rotation on the plane $\mathbf{R}^2$ with the angle of rotation θ not an integer multiple of π. Then T is a rigid motion of $\mathbf{R}^2$, therefore, a unitary (orthogonal) linear transformation of $\mathbf{R}^2$. It is clear that T can not be diagonalized.* ∎

Exercises

(1) Let $A \in FL(3, \mathbf{R})$ be as follows,

$$A = \begin{pmatrix} 8 & -2 & 2 \\ -2 & 5 & 4 \\ 2 & 4 & 5 \end{pmatrix}$$

Find a change of basis such that A is diagonalized.
In the following we will assume the field is complex $\mathbf{C}$ and fix an inner product $[\ ,\ ]$.

(2) Let $\mathbf{V}$ be finite dimensional and T a linear transformation from $\mathbf{V}$ to $\mathbf{V}$. Suppose that T is self-adjoint. Show that $I + iT$ is invertible and $(I - iT)(I + iT)^{-1}$ is unitary.

(3) A linear transformation T is said to be *normal* if $TT^* = T^*T$. Prove
 (i) Self-adjoint linear transformations and unitary linear transformations are normal.
 (ii) Let T be normal. Then λ is a characteristic value of $T \Leftrightarrow \lambda$ is a characteristic value of T^*.
 (iii) Let $\lambda_1 \neq \lambda_2$ be characteristic values of a normal linear transformation T. Then the characteristic spaces $\mathbf{V}_{\lambda_1}$ and $\mathbf{V}_{\lambda_2}$ are perpendicular to each other.
 (iv) State and prove a spectral theorem for normal linear transformations.

(4) Is the linear transformation T induced by the following matrix A normal?

$$A = \begin{pmatrix} 1 & i \\ i & 1 \end{pmatrix}$$

(5) Let **V** be finite dimensional and T a linear transformation from **V** to **V**. Suppose that T is normal. Show that the adjoint T^* of T is a polynomial in T.

(6) Let **V** be finite dimensional and T_1, T_2 linear transformations from **V** to **V**. Suppose that T_1, T_2 are normal and $T_1T_2 = T_2T_1$. Show that T_1T_2 is normal.

(7) Let **V** be finite dimensional and T a linear transformation from **V** to **V** with $T^2 = T$. Show that T is self-adjoint if and only if $TT^* = T^*T$.

(8) Let **V** be finite dimensional and T a linear transformation from **V** to **V**. Show that T is self-adjoint if and only if $[T(v), v]$ is real for all $v \in \mathbf{V}$.

(9) Let T be a self-adjoint linear transformation. If we have $[T(u), u] > 0$ for all $u \in \mathbf{V}$, then T is said to be ***positive definite.*** Show that T is positive definite $\Leftrightarrow$ all characteristic values of T are positive.

(10) Let T be an invertible self-adjoint linear transformation. Show that T^2 is positive definite.

(11) Let T be an invertible self-adjoint linear transformation. Show that TT^* is positive definite.

(12) Let A be a $n \times n$ positive definite hermitian matrix. Show that there is an invertible $n \times n$ matrix G such that $A = \overline{G}G^t$.

(13) Let T be a self-adjoint linear transformation, and λ a characteristic value of T. Show that the geometric multiplicity of $\lambda =$ the algebraic multiplicity of λ.

(14) Let T be a linear transformation with $T^* = -T$. Show that all characteristic values of T are either 0 or pure imaginary.

CHAPTER V

Polynomials in One Variable and Field Theory

§1 Algebraically Closed Field

Let $(x_1, x_2, \cdots, x_n)$ be unknowns whose values we are searching for. By experiments or reasonings, sometimes we may find a system of equations which restricts the possible values of $(x_1, x_2, \cdots, x_n)$. Due to the nature of the unknown $(x_1, x_2, \cdots, x_n)$, and the methods employed, sometimes we have polynomial equations, analytic equations, differential equations, integral equations, etc. The most elementary and fundamental one is the study of a system of polynomial equations. We shall study the simplest case of one polynomial equation in one variable x_1 which will be denoted by x in this Chapter.

Recall the following statement:

Corollary 2., Theorem 3.15. *If* $\mathbf{R}$ *is an integral domain,* $f(x) \in \mathbf{R}[x]$*, and deg* $f(x) = n$*, then* $f(x)$ *has at most* n *roots.* ∎

The condition that $\mathbf{R}$ is an integral domain is necessary. Let us look at the following example.

Example 1. *(1) Let us consider the group* $\mathbf{K}_4 = \{e, (1,2)(3,4), (1,3)(2,4), (1,4)(2,3)\}$*. Let us consider the following equation*

$$x^2 = e$$

Then all elements in $\mathbf{K}_4$ *satisfy the above equation. Hence it has 4 roots which is bigger than 2 = degree of the equation.*

(2) Let us consider $FL(n, \mathbf{K})$ *where* $\mathbf{K}$ *is a field. We may map* $\mathbf{K}$ *into* $FL(n, \mathbf{K})$ *by sending* $k \in \mathbf{K}$ *to* kI *where* I *is the identity matrix. It is easy to see the preceding map*

is an injective map. Let $A \in FL(n, \mathbf{K})$, and det $(xI - A) \in \mathbf{K}[x]$ be the characteristic polynomial of A. We may seek all solutions of det $(xI - A) = 0$ in $FL(n, \mathbf{K})$. We know that all matrices which are similar to A will satisfy the same characteristic polynomial. Therefore, in general one characteristic polynomial will have infinitely many solutions (cf Problem 1). ∎

Let $\mathbf{K}$ be a field. Then any polynomial equation $f(x) = 0$ has at most n roots in $\mathbf{K}$ where $n = deg\ f(x)$. There are some special fields such that the above equation has precisely n solutions. We shall define,

Definition 5.1. *Let $\mathbf{K}$ be a field. If every irreducible polynomial in the ring $\mathbf{K}[x]$ is linear, then $\mathbf{K}$ is said to be algebraically closed.* ∎

The following theorem establishes equivalent definitions for algebraic closeness.

Theorem 5.1. *Let $\mathbf{K}$ be a field. Then the following three conditions are equivalent,*

(i) *$\mathbf{K}$ is algebraically closed.*
(ii) *Every polynomial $f(x) \in \mathbf{K}[x]$ of positive degree can be factored into product of linear polynomials.*
(iii) *Every polynomial $f(x) \in \mathbf{K}[x]$ of positive degree has at least one root in $\mathbf{K}$.*

Proof: ((i)⇒(ii)) Since $\mathbf{K}[x]$ is a U.F.D., then $f(x)$ can be factored into product of irreducible polynomials which must be linear.

((ii)⇒(iii)) Let $f(x)$ be factored into product of linear polynomials as follows,

$$f(x) = \delta \prod_{i=1}^{n} (a_i x - b_i), \quad 0 \neq \delta \in \mathbf{K}, 0 \neq a_i, b_i \in \mathbf{K}.$$

Let $c_1 = b_1/a_1 \in \mathbf{K}$, since $\mathbf{K}$ is a field. Then it is easy to check that $f(c_1) = 0$.

((iii)⇒(i)) Let $f(x)$ be an irreducible polynomial. Then $f(x)$ must be of positive degree. Let c be a root of $f(x) = 0$. It follows from the Euclidean Algorithm that

$$f(x) = d(x)(x - c) + k, \qquad \text{where } deg\ k < deg\ (x - c) = 1.$$

Note that $k \in \mathbf{K}$. Replacing x by c in the above equation, we conclude that $k = 0$ and $f(x)$ is divisible by $(x - c)$. Since $f(x)$ is given to be irreducible, then we conclude that $f(x)$ is associated with $(x - c)$ and is linear. ∎

Example 2. *(1) The field of rational numbers $\mathbf{Q}$ is not algebraically closed. We will show that the following equation has no solution in $\mathbf{Q}$,*

$$f(x) = x^2 - 2 = 0 \tag{1}$$

Suppose that $f(x)$ has a root c. We may write c in its reduced form as follows,

$$c = \frac{a}{b}, \quad a, b \in \mathbf{Z}, \quad a, b \text{ have no common divisor.}$$

Then the above Eq (1) can be rewritten as an equation in $\mathbf{Z}$ as follows,

$$a^2 - 2b^2 = 0$$

Therefore, a is even. Let $a = 2c$ with c an integer. The above Eq (1) becomes

$$2c^2 - b^2 = 0$$

Henceforth, b is even, and 2 is a common divisor of a, b. Thie contradicts the assumption that a, b have no common divisor.

(2) The field of real numbers $\mathbf{R}$ is not algebraically closed. We will show that the following equation has no solution in $\mathbf{R}$,

$$(1) \qquad f(x) = x^2 + 1 = 0$$

Note that $r^2 \geq 0$ for any $r \in \mathbf{R}$. Therefore, we have $r^2 + 1 \geq 1$ and $r^2 + 1 \neq 0$. ∎

Later on we will see that inside C there is a countable field which is algebraically closed (cf §3). Due to the importance of the real field **R** and the complex field C in Analysis, Geometry, topology and Science in general, we will prove that C is algebraically closed (cf Theorem 5.2. below). The original proof was due to Gauss. Now there are many simple proofs, we will list three proofs.

Theorem 5.2. (The Fundamental Theorem of Algebra). [1] *The complex field C is algebraically closed.*

Proof: (1st proof) Assume *Liouville's Theorem*: 'Every bounded analytic function on C is a constant function'. Suppose that a polynomial $f(x) \in \mathbf{C}[x]$ of positive degree is never zero on C. Let

$$g(x) = \frac{1}{f(x)}$$
$$D_M = \{x : |x| \leq M\}$$

Note that

$$\lim_{x \to \infty} g(x) = 0$$

Therefore, if M is large enough, then we have

$$|g(x)| < 1, \qquad \forall\, x \notin D_M$$

[1] Due to Gauss

The function $g(x)$ is analytic and hence continuous on C. On the closed disc D_M, $g(x)$ is bounded. Therefore, $g(x)$ is a bounded analytic function on C. It follows from Liouville's Theorem that $g(x)$ is constant, and its inverse $f(x)$ is a constant function. This contradicts $f(x)$ being of positive degree.

(2nd proof) We assume that *minimal modular principle* of complex analysis: 'A non-constant function can not have a non-zero minimal absolute value in its domain of definition.' Let $f(x) \in C[x]$ be a polynomial of positive degree. Clearly, $f(x)$ is defined over the whole complex plane C. Suppose that $f(x)$ is never zero on C. Note that

$$\lim_{x \to \infty} g(x) = \infty$$

Therefore, there exists large number M such that

$$|f(x)| > f(0), \quad \forall\, x \notin D_M$$

The function $|f(x)|$ is continuous on the closed disc D_M. It has a minimal value in D_M. This value is clearly the minimal absolute value of $f(x)$ on the whole complex plane C. A contradiction.

(3rd proof) We shall prove the *minimal modular principle* for polynomials directly, and then apply the arguments of the 2nd proof. Let $f(x)$ be a non-constant polynomial. Suppose that $f(x)$ assume the non-zero minimal absolute value at a point $x = a$. Replacing x by $x - a$, we may assume that $a = 0$. Let $f(0) = c \neq 0$. Replacing $f(x)$ by $c^{-1}f(x)$, we may assume that $f(0) = 1$. We will expand $f(x)$ as follows,

$$\begin{aligned} f(x) &= 1 + a_m x^m + a_{m+1}x^{m+1} + \cdots + a_n x^n, \quad a_m \neq 0. \\ &= 1 + a_m x^m(1 + b_1 x + \cdots + b_{n-m}x^{n-m}). \end{aligned}$$

Note that every complex number c can be written as $re^{i\theta}$ in the polar coordinate system where r is a positive real number. Hence, every complex number c has an m-th root $r^{1/m}e^{i\theta/m}$. Therefore, there exists a satisfying the following equation,

$$a = \frac{-1}{\sqrt[m]{a_m}}$$

Replacing x by at where t approaching 0 along the positive real axis, we have

$$f(at) = 1 - t^n(1 + \epsilon(t))$$

with

$$\lim_{t \to 0} \epsilon(t) = 0$$

For t small enough, we have

$$|f(at)| = |1 - t^n + t^n\epsilon(t)| \leq |1 - t^n| + t^n|\epsilon(t)| < 1$$

Contradict to the assumption that $|f(0)| = 1$ is the minimal. ∎

The importance of the previous theorem is that many polynomials $f(x)$ considered by us are in the ring $\mathbf{C}[x]$, it follows $f(x)$ has all roots in $\mathbf{C}$. However, sometimes the polynomial $f(x) \in \mathbf{Q}[x]$. Its roots may be in a smaller field, which is indeed the case (cf §2, §3 below). We have the following definition,

Definition 5.2. *If a complex number c satisfies a non-zero polynomial $f(x) \in \mathbf{Q}[x]$, then c is called an **algebraic number**, otherwise, a **transcendental number**.* ∎

Later on (cf §2, §3 below), we will prove that all algebraic numbers form an algebraically closed field. Let us count the total number of all algebraic numbers.

Lemma. *The set $\mathbf{Q}[x]$ is a countable set.*

Proof: Let P_n be defined as usual,

$$P_n = \{f(x) : f(x) \in \mathbf{Q}[x], deg\ f(x) < n\}$$

Then we know that P_n is an n-dimensional vector space over $\mathbf{Q}$. Therefore, $P_n \approx \prod_{i=1}^{n} \mathbf{Q}$. It follows from Corollary of Theorem 1.1. that P_n is countable. Furthermore, we have

$$\mathbf{Q}[x] = \cup_{i=1}^{\infty} P_i$$

It follows from Corollary of Theorem 1.1. that $\mathbf{Q}[x]$ is countable. ∎

Theorem 5.3. *The set of all algebraic numbers is countable.*

Proof: Let us arrange all polynomials in $\mathbf{Q}[x]$ in a row as $f_1(x), f_2(x), \cdots, f_i(x), \cdots$. Then we have

$$\text{The set of all algebraic numbers} = \cup_{i=1}^{\infty}\{a : a \text{ is a root of } f_i(x)\}$$

It follows from Corollary of Theorem 1.1. that the set is countable. ∎

It is common knowledge from *measure theory* that every countable set is of measure zero. Since we have

$$\mathbf{C} = \{\text{ all algebraic numbers}\} \cup \{\text{ all transcendental numbers}\}$$

It follows that any number has a probability zero to be algebraic. However, it is a hard problem to decide if any given number is indeed non-algebraic. For instance, it is well-known that e, π[2] are transcendental, and it is an open problem to decide if $e \pm \pi, e\pi, e/\pi$ are transcendental.

[2] Euler proved the irrationality of e in 1744, Hermite proved the transcendence of e in 1873. Lambert proved the irrationality of π in 1761, Lindemann proved the transcendence of π in 1882

Exercises

(1) Find $f(x) \in \mathbb{C}[x]$ with $deg\ f(x) = n \geq 2$ such that there are non-similar matrices A, B with $f(x)$ as their characteristic polynomials.

(2) Let $\mathbf{R}$ be an integral domain which contains $\mathbb{C}$ and is a finite dimensional vector space over $\mathbb{C}$ with respect to its addition and multiplication. Prove that $\mathbf{R}$=$\mathbb{C}$.

(3) Let $f(x) = a_0x^n + a_1x^{n-1} + \cdots + a_{n-1}x + a_n \in \mathbb{Z}[x]$. Suppose that $f(0), f(1)$ are odd integers. Prove that $f(x)$ has no integer root.

(4) Let $f(x) = x^n + a_1x^{n-1} + \cdots + a_{n-1}x + a_n \in \mathbb{Z}[x]$. Suppose that all roots of $f(x)$ are in the closed disc $D_1 = \{c : |c| \leq 1\}$. Show that all roots of $f(x)$ must have absolute value 1.

(5) Let $f(x) = a_0x^n + a_1x^{n-1} + \cdots + a_{n-1}x + a_n \in \mathbb{R}[x]$ with all roots real numbers. Prove that the roots of $g(x)$ are all real, where $g(x)$ is given as

$$g(x) = \sum_{i=1}^{n} \binom{n}{j} a_j x^{n-j}$$

§2 Algebraic Extension

We shall lay a theoretic foundation for the discussion of algebraic numbers. We shall generalize Definition 5.2. to the following,

Definition 5.3. *(1) Given two fields* $\mathbf{K}$, $\mathbf{L}$, *if* $\mathbf{K}$ *is a subfield of* $\mathbf{L}$, *then* $\mathbf{L}$ *is said to be an extension of* $\mathbf{K}$.

(2) Let $\alpha \in \mathbf{L}$. *If* α *is a root of a non-zero polynomial* $f(x) \in \mathbf{K}[x]$, *then* α *is called an algebraic element over* $\mathbf{K}$, *otherwise, a transcendental element over* $\mathbf{K}$.

(3) If all elements of $\mathbf{L}$ *are algebraic over* $\mathbf{K}$, *then* $\mathbf{L}$ *is said to be an algebraic extension of* $\mathbf{K}$.

(4) Given two fields $\mathbf{L}$, $\mathbf{L'}$ *which both contain a common subfield* $\mathbf{K}$. *Let* $\sigma : \mathbf{L} \to \mathbf{L'}$ *be a map. If* $\sigma(k) = k\ \forall\ k \in \mathbf{K}$, *then* σ *is called a* $\mathbf{K}$*-map. If* σ *is an injection, then* σ *is called a* $\mathbf{K}$*-injection. If* σ *is an isomorphism. the* σ *is called a* $\mathbf{K}$*-isomorphism.*

We have the following criteria for algebraic elements,

Theorem 5.4. *Let* $\mathbf{L}$ *be an extension of* $\mathbf{K}$, *and* $\alpha \in \mathbf{L}$. *Then the following four conditions are equivalent*

(1) *α is algebraic over* $\mathbf{K}$.

(2) *Let the ring homomorphism* $\rho : \mathbf{K}[x] \to \mathbf{K}[\alpha]$ *be defined as* $\rho(x) = \alpha$. *Then* $\rho^{-1}(0) \neq (0)$. *Note that if we let* $\rho^{-1}(0) = (f(x))$, *then* $\mathbf{K}[x]/(f(x))$ *is canonically isomorphic to* $\mathbf{K}[\alpha]$.

(3) $\mathbf{K}[\alpha]$ *is naturally a finite dimensional vector space over* $\mathbf{K}$.

(4) $\mathbf{K}[\alpha]$ *is a field.*

Proof: We shall establish (1) $\Rightarrow$ (2) $\Rightarrow$ (3) $\Rightarrow$ (4) $\Rightarrow$ (1).

(1) $\Rightarrow$ (2). It is clear that ρ is surjective. We assume that α is algebraic. Therefore, there is a non-zero polynomial $h(x) \in \mathbf{K}[x]$ such that $h(\alpha) = 0$. Henceforth, $h(x) \in \rho^{-1}(0)$ and $\rho^{-1}(0) \neq (0)$.

(2) $\Rightarrow$ (3). Since $\rho^{-1}(0) \neq (0)$, and $\mathbf{K}[x]$ is a P.I.D., then there is a non-zero polynomial $f(x)$ such that $(f(x)) = \rho^{-1}(0)$. Clearly, $f(x)$ can not be a non-zero constant. It suffices to show that $\mathbf{K}[x]/(f(x))$ is a finite dimensional vector space over $\mathbf{K}$. Let $deg\ f(x) = n > 0$. We claim that the image $\{[1], [x], [x^2], \cdots, [x^{n-1}]\}$ of $\{1, x, x^2, \cdots, x^{n-1}\}$ under the canonical map ρ generates $\mathbf{K}[x]/(f(x))$.

Given any $[g(x)] \in \mathbf{K}[x]/(f(x))$, we have the following by the Euclidean Algorithm,

$$g(x) = d(x)f(x) + r(x), \quad deg\ r(x) < deg\ f(x) = n$$

Therefore, $[g(x)] = [r(x)]$, and $\{[1], \cdots, [x^{n-1}]\}$ generates.

(3) $\Rightarrow$ (4). Since $\mathbf{K}[\alpha] \subset \mathbf{L}$, then $\mathbf{K}[\alpha]$ is an integral domain. To prove it is a field, it suffices to show that every non-zero element $h(\alpha)$ has an inverse. Let us assume that $dim\ \mathbf{K}[\alpha] = n$. Then $\{1, h(\alpha), h(\alpha)^2, \cdots, h(\alpha)^n\}$ must be a linearly dependent set. Therefore, there must be a non-trivial relation of the following form and with m minimal

$$a_0 + a_1 h(\alpha) + \cdots + a_m h(\alpha)^m = 0$$

Note that $a_0 \neq 0$. Otherwise, we may factor out $h(\alpha)$, which is possible because $\mathbf{L}$ is a field and $h(\alpha) \neq 0$, and reduce the number m. The above equation can be rewritten as

$$1 = -\left(\frac{a_1}{a_0} + \cdots + \frac{a_m}{a_0} h(\alpha)^{m-1}\right) h(\alpha)$$

Therefore, $h(\alpha)$ has an inverse and $\mathbf{K}[\alpha]$ is a field.

(4) $\Rightarrow$ (1). If $\alpha = 0$, then it satisfies $x = 0$, and 0 is an algebraic element. If $\alpha \neq 0$, since $\mathbf{K}[\alpha]$ is a field, α will have an inverse in $\mathbf{K}[\alpha]$. We have

$$\alpha\left(\sum_{i=0}^{m} a_i \alpha^i\right) = 1$$

Therefore, α satisfies $x(\sum_{i=0}^{m} a_i x^i) - 1 = 0$, and α is algebraic over $\mathbf{K}$. ∎

Corollary. *Let* $\mathbf{L}$ *be an extension of* $\mathbf{K}$, *and* $\alpha \in \mathbf{L}$. *Then the following four conditions are equivalent*

(1) *α is transcendental over* $\mathbf{K}$.

(2) *Let the ring homomorphism* $\rho : \mathbf{K}[x] \to \mathbf{K}[\alpha]$ *be defined as* $\rho(x) = \alpha$. *Then* $\rho^{-1}(0) = (0)$. *Note that* $\mathbf{K}[x]$ *is canonically isomorphic to* $\mathbf{K}[\alpha]$.

(3) $\mathbf{K}[\alpha]$ *is naturally an infinite dimensional vector space over* $\mathbf{K}$.

(4) $\mathbf{K}[\alpha]$ *is not a field.*

Proof: The statements of the Corollary complement the statements of the Theorem. ∎

Theorem 5.5. *Let* K *be a subfield of* L, *and* α *an element in* L *which is algebraic over* K. *Then the following conditions are equivalent for a polynomial* $f(x) \in \mathbf{K}[x]$,

(1) $f(\alpha) = 0$ *and* $f(x)$ *is irreducible in* $\mathbf{K}[x]$.
(2) $f(\alpha) = 0$ and $(f(x)) = \{g(x) : g(x) \in \mathbf{K}[x], g(\alpha) = 0\}$

Furthermore, if $f(x)$ *is monic, then* $f(x)$ *is called the minimal polynomial of* α, *and its degree is called the algebraic degree of* α *over* K.

Proof: (1) $\Rightarrow$ (2). The ideal $\{g(x) : g(x) \in \mathbf{K}[x], g(\alpha) = 0\}$ is principal and equals $(h(x))$ for some suitable $h(x) \notin \mathbf{K}$. Clearly, we have $f(x) \in (h(x))$, and

$$h(x) \mid f(x)$$

Since we assume that $f(x)$ is irreducible, then we have $f(x) \sim h(x)$, and $(f(x)) = (h(x))$.
(2) $\Rightarrow$ (1). It follows from the condition (2) of Theorem 5.4. that

$$\mathbf{K}[\alpha] \approx \mathbf{K}[x]/(f(x))$$

Since $\mathbf{K}[x] \subset \mathbf{L}$, then it is an integral domain. It follows from Theorem 3.24. that $(f(x))$ is a prime ideal, therefore, $f(x)$ is a prime element, and hence irreducible. ∎

Corollary. *The dimension of* $\mathbf{K}[\alpha]$ *as a vector space over* K *is the algebraic degree of* α.

Proof: It follows from the condition (2) of Theorem 5.4. that

$$\mathbf{K}[\alpha] \approx \mathbf{K}[x]/(f(x)), \quad n = deg\ f(x)$$

Let $[x^i]$ be the image of x^i under the canonical map $\mathbf{K}[x] \to \mathbf{K}\ [x]/(f(x))$. Then $\{[1], [x], \cdots, [x^{n-1}]\}$ is a basis of $\mathbf{K}[x]/(f(x))$. ∎

Definition 5.4. *Let* K *be a subfield of* L. *We use [L:K] to denote the vector space dimension of* L *over* K. *If [L:K]*$< \infty$, *then* L *is called a finite extension of* K. ∎

Discussion
(1) If L is a finite extension of K, and $\alpha \in \mathbf{L}$, then $\mathbf{K}[\alpha]$ is a subspace of L, and therefore a finite dimensional vector space over K. It follows that every α is algebraic over K. ∎

Theorem 5.6. *Let* K *be a subfield of* L, *and* L *be a subfield of* S. *Then we have*

$$[\mathbf{S} : \mathbf{K}] = [\mathbf{S} : \mathbf{L}][\mathbf{L} : \mathbf{K}]$$

Especially, if S *is a finite extension of* L *and* L *is a finite extension of* K, *then* S *is a finite extension of* K.

Proof: Let $\{s_i\}_{i\in I}$ be a basis of S over L, and $\{\ell_j\}_{j\in J}$ be a basis of L over K. It suffices to prove $\{s_i\ell_j\}_{(i,j)\in I\times J}$ is a basis of S over K.

(i) We claim that $\{s_i\ell_j\}_{(i,j)\in I\times J}$ is linearly independent over K. Suppose the converse. Let the following be a non-trivial relation among them,

$$\sum_{finite} k_{ij}s_i\ell_j = 0$$

Then we have

$$\sum_i(\sum_j k_{ij}\ell_j)s_i = 0, \quad \sum_j k_{ij}\ell_j \in \mathbf{L}$$

Since $\{s_i\}$ is a linearly independent set over L, then we must have

$$\sum_j k_{ij}\ell_j = 0, \quad \forall\, i$$

Since $\{\ell_j\}$ is a linearly independent set over K, then we must have

$$k_{ij} = 0, \quad \forall\, i,j$$

A contradiction.

(ii) We claim that $\{s_i\ell_j\}_{(i,j)\in I\times J}$ generates S over K. For any $s \in$S, since $\{s_i\}$ generates S over L, then we have,

$$s = \sum_{finite} a_is_i, \qquad \text{where } a_i \in \mathbf{L}$$

Since $\{\ell_j\}$ generates L over K, then we have

$$a_i = \sum_{finite} k_{ij}\ell_j, \qquad \text{where } k_{ij} \in \mathbf{K}$$

Therefore, we have

$$s = \sum_{finite} k_{ij}s_i\ell_j$$

∎

Theorem 5.7. *(1) Let* L *be an extension of* K, *and* $\alpha,\beta \in$ L *be algebraic elements over* K. *Then* K$[\alpha,\beta]$=K$[\alpha][\beta]$ *is a finite extension of* K. *Therefore, all elements are of the form* $\alpha \pm \beta, \alpha\beta$, $\alpha\beta^{-1}$ *where* $\beta \neq 0$ *are algebraic over* K.

(2) The set of all algebraic elements over **K** *in* **L** *is an extension of* **K**, *the **algebraic closure**,* $\mathbf{K}_L^C$, *of* **K** *in* **L**. *If* $\mathbf{K}=\mathbf{K}_L^C$, *in other words, if any element in* **L** *which is algebraic over* **K** *must be in* **K**, *then we say that* **K** *is **algebraically closed** in* **L**.

(3) If **S** *is an algebraic extension of* **L**, *and* **L** *is an algebraic extension of* **K**, *then* **S** *is an algebraic extension of* **K**.

Proof: (1) Let β be algebraic over **K** satisfying the polynomial equation $f(x) = 0$. Clearly, we have $f(x) \in \mathbf{K}[x] \subset \mathbf{K}[\alpha][x]$. Therefore, β is algebraic over $\mathbf{K}[\alpha]$. It follows from the preceding theorem that

$$[\mathbf{K}[\alpha,\beta] : \mathbf{K}] = [\mathbf{K}[\alpha,\beta] : \mathbf{K}[\alpha]][\mathbf{K}[\alpha] : \mathbf{K}]$$

The two numbers on the right hand side of the above equation are finite. We conclude that $\mathbf{K}[\alpha,\beta]$ is a finite extension of **K**. Therefore, all elements in it are algebraic over **K**.

(2) It follows from (1) that $\mathbf{K}_L^C$ is a field.

(3) Let s be any element in **S**, and let s be a root of $f(x) \in \mathbf{L}[x]$ as follows,

$$f(x) = \sum_{i=0}^{n} \ell_i x^i$$

Note that all ℓ_i are algebraic over **K**. We have the following chain of fields,

$$\mathbf{K} \subset \mathbf{K}[\ell_0] \subset \mathbf{K}[\ell_0,\ell_1] \subset \cdots \subset \mathbf{K}[\ell_0,\cdots,\ell_n] \subset \mathbf{K}[\ell_0,\cdots,\ell_n,s]$$

Each field in the above chain is a finite extension of the preceding one. After applying Theorem 5.6. finitely many times, we conclude that the last field is a finite extension of the first one. Therefore, s is algebraic over **K**. ∎

Corollary. *(1)* $\mathbf{K}_L^C$ *is algebraically closed in* **L**. *(2) If* **L** *is algebraically closed, then* $\mathbf{K}_L^C$ *is algebraically closed.*

Proof: (1) It follows from the Theorem. (2) Let $f(x) \in \mathbf{K}_L^C$ be a non-constant polynomial. Then we have $f(x) \in \mathbf{L}[x]$. Since **L** is algebraically closed, there is a root $\alpha \in \mathbf{L}$. Note that α is algebraic over **K**. It follows from the preceding Theorem that $\alpha \in \mathbf{K}_L^C$, and $\mathbf{K}_L^C$ is algebraically closed. ∎

Example 3. *It follows from the preceding Corollary that the algebraic closure of* **Q** *in* **C**, $\mathbf{Q}_{\mathbf{C}}^C$, *is algebraically closed. Note that* $\mathbf{Q}_{\mathbf{C}}^C$ *is the collection of all algebraic numbers, and therefore, a countable set.*

If we start with polynomial equations with rational coefficients, and solve these equations, we then get algebraic numbers as roots. Successively, we use those algebraic numbers created by solving equations as coefficients of new polynomial equations, and solve the new equations to create new roots. What we get are still algebraic numbers. In the way of solving polynomial equations, we will stay in the field of algebraic numbers, and can no reach the field of complex numbers. To reach the complex field **C**, *we first create the rea field* **R** *by the process of completion (cf Section 6, Chapter I), and then solve equation with real coefficients.*

As pointed out in Theorem 5.5., the minimal polynomial $f(x)$ of an algebraic element α is irreducible. In general it is hard to tell if a polynomial $f(x)$ is irreducible in the ring $\mathbf{K}[x]$. The following criterion is useful,

Theorem 5.8. (Eisenstein Criterion). *Let* **D** *be a U.F.D., and* **K** *its quotient field. Let $f(x) \in \mathbf{D}[x] \subset \mathbf{K}[x]$ be a polynomial of positive degree and with the following expansion,*

$$f(x) = a_0x^n + a_1x^{n-1} + \cdots + a_{n-1}x + a_n$$

If there is a prime element $p \in \mathbf{D}$ such that

$$p \nmid a_0, \quad p \mid a_i, \text{ for } i = 1, 2, \cdots, n, \quad p^2 \nmid a_n$$

Then $f(x)$ is an irreducible polynomial in $\mathbf{K}[x]$.

Proof: (1) We shall reduce the proof to the case that $f(x)$ is a *primitive* polynomial (cf Definition 3.12.). Let d be a greatest common divisor of all coefficients $\{a_0, a_1, \cdots, a_n\}$. Let $f(x) = df^*(x)$. Then $f(x)$ is irreducible in $\mathbf{K}[x] \Leftrightarrow f^*(x)$ is irreducible in $\mathbf{K}[x]$. Note that $f^*(x)$ is primitive with the following expansion,

$$f^*(x) = a_0^*x^n + a_1^*x^{n-1} + \cdots + a_{n-1}^*x + a_n^*$$

We have

$$p \nmid a_0^*, \quad p \mid a_i^*, \text{ for } i = 1, 2, \cdots, n, \quad p^2 \nmid a_n^*$$

(2) Let us assume that $f(x)$ is a primitive polynomial. It follows from Theorem 3.12. that we have to show $f(x)$ is irreducible in $\mathbf{D}[x]$. Suppose the converse. Let the following be a factorization in $\mathbf{D}[x]$ with two factors $g(x), h(x)$ positive degrees,

$$f(x) = g(x)h(x)$$

Let $\rho : \mathbf{D}[x] \to (\mathbf{D}/(p))[x]$ be the canonical map. Let $\mathbf{R}$ be the quotient field of $\mathbf{D}/(\mathrm{p})$. Then we have

$$\rho(f(x)) = \rho(a_0)x^n = \rho(g(x))\rho(h(x))$$

Comparing the degrees on all sides of the above equation, and noting that $\mathbf{R}[x]$ is a U.F.D., we conclude that

$$\rho(g(x)) = b_1x^m, \qquad \rho(h(x)) = b_2x^{n-m}$$

Thus we have

$$\rho(g(0)) = 0, \qquad \rho(h(0)) = 0$$

which means

$$p \mid g(0), \quad p \mid h(0) \Rightarrow p^2 \mid g(0)h(0) = a_n$$

A contradiction. ∎

Example 4. (Existence of Irrational Number). *The ancient Greeks believed that all numbers are rational, largely due to the influence of Pythagoras. However, Hippasus, a member of Pythagoras school, showed that the* golden mean *was not rational, which caused a chaos in the ancient Greek mathematics.*

We shall use the above theorem to show that $\sqrt{2}$ is not rational. Note that $\sqrt{2}$ is a root of $x^2 - 2 \in \mathbb{Z}[x] \subset \mathbb{Q}[x]$. We claim that $x^2 - 2$ is irreducible in $\mathbb{Q}[x]$. Note that $\mathbb{Z}$ is U.F.D., and 2 is a prime number, and the conditions of Theorem 4.8. are satisfied for $f(x) = x^2 - 2$, $p = 2$. Therefore, $x^2 - 2$ is irreducible in $\mathbb{Q}[x]$. It follows from Theorem 5.5. and its Corollary that

$$[\mathbb{Q}[\sqrt{2}] : \mathbb{Q}] = 2$$

Therefore, $\sqrt{2} \notin \mathbb{Q}$. ∎

Example 5. (Cyclotomic Polynomials). *(1) Let $\alpha \in \mathbb{C}$ with $\alpha^n = 1$ for some positive integer n. Then α is called an n-th root of unity. It is easy to see that*

$$\alpha = e^{\frac{k2\pi i}{n}}, \qquad \text{for some } k \text{ with } 0 \le k \le n-1$$

Note that all possible n-th roots of unity form the vertices of a regular n-gon on the unit circle. If n is the minimal positive integer such that $\alpha^n = 1$, then α is called a **primitive n-th root of unity.**

(2) The minimal polynomial, $\varphi_n(x)$, of a primitive n-th root of unity over $\mathbb{Q}$ is called the **n-th cyclotomic polynomial.** *Later on (cf Example 9., §4) we will show that all primitive n-th roots of unity will share the same minimal polynomial $\varphi_n(x)$.*

(3) Let us consider the p-th **cyclotomic polynomial**, *where p is a prime number. Let ζ be a primitive p-th root of unity. Then $\zeta \neq 1$, and it must be a root of the following polynomial,*

$$f(x) = \frac{x^p - 1}{x - 1} = x^{p-1} + x^{p-2} + \cdots + x + 1$$

Let us prove that the above polynomial is irreducible. Note that it then follows from Theorem 5.5. that the above polynomial is the minimal polynomial of ζ, and is $\varphi_p(x)$. Further note that all primitive p-th roots of unity will be roots of the same polynomial as we claim in (2).

Let $y = x - 1, x = y + 1$. Then we have

$$f(x) = f(y+1) = \frac{(y+1)^p - 1}{y} = y^{p-1} + \binom{p}{1} y^{p-2} + \cdots + \binom{p}{i} y^{p-i-1} + \cdots + p$$

It is clear that the conditions of Theorem 5.8. are satisfied. Therefore, $f(x) = f(y+1)$ is irreducible. We conclude that

$$\varphi_p(x) = x^{p-1} + x^{p-2} + \cdots + x + 1$$
$$[\mathbf{Q}[\zeta] : \mathbf{Q}] = p - 1$$

∎

Example 6. (Construction with Straight-Edge and Compass). *The ancient Greeks believed that the straight lines and circles were perfect, therefore, they preferred to use only straight-edge and compass in their geometric construction on the plane. The rules were as follows,*

(1) *Pick up any two distinct points on the plane.*
(2) *Draw the line passing the two points.*
(3) *Use one point as the center and the distance between the two points as the radius to draw a circle. It intersects the line at a new point.*
(4) *Lines may be drawn passing any two points created previously, circles may be drawn with any point created previously as center, and the distance between any two points created previously as radius.*
(5) *Repeating the above process finitely many times.*

The question is: what new points can be created this way? what problem can this method solve?

*We shall use algebra to understand the above procedure. Let us use the complex numbers to represent the plane, and assume the two points picked up are the origin and the point 1 on the real line. Note that we know how to construct a line passing through the origin and perpendicular to the real line from plane geometry. Therefore, we have the imaginary axis. Any number which can be constructed with straight-edge and compass will be called a **constructible number**.*

Let us consider the real numbers which can be constructed with straight-edge and compass. The following two diagrams show that if α, β can be constructed, then $\alpha \pm \beta$, $\alpha\beta$, β^{-1} (if $\beta \neq 0$) can be constructed,

The left diagram shows how to construct $\alpha \pm \beta$. In the right diagram, we know that $\Delta oab \sim \Delta cad$. If we let $\overline{oa} = \alpha$, $\overline{ca} = 1$, then we have the product $\overline{ob} = \alpha\beta$. If we let $\overline{ob} = 1 = \overline{ca}, \overline{oa} = \beta$, then we have $\overline{cd} = \beta^{-1}$. We conclude that we may form arithmetical operations among the constructible real numbers, and they form a field.

It is similar among complex numbers. Let the complex number $c = \alpha + \beta i$ correspond to the point (α, β). Then we have

$$c = \alpha + \beta i \text{ is constructible.} \Leftrightarrow \alpha, \beta \text{ are constructible.}$$

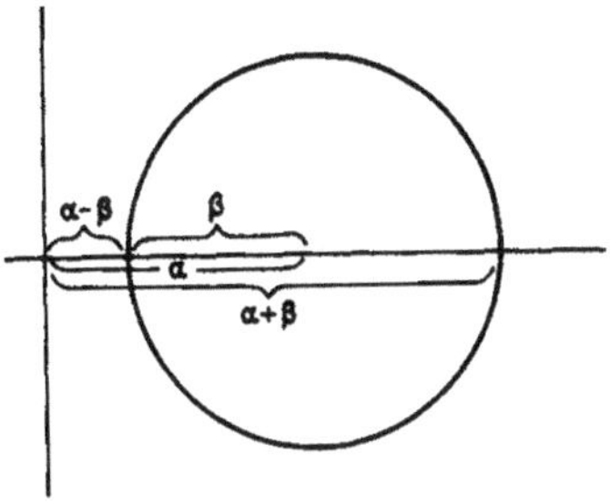

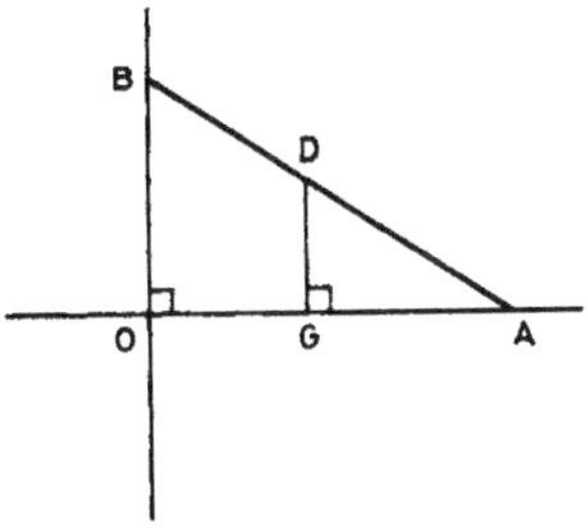

Let $c_1 = \alpha_1 + \beta_1 i, c_2 = \alpha_2 + \beta_2 i$. Then we have

$$c_1 \pm c_2 = (\alpha_1 \pm \alpha_2) + (\beta_1 \pm \beta_2)i$$
$$c_1 c_2 = (\alpha_1\alpha_2 - \beta_1\beta_2) + (\alpha_1\beta_2 + \alpha_2\beta_1)i$$
$$c_1^{-1} = \frac{\alpha_1 - \beta_1 i}{\alpha_1^2 + \beta_1^2}$$

We conclude that we may form arithmetical operations among the constructible complex numbers, and they form a field.

Let us study the procedure of constructions again. In finite steps, we have constructed numbers $\alpha_1, \alpha_2, \cdots, \alpha_n, \beta_1, \cdots, \beta_m$. We have the field $\mathbf{K}_i = \mathbf{Q}[\alpha_1, \cdots, \alpha_n, \beta_1, \cdots \beta_m]$. We may intersect

(i) two lines, one passes two points with coordinates $(\alpha_1, \beta_1), (\alpha_2, \beta_2)$, other passes two points with coordinates $(\alpha_3, \beta_3), (\alpha_4, \beta_4)$.

(ii) a line passing through points (α_1, β_1), (α_2, β_2) and a circle with point (α_3, β_3) as the center and α_4 as the radius.

(iii) two circles with points $(\alpha_1, \beta_1), (\alpha_2, \beta_2)$ as centers and α_3, β_3 as radius.

Case (i). Let the equations of the two lines be

$$\begin{cases} (\alpha_1 - \alpha_2)(y - \beta_2) = (\beta_1 - \beta_2)(x - \alpha_1) \\ (\alpha_3 - \alpha_4)(y - \beta_4) = (\beta_3 - \beta_4)(x - \alpha_3) \end{cases}$$

It is easy to see that the intersection point is with coordinates in the field $\mathbf{K}_i$.

Case (ii). Let the equation of the line and the circle be the following,

$$\begin{cases} (\alpha_1 - \alpha_2)(y - \beta_2) = (\beta_1 - \beta_2)(x - \alpha_1) \\ (x - \alpha_3)^2 + (y - \beta_3)^2 = \alpha_4^2 \end{cases}$$

Solving the first equation to express one variable, say y, in term of the other, say x, and then substituting it into the second equation, we get a quadratic equation in one variable, y. This quadratic equation may have two real roots, or one real root, or no real root, corresponding to the line and the circle intersecting at two points, or one point, or no point. It is clear that the other x is linear in y. Let one of the new points be (α_s, β_s). Therefore, we have $\mathbf{K}_i[\alpha_s, \beta_s] = \mathbf{K}_i[\beta_s]$, and

$$[\mathbf{K}_i[\beta_s] : \mathbf{K}_i] \leq 2$$

When we consider the new distance produced by measuring between one old point and the new point, it follows from the Pythagonean Theorem that the field extension is at most quadratic. We conclude that if the new points and the new distance are added to the field $\mathbf{K}_i$, we have a finite sequence of quadratic extensions.

Case (iii). Let the equations of the circles be

$$\begin{cases} (x-\alpha_1)^2 + (y-\beta_1)^2 = \alpha_3^2 \\ (x-\alpha_2)^2 + (y-\beta_2)^2 = \beta_3^2 \end{cases}$$

subtracting one from other, we may assume one of the above equations is linear. Case (iii) is reduced to Case (ii).

To summarize the above discussion, a finite step construction with straight-edge and compass only produces elements in a field $\mathbf{K}$ which can be connected to the rational field $\mathbf{Q}$ by a chain of intermediate fields such that every one is a quadratic extension of the preceding one as follows,

$$\mathbf{Q} \overset{2}{\subset} \mathbf{K}_1 \overset{2}{\subset} \cdots \overset{2}{\subset} \mathbf{K}_n = \mathbf{K} \tag{1}$$

Conversely, given any chain of quadratic extensions as above, then every elements in $\mathbf{K}$ can be constructed. To prove the preceding statement, it suffices to show that if all elements in $\mathbf{K}_{j-1}$ can be constructed, then all elements in $\mathbf{K}_j$ can be constructed. Let $k \in \mathbf{K}_j$. If $k \in \mathbf{K}_{j-1}$, then certainly k can be constructed. Let us assume that $k \notin \mathbf{K}_{j-1}$. Then we have $[\mathbf{K}_{j-1}[k] : \mathbf{K}_{j-1}] = 2$. Let the following be the minimal polynomial of k over $\mathbf{K}_{j-1}$,

$$f(x) = x^2 - 2ax + b$$

The solutions of the above equation are as follows,

$$k = a \pm \sqrt{a^2 - b}$$

It is clear that

$$k \text{ is constructible} \Leftrightarrow \sqrt{a^2-b} \text{ is constructible.}$$

Note that if $a^2 - b < 0$, then $\sqrt{a^2 - b}$ is pure imaginary and with the same length as $\sqrt{b - a^2}$. Therefore, we may assume that $a^2 - b > 0$. The following diagram exhibit a way to construct $\sqrt{a^2 - b}$. Our proof is finished.

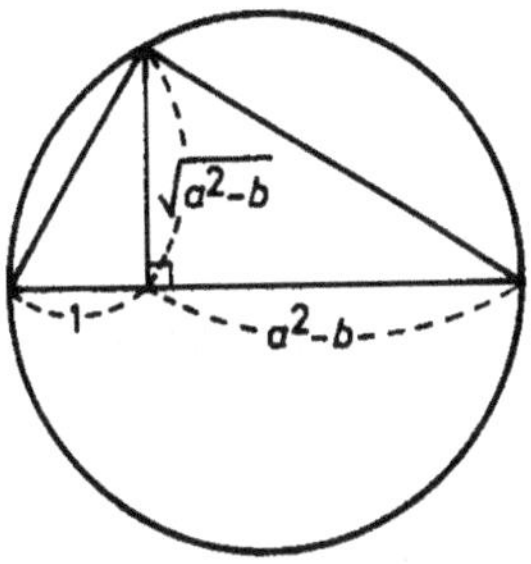

∎

Example 7. *The ancient Greeks had three problems: (1) Can any constructed angle be trisected in equal parts? (2) Can a cube which doubles the volume of a given cube be constructed? (3) Can one find the quadrature of a circle? In other words, can a square with the area of a given constructed circle be constructed? Here the constructions are limited with straight-edge and compass only.*

In fact, all those problems can be discussed in the field theory. We have the following,

(1) The answer to the first problem is no. We have the following equation among angles,

(1) $$4\cos^3\theta - 3\cos\theta = \cos 3\theta$$

Let $\theta = 20^o, 3\theta = 60^o$. It is well known that regular triangle can be constructed, hence the angle 60° can be constructed. We claim that 20° can not be constructed. In other words, the point $(\sin 20^o, \cos 20^o)$, which is the intersection of the line with 20° inclination against the real line and the unit circle, can not be constructed. Note that $\cos 60^o = 1/2$. Hence, $\cos 20^o$ is a root of the following polynomial,

(2) $$4x^3 - 3x - \frac{1}{2} = 0$$

We will prove that the above polynomial is irreducible. Let

$$x = \frac{1}{2}(y+1)$$

Then Eq (ii) can be simplified to the following

(3) $$y^3 + 3y^2 - 3 = 0$$

It follows from the Eisenstein Criterion that let $\mathbf{D}=\mathbf{Z}$, $p=3$, *then Eq (iii) is irreducible. Henceforth, Eq (ii) is irreducible, and it follows from Corollary of Theorem 5.5. that*

$$[\mathbf{Q}[\cos 20^\circ] : \mathbf{Q}] = 3 \tag{4}$$

If $\cos 20^\circ$ *is constructible, then there exists a chain of quadratic extensions as follows*

$$\mathbf{Q} \overset{2}{\subset} \mathbf{K}_1 \overset{2}{\subset} \cdots \overset{2}{\subset} \mathbf{K}_n, \quad \cos 20^\circ \in \mathbf{K}_n$$

We have

$$2^n = [\mathbf{K}_n : \mathbf{Q}] = [\mathbf{K}_n : \mathbf{Q}[\cos 20^\circ]][\mathbf{Q}[\cos 20^\circ] : \mathbf{Q}] = \ell \cdot 3$$

This is impossible, since 2^n *is not divisible by 3. Therefore,* $\cos 20^\circ$ *is not constructible.*

(2) The answer to the second problem is no. Let the volume of the doubling cube be $2a^3$, *and the length of an edge of the doubling cube be* $\sqrt[3]{2}a$, *where* a *is constructible. It is easy to see that the doubling cube can be constructed if and only if* $\sqrt[3]{2}$ *can be constructed. Note that* $\sqrt[3]{2}$ *is a root of the following polynomial*

$$x^3 - 2 = 0$$

It follows from the Eisenstein Criterion that the above polynomial is irreducible. Therefore, we have

$$[\mathbf{Q}[\sqrt[3]{2}] : \mathbf{Q}] = 3$$

The arguments of (1) may be copied to show that $\sqrt[3]{2}$ *is not constructible.*

(3) The answer to the third problem is no. Let the radius of the given circle be r, *where* r *is constructible. Then we want to find a square with area* πr^2, *or an edge with length* $r\sqrt{\pi}$. *Clearly, it is equivalent to construct* $\sqrt{\pi}$. *It is easy to see that* $\pi \in \mathbf{Q}[\sqrt{\pi}]$, *and if* $\sqrt{\pi}$ *is algebraic over* $\mathbf{Q}$, *then* π *is algebraic over* $\mathbf{Q}$. *In 1882, Lindemann proved that* π *is transcendental over* $\mathbf{Q}$. *Therefore,* $\sqrt{\pi}$ *is not algebraic over* $\mathbf{Q}$, *and non-constructible. The proof of the transcendence of* π *is outside the scope of this book, and will not be discussed.* ∎

Example 8. (Construction of Regular Polygon)[3]. *Let* $p > 2$ *be a prime number. Can we construct (cf Example 6.) a regular p-gon? It is equivalent to construct a regular p-gon inscribing the unit circle. We may assume that one of the vertices is 1 on the real line. For the time being, we shall look for a necessary numerical condition. The general discussions will be postponed to Examples 10. 16. & 17..*

The problem of construction of a regular p-gon is equivalent to the construction of the following ζ,

$$\zeta = e^{\frac{2\pi i}{p}}$$

---

[3] Please see the related Examples 16. & 17.. Those Examples were due to Gauss

We have (cf Example 5.)

$$[\mathbf{Q}[\zeta] : \mathbf{Q}] = p - 1$$

It follows from Example 6. that zeta can be constructed if and only if this is a chain of quadratic extensions as follows,

$$\mathbf{Q} \overset{2}{\subset} \mathbf{K}_1 \overset{2}{\subset} \cdots \overset{2}{\subset} \mathbf{K}_n, \quad \zeta \in \mathbf{K}_n$$

Therefore, we have

$$2^n = [\mathbf{K}_n : \mathbf{Q}] = [\mathbf{K}_n : \mathbf{Q}[\zeta]][\mathbf{Q}[\zeta] : \mathbf{Q}] = \ell \cdot (p-1)$$

Therefore, we have

$$p - 1 = 2^m, \qquad p = 2^m + 1$$

If m has an odd factor, let $m = rs, r > 1, s > 1$ and s odd, then we have

$$p = 2^m + 1 = (2^r)^s + 1 = (2^r + 1)[(2^r)^{s-1} - (2^r)^{s-2} + \cdots + 1]$$

Therefore, $(2^r + 1) \mid p$ and p is not prime. Henceforth, we have $m = 2^q$ and

$$p = 2^{2^q} + 1 = F_q$$

*The numbers F_q of the above form are called **Fermat numbers**. We conclude that a regular p-gon, where p is a prime number, can be constructed only if p is a Fermat number. The first five Fermat numbers are*

$$F_0 = 3, \quad F_1 = 5, \quad F_2 = 17, \quad F_3 = 257, F_4 = 65537$$

They are all primes. Fermat conjectured that all Fermat numbers are primes. In 1732, Euler showed that

$$F_5 = 641 \times 6700417$$

It has been shown that for $q = 6, 7, 8, 9, 11, 12, 18, 23, 36, 38, 73$, F_q is not prime. No one has found any other prime Fermat number. ∎

Exercises

(1) Let $\mathbf{K}$ be a field, and $f(x), g(x) \in \mathbf{K}[x]$ be two co-prime polynomials, namely $(f(x), g(x)) = 1$. Let

$$max(deg\ f(x), deg\ g(x)) = n$$

Prove that

$$[\mathbf{K}(x) : \mathbf{K}(\frac{f(x)}{g(x)})] = n$$

(2) Let $\mathbf{K}$ be a field, and

$$y = \frac{x^3}{x+1}$$

Find the minimal polynomial of x over $\mathbf{K}(y)$.

(3) Let $\mathbf{K}$ be a field, and x, y two variables. Show that the following polynomial is irreducible in $\mathbf{K}[x, y]$,

$$f(x, y) = x^3 + 3x^2y^2 + 2x^2y + y^4 + 7y + y^2$$

(4) Let ω be a root of $x^4 + 1$. Factor $x^4 + 1$ in $\mathbf{Q}(\omega)$.

(5) Find the degree of field extension $[\mathbf{Q}[i, \sqrt{2}] : \mathbf{Q}]$.

(6) Let α be a root of $x^3 - 2$. Does $x^3 - 2$ factor into linear polynomials in $\mathbf{Q}[\alpha][x]$?

(7) Prove that $[\mathbf{Q}[e^{2\pi i/8}] : \mathbf{Q}] = 2$.

(8) Let α be a root of the polynomial $x^3 - x^2 + x + 2$. Express $(\alpha^2 + \alpha + 1)(\alpha^2 - \alpha)$ and $(\alpha - 1)^{-1}$ in the following form,

$$a\alpha^2 + b\alpha + c, \qquad a, b, c \in \mathbf{Q}$$

(9) Let

$$\alpha = cos\ \frac{\pi}{6} + i\ sin\ \frac{\pi}{6}$$

Prove $[\mathbf{Q}[\alpha] : \mathbf{Q}] = 4$, and find the minimal polynomial of α over $\mathbf{Q}$.

(10) Let α be a root of an irreducible polynomial of odd degree in $\mathbf{K}[x]$. Show that $\mathbf{K}[\alpha] = \mathbf{K}[\alpha^2]$.

(11) Suppose that $x^n - a$ is an irreducible polynomial in $\mathbf{K}[x]$, where $\mathbf{K}$ is a field. Let α be a root of $x^n - a$, and $n = mr$. Show that α^r satisfies an irreducible polynomial of degree m over $\mathbf{K}$.

(12) Let $\mathbf{L}$ be an algebraic extension of $\mathbf{K}$, and $\mathbf{L} \supset \mathbf{R} \supset \mathbf{K}$, where $\mathbf{R}$ is a ring. Show that $\mathbf{R}$ is a field.

(13) Let us define *algebraic integer* as an algebraic number which satisfies a monic polynomial with integer coefficients. Prove that if α is an algebraic number, then there exists an integer n such that $n\alpha$ is an algebraic integer.

(14) Let α be an algebraic integer (see Problem (13)). Then the minimal polynomial is with integer coefficients.

(15) Let $\mathbf{L}=\mathbf{K}(\alpha)$, and α be a transcendental element over $\mathbf{K}$. Let $\mathbf{S}$ be a subfield of $\mathbf{L}$ with $\mathbf{S}\subset \mathbf{K}$ and $\mathbf{S}\neq\mathbf{K}$. Show that α is algebraic over $\mathbf{S}$.

(16) Suppose that $\mathbf{L}\subset \mathbf{S}_1$, $\mathbf{S}_2 \subset \mathbf{K}$. Let $\alpha \in \mathbf{L}$. If every element in $\mathbf{S}_1$ is algebraic over $\mathbf{S}_2$, then every element in $\mathbf{S}_1(\alpha)$ is algebraic over $\mathbf{S}_2(\alpha)$.

(17) Let $\mathbf{K}$ be a field, and $\mathbf{L}_i$ a chain of field extensions as follows,

$$\mathbf{K} \subset \mathbf{L}_1 \subset \cdots \subset \mathbf{L}_i \subset \cdots$$

Prove that $\cup_{i=1}^{\infty} \mathbf{L}_i$ is a field.

§3 Algebraic Closure

We shall discuss the existence and uniqueness up to isomorphism of an *algebraic closure* (cf Definition 5.5. below) of any field $\mathbf{K}$. The theorem proved in this section is abstract and non-constructive. In many interesting cases, it is interesting to realize the algebraic closure in a concrete way. For instance, it is more than the abstract existence aspect to notice that the complex field $\mathbf{C}$ is an algebraic closure of the real field $\mathbf{R}$. The proof of Theorem 5.11. may be skipped if the reader is concerned with only the complex field $\mathbf{C}$ and the subfields of $\mathbf{C}$.

Definition 5.5. *Let $\mathbf{L}$ be an extension of a field $\mathbf{K}$. If all elements in $\mathbf{L}$ are algebraic over $\mathbf{K}$, and $\mathbf{L}$ is algebraically closed (cf Definition 5.1.), then $\mathbf{L}$ is called an* ***algebraic closure*** *of $\mathbf{K}$.* ∎

Discussion

(1) Let $\mathbf{K}$ be any subfield of $\mathbf{C}$, Then $\mathbf{K}_{\mathbf{C}}^{c}$ is an algebraic closure of $\mathbf{K}$. ∎

Theorem 5.9. *Let $\mathbf{K}$ be a field, and $f(x)$ a non-constant, irreducible polynomial in $\mathbf{K}[x]$. Let $\rho : \mathbf{K}[x] \to \mathbf{K}[x]/(f(x))$ be the canonical map, and $\bar{x} = \rho(x)$. Then we have,*

(1) *$\mathbf{K}[x]/(f(x))$ is an algebraic extension of $\mathbf{K}$, and $[\mathbf{K}[\bar{x}] : \mathbf{K}] = deg\ f(x)$.*

(2) *$f(\bar{x}) = 0$, namely $f(x)$ has a root $\bar{x}$ in $\mathbf{K}[\bar{x}]$.*

Proof: (1) Since $f(x)$ is irreducible, then the ideal $(f(x))$ is prime, and $\mathbf{K}[x]/(f(x)) \approx \mathbf{K}[\bar{x}]$ is an integral domain. Since we have,

$$dim\ \mathbf{K}[\bar{x}] = deg\ f(x) < \infty$$

then it follows from Theorem 5.4. that $\mathbf{K}[\overline{x}]$ is an algebraic extension of **K**, and $[\mathbf{K}[\overline{x}] : \mathbf{K}] = deg\ f(x)$.

(2) It follows trivially that

$$\rho(f(x)) = f(\overline{x}) = 0$$

∎

Theorem 5.10. *Let* **K**, **K'** *be two isomorphic fields, and* $\sigma : \mathbf{K} \to \mathbf{K'}$ *be the isomorphism. Then* σ *can be extended to an isomorphism from* $\mathbf{K}[x]$ *to* $\mathbf{K'}[x]$ *as follows,*

$$\sigma(\sum_i a_i x^i) = \sum_i \sigma(a_i) x^i$$

Assume **L** *is an extension of* **K** *such that there is a root* α *of* $f(x)$ *in* **L**. *Let* **L'** *be an extension of* **K'** *such that* $\sigma(f(x))$ *has a root* $\beta \in \mathbf{L'}$. *Then* σ *can be extended to a map* $\overline{\sigma} : \mathbf{K}[\alpha] \to \mathbf{K'}[\beta]$, *such that* $\overline{\sigma}(\alpha) = \beta$.

Proof: It is easy to see that σ can be extended to an isomorphism from $\mathbf{K}[x]$ to $\mathbf{K'}[x]$. We claim that $\sigma(f(x))$ is irreducible. Suppose not. Then there are two polynomials of positive degrees, $g(x), h(x)$, in $\mathbf{K'}[x]$ such that

$$\sigma(f(x)) = g(x)h(x)$$

Applying σ^{-1} to the above equation, we get

$$f(x) = \sigma^{-1}(g(x))\sigma^{-1}(h(x))$$

contradicting our assumption that $f(x)$ is irreducible.

It follows from Theorem 5.4. that

$$\begin{array}{ccccccc} \mathbf{K}[\alpha] & \approx & \mathbf{K}[x]/(f(x)) & \approx & \mathbf{K'}[x]/(\sigma(f(x))) & \approx & \mathbf{K'}[\beta] \\ \alpha & \leftrightarrow & \overline{x} & \leftrightarrow & \overline{x} & \leftrightarrow & \beta \end{array}$$

∎

Now we can prove the existence and uniqueness of the algebraic closure of a given field **K**.

Theorem 5.11. *Let* **K** *be a field. Then there exists an algebraic closure* Ω *of* **K**. *Moreover, let* Ω' *be another algebraic closure of* **K**, *then there is a* **K**-*isomorphism* $\sigma : \Omega \to \Omega'$, *namely* $\sigma(k) = k$ *for all* $k \in \mathbf{K}$.

Proof: We shall use Zorn's Lemma. Let

$$\mathsf{F} = \{(R, \sigma, R') : R, R' \text{ are algebraic extensions of } \mathbf{K}, \sigma : R \to R' \text{ is a } \mathbf{K}\text{-injection.}\}.$$

Note that $(\mathbf{K}, identity, \mathbf{K}) \in \mathsf{F}$. Therefore, $\mathsf{F} \neq \emptyset$.

Let us define a partial ordering "$\leq$" as follows,

$$(R_1,\sigma_1,R_1') \leq (R_2,\sigma_2,R_2')$$
$$\Leftrightarrow R_1 \subset R_2, R_1' \subset R_2', \sigma_2(r_1) = \sigma_1(r_1), \quad \forall\, r_1 \in R_1$$

It is easy to see that "$\leq$" is indeed a partial ordering.

Let us verify the hypothesis of Zorn's Lemma. Let $\{(R_i,\sigma_i,R_i') : i \in I\}$ be a chain in F. Let $S = \cup R_i$, $S' = \cup R_i'$, and $\rho : S \to S'$ be defined as follows, for any $s \in S = \cup R_i$i, say $s \in R_j$,

$$\rho(s) = \sigma_j(s)$$

to verify the above definition is well-defined, let $s \in R_k$. We may assume that $(R_j,\sigma_j,R_j') \leq (R_k,\sigma_k,R_k')$. Then we have

$$\sigma_j(s) = \sigma_k(s)$$

Therefore, the definition of $\rho(s)$ is independent of the index j. It is easy to see that $(S,\rho,S') \in$ F and is an upper bound of the given chain. Henceforth, the hypothesis of Zorn's Lemma has been verified.

It follows from Zorn's Lemma that there is a maximal element (L,σ,L') in F. We claim that L' is an algebraic closure of **K**.

Certainly, L' is an algebraic extension of **K**. It suffices to show that L' is algebraically closed. Suppose not. Then there is an irreducible polynomial $f(x) \in L'[x]$ of degree ≥ 2. It follows from Theorem 5.7. that $L'[x]/(f(x)) = \overline{L}$ is an algebraic extension of L' and **K**, and $\overline{L} \neq L'$. It is clear that the following map

$$\sigma : L \to L' \subset \overline{L}$$

is a ring injection. Therefore, $(L,\sigma,\overline{L}) \in$ F, and $(L,\sigma,L') < (L,\sigma,\overline{L})$. This contradicts the maximality of $(L.\sigma,L')$. Thus we established that L' is an algebraic closure of **K**.

Let Ω be an algebraic closure of **K**. Suppose that Ω' is another algebraic closure of **K**. We shall rearrange the above arguments slightly. Let

$$\mathsf{F}' = \{(R,\sigma,R') : (R,\sigma,R') \in \mathsf{F}, R \subset \Omega, R' \subset \Omega'.\}$$

We define the partial ordering as before. The hypothesis of Zorn's Lemma can be verified similarly. We conclude that there is a maximal element $(L,\sigma,L') \in$ F$'$. A similar argument as before proves that L' is an algebraic closure of **K**. Since Ω' is algebraic extension of **K**, then it is an algebraic extension of L' which is algebraically closed. Therefore, we have $L' = \Omega'$.

We claim $L = \Omega$. Note that it suffices to show that L is algebraically closed. Suppose not. Then there exists an irreducible polynomial $f(x) \in L[x]$ with *deg* $f(x) \geq 2$. We shall apply Theorem 5.10.. The map $\sigma : L \to \sigma(L)$ is an isomorphism between fields. Since Ω and Ω' are both algebraically closed, then $f(x)$ will have a root α in Ω and $\sigma(f(x))$ will have a root β in Ω'. It follows from Theorem 5.10. that the map σ can be extended to $\overline{\sigma} : L[\alpha] \to L'[\beta]$. Therefore, $(L[\alpha], \overline{\sigma}, L'[\beta]) \in \mathsf{F}'$, and

$$(L, \sigma, L') < (L[\alpha], \overline{\sigma}, L'[\beta])$$

This contradicts the maximality of (L, σ, L'). Henceforth, we have $L = \Omega$. Moreover, $\sigma(L) = \sigma(\Omega) \subset \Omega'$ are all algebraic closures of $\mathbf{K}$. Therefore, we must have $\sigma(\Omega) = \Omega'$, and Ω, Ω' isomorphic fields. ∎

Corollary. *Let Ω be an algebraic closure of* $\mathbf{K}$. *Let* $\mathbf{L}$ *be an algebraic extension of* $\mathbf{K}$. *Then there is a* $\mathbf{K}$*-isomorphism which maps* $\mathbf{L}$ *into* Ω.

Proof: Let Ω' be an algebraic closure of $\mathbf{L}$. Then Ω' is an algebraic closure of $\mathbf{K}$. It follows from the Theorem that there is a $\mathbf{K}$-isomorphism which maps Ω' to Ω. It is the map we look for. ∎

Exercises

(1) Prove that $\mathbf{C}$ is not an algebraic closure of $\mathbf{Q}$.
(2) Prove that an algebraic closure of $\mathbf{Q}$ is not a finite algebraic extension of $\mathbf{Q}$.
(3) Show that an algebraic closure of $\mathbf{Q}$ is $\mathbf{Q}$-isomorphic to an algebraic closure of $\mathbf{Q}[i]$.
(4) Show that $\mathbf{C}$ is an algebraic closure of $\mathbf{R}$, and $\mathbf{C}$ is a quadratic extension of $\mathbf{R}$.
(5) Let p be a prime number. Find an algebraic extension of the p-adic number field $\mathbf{Q}_p$ such that the polynomial $x^2 + 1$ has a root.
(6) Find an algebraic extension of $\mathbf{Q}(x)$ such that the polynomial $f(y) = y^2 - g(x) \in \mathbf{Q}(x)[y]$ has a root, where

$$g(x) = \frac{x^3}{x^2 + 1}$$

§4 Characteristic and Finite Field

Classically, we start with the rational field $\mathbf{Q}$. All fields which contain the rational field $\mathbf{Q}$ are infinite fields. The theory of finite fields, which was invented by Galois, was considered as a fancy of mathematicians. Nowadays, there are many applications of finite field theory to Computer Sciences (cf Example 2, Chapter III). In fact, there are many practical values of the finite field theory.

To begin with, let $\mathbf{K}$ be any field, finite or infinite, with the multiplicative identity e. Let us define a map $\sigma : \mathbf{Z} \to \mathbf{K}$ as follows,

$$\sigma(0) = 0, \quad \sigma(1) = e, \quad \sigma(n \cdot 1) = n\sigma(1) = ne$$

Let $ker(\sigma) = \sigma^{-1}(0) = (p)$, where $p \geq 0$. We define,

Definition 5.6. *(1) The number p defined above is called the **characteristic** of the field $\mathbf{K}$.*

*(2) A finite field is sometimes called a **Galois field**.*

*(3) The rational field $\mathbf{Q}$ and the fields $\mathbf{Z}/(p) = \mathbf{Z}_p$, where p is a prime number, are called the **prime fields**.* ∎

Theorem 5.12. *(1) The characteristic of a field is either 0 or a prime number.*

(2) If the characteristic of a field $\mathbf{K}$ is 0, then $\mathbf{K}$ contains a copy of $\mathbf{Q}$. If the characteristic of a field $\mathbf{K}$ is $p > 0$, then $\mathbf{K}$ contains a copy of $\mathbf{Z}_p$.

Proof: (1) If the characteristic n is not a prime, then let $n = mr, n > m > 1, n > r > 1$. We have,

$$0 = mre = (me)(re) \Rightarrow me = 0, \text{ or } re = 0$$

In either case, n is not the generator of $\sigma^{-1}(0)$. Contradiction.

(2) Suppose that characteristic of the field is 0. Then σ is an injection, and there is a copy of $\mathbf{Z}$ in $\mathbf{K}$. We shall identify this copy of $\mathbf{Z}$ with $\mathbf{Z}$. Since $\mathbf{K}$ is a field, then it contains the quotient field of $\mathbf{Z}$ which is (a copy of) $\mathbf{Q}$. Suppose that the characteristic of the field is a prime number p. It follows from the First Isomorphism Theorem of rings that there is a canonical map which send $\mathbf{Z}_p$ injectively into $\mathbf{K}$. We shall identify this copy of $\mathbf{Z}_p$ with $\mathbf{Z}_p$. ∎

Discussion

(1) Every field $\mathbf{K}$ is an extension of a unique prime field. Therefore, in field theory we may consider only extensions from a given prime field. Furthermore, it follows from Corollary of Theorem 5.11. that any field $\mathbf{L}$ which is an algebraic extension of another field $\mathbf{K}$ can be considered as a subfield of a given algebraic closure, Ω, of $\mathbf{K}$. Henceforth, we may fix the algebraic closure of $\mathbf{K}$, Ω, at the beginning, if we only consider algebraic extensions of $\mathbf{K}$.

(2) Let $\mathbf{K}$ be a finite field of characteristic p. Then it is an extension of the prime field $\mathbf{Z}_p$ of a finite algebraic degree n. Since $\mathbf{K}$ is a vector space of dimension n over $\mathbf{Z}_p$. We conclude at once that $Card(\mathbf{K}) = p^n$. ∎

Suppose the characteristic of a field $\mathbf{K}$, finite or infinite, is p. Then we have the following rules of arithmetic,

$$(\alpha + \beta)^p = \alpha^p + \cdots + \binom{p}{i}\alpha^{p-i}\beta^i + \cdots + \beta^p$$

Note that

$$p \mid \binom{p}{i}, \quad \text{for } 1 \leq i < p$$

Therefore, we have the following simple rule

(1) $$(\alpha \pm \beta)^p = \alpha^p \pm \beta^p$$

As usual, we have

(2) $$(\alpha\beta)^p = \alpha^p \beta^p, \quad (\beta^{-1})^p = (\beta^p)^{-1}.$$

Therefore, the map ρ defined as follows,

$$\rho(\alpha) = \alpha^p$$

is a field injection from **K** to **K**. This map is very important in the theory of fields of positive characteristic. We have the following definition,

Definition 5.7. *Let the characteristic of the field* **K** *be* $p > 0$. *Then the field injection* $\rho : \mathbf{K} \to \mathbf{K}$ *with* $\rho(\alpha) = \alpha^p$ *is called the* ***Frobinus map*** *of the field. In case the field* **K** *is finite,* ρ *is an isomorphism.* ∎

Discussion

(1) The last part of the definition is a consequence of the Pigeon Hole Principle: for a finite set **K**, an injection from it to itself must be a bijection. ∎

We have the following theorem which gives the structure of the multiplicative group of a finite field,

Theorem 5.13. *Let* **K** *be any field, and* **G** *a finite subgroup of the multiplicative group* $\mathbf{K}^* = (\mathbf{K} \backslash \{0\})$. *Then* **G** *is a cyclic group.*

Proof: It follows from Corollary, Theorem 4.18. that it suffices to show the primes $\{p_1, p_2, \cdots, p_q\}$ deduced from the set of the elementary divisors $\{p_1^{s_1}, p_2^{s_2}, \cdots, p_q^{s_q}\}$ are all distinct. Suppose that $p_1 = p_2$. Then there are subgroups H_1, H_2 such that ord $H_i = p_i^{s_i}$ and $H_1 \cap H_2 = \{e\}$. Let $s_1 \leq s_2$. Then all elements in $H_1 \cup H_2$ will satisfy the following equation,

$$x^{p^{s_2}} = e$$

Note that $Card(H_1 \cup H_2) = p^{s_1} + p^{s_2} - 1 > p^{s_2}$. The above equation of degree p^{s_2} can not have more than p^{s_2} roots in the field **K**. Therefore, the hypothesis of Corollary, Theorem 4.18. is satisfied, and **G** is cyclic. ∎

Corollary. *The multiplicative group* $\mathbf{K}^*$ *of a finite field* **K** *is cyclic.*

Proof: The above theorem. ∎

Discussion

(1) Any finite subgroup of the multiplicative group of the unit circle $\{e^{\theta i}\} \subset \mathbf{C}$ is cyclic.

(2) Let $\mathbf{K}$ be a finite field. The additive group of $\mathbf{K}$ is simply a vector space over the prime field, and the multiplicative group $\mathbf{K}^*$ is a cyclic group. Both are very simple. However, the relation between those two operations are very complicated.

(3) Let $\mathbf{K}$ be a finite field., and $0 \neq \alpha \in \mathbf{K}$. The *order* of α, *ord* (α), is defined be the order of the multiplicative subgroup generated by α. An element α is called a ***primitive element*** of $\mathbf{K}$, if α generates the multiplicative group $\mathbf{K}^*$, namely if $ord\ \alpha = p^n - 1$, where n is the algebraic degree of $\mathbf{K}$ over the prime field $\mathbf{Z}_p$.

(4) Let α be a primitive of a finite field $\mathbf{K}$. Then the minimal polynomial of α over the prime field is called a ***primitive polynomial.*** ∎

We shall construct all finite fields by solving equations.

Theorem 5.14. *Let* $\mathbf{K}$ *be a finite field with characteristic* $p > 0$ *which is a degree* n *extension of the prime field* $\mathbf{Z}_p$. *Then we have,*

(1) *All elements of* $\mathbf{K}$ *satisfy the following equation,*

$$x^{p^n} - x = 0$$

(2) *The isomorphism* ρ *has an order* n.

Proof: (1) It is clear that 0 is a root of the above equation. Let $\alpha \neq 0$, then $\alpha \in \mathbf{K}^*$, a cyclic group of order $p^n - 1$. Therefore, we have

$$\alpha^{p^n-1} = e, \quad \alpha^{p^n} - \alpha = 0$$

Therefore all elements in $\mathbf{K}$ satisfy said equation.

(2) Let r be any element in $\mathbf{K}$. Then we have

$$\rho^n(r) = \rho^{n-1}(\rho(r)) = \rho^{n-1}(r^p) = r^{p^n} = r$$

On the other hand, let $0 < m < n$. If $\rho^m = I$ the identity map, then we have,

$$\rho^m(r) = r^{p^m} = r, \quad \forall\, r \in \mathbf{K}$$

and any element r will satisfy the following equation,

$$x^{p^m} - x = 0$$

Note that the above equation is of degree $p^m < p^n$. Therefore, it can not have p^n roots. A contradiction. We conclude that the order of ρ is n. ∎

Theorem 5.15. *Let Ω be an algebraic closure of $\mathbb{Z}_p$. Let F_n, or $GF(p^n)$, be the set the solutions of the following equation in Ω,*

(1)
$$x^{p^n} - x = 0$$

Then we have,

(1) *F_n is a finite field of cardinality p^n.*
(2) *F_n is the only finite subfield of cardinality p^n in Ω.*
(3) *$F_n \subset N_m \Leftrightarrow n \mid m$.*
(4) *Suppose that $F_n \subset F_m$. Then ρ^n is a F_n-isomorphism of F_m to itself with order m/n, namely ρ^n is an isomorphism of F_m to itself and*

$$\rho^n(r) = r, \quad \forall\, r \in F_n$$

Proof: (1) Note that $0, 1 \in F_n$. Therefore, $F_n \neq \emptyset$. Let $\alpha, \beta \in F_n$. Then we have

$$\begin{aligned}
&\alpha^{p^n} = \alpha, \quad \beta^{p^n} = \beta \\
&\Rightarrow \begin{cases} (\alpha \pm \beta)^{p^n} = \alpha^{p^n} \pm \beta^{p^n} = \alpha \pm \beta \\ (\alpha\beta)^{p^n} = \alpha^{p^n}\beta^{p^n} = \alpha\beta \end{cases} \\
&\Rightarrow \begin{cases} \alpha \pm \beta \in F_n \\ \alpha\beta \in F_n \end{cases} \\
&(\beta^{-1})^{p^n} = (\beta^{p^n})^{-1} = (\beta)^{-1}, \quad \text{if } \beta \neq 0 \\
&\Rightarrow \beta^{-1} \in F_n
\end{aligned}$$

We conclude that F_n is a field. Let us show that it has precisely p^n elements. Since Ω is algebraically closed, then Eq (1) has all roots in Ω. Therefore, it suffices to show that Eq (1) has no multiple root. We shall use derivative arguments. The derivative of $x^{p^n} - x$ is $p^n x^{p^n - 1} - 1 = -1$ which has no common root with $x^{p^n} - x$. It follows from Theorem 3.20. that Eq (1) has no multiple root. We establish that F_n is a field of p^n elements.

(2) Let F'_n be any other field in Ω with p^n elements. It follows from the preceding theorem that all elements of F'_n must satisfy Eq (1). Therefore, $F'_n \subset F_n$. Now the cardinalities force those two to be equal.

(3) Suppose that $F_n \subset F_m$. Let $[F_m : F_n] = \ell$. Then F_m is a vector space of dimension ℓ over F_n, and

$$p^m = (p^n)^\ell = p^{n\ell}, \quad m = n\ell$$

Conversely, suppose that $n \mid m$, $m = n\ell$. Clearly we have

$$x^{p^m} - x = x(x^{p^{n\ell}-1} - 1) = x(x^{p^n - 1} - 1)(x^{(p^{n\ell}-1)/(p^n-1)} + \cdots + 1)$$

and

$$(x^{p^n} - x) \mid (x^{p^m} - x)$$
$$F_n = \{ \text{ roots of } x^{p^n} - x\} \subset \{ \text{ roots of } x^{p^m} - x\} = F_m$$

(4) Obvious. ∎

Discussion

(1) In §6, we will show that all F_n-isomorphisms of F_m are of the form ρ^{ni} for $i = 0, 1, \cdots, m/n - 1$. There are precisely m/n F_n-isomorphisms of F_m. ∎

Example 9. (Cyclotomic Polynomials). *We shall apply the preceding discussions of the finite fields to the infinite fields of characteristic* 0. *Let us discuss the roots of unity in* **C**. *The materials of Example 5. will be assumed. We have the following theorem,*

Theorem. *Let ζ be a primitive n-th root of unity, and $\varphi_n(x) \in$ **Q**$[x]$ be the minimal polynomial of ζ with respect to* **Q**. *Then we have*

(1) $\varphi_n(x) \in \mathbf{Z}[x]$.
(2) *$\varphi_n(x)$ has no multiple root. The set of roots of $\varphi_n(x)$ is the set of all primitive n-th roots of unity.*
(3) *deg $\varphi_n(x) = \varphi(n)$ the Euler number.*

Proof. Note that above had been proved in Example 5. for any prime number n.

(1) Note that $x^n - 1$ is a primitive polynomial, and $\varphi_n(x) \mid x^n - 1$. It follows from Theorem 3.12. that $x^n - 1$ is a product of irreducible polynomials in **Q**$[x]$ which are primitive polynomials in $\mathbf{Z}[x]$ as follows,

$$x^n - 1 = \prod_i g_i(x)$$

Since $\varphi_n(x)$ is irreducible in **Q**$[x]$, then there exists an integer s such that $s\varphi_n(x) = g_j(x)$. Comparing the coefficients of x^n on both sides of the above equation, we conclude $s = \pm 1$. Therefore $f(x) \in \mathbf{Z}[x]$.

(2) Since $x^n - 1$ has no multiple root, then a factor of it, $\varphi_n(x)$, has no multiple root. First, we will show that every root of $\varphi_n(x)$ is a primitive n-th root of unity. Let η be any root of $\varphi_n(x)$. Then we have

$$\mathbf{Q}[\eta] \approx \mathbf{Q}[x]/(\varphi_n(x)) \approx \mathbf{Q}[\zeta]$$
$$\eta^m = 1 \Leftrightarrow \zeta^m = 1$$

Therefore, η is a primitive n-th root of unity.

Secondly, we will show that every primitive n-th root of unity is a root of $\varphi_n(x)$. We will use characteristic $p > 0$ arguments.

Let ξ be any n-th primitive root of unity. Since ζ generates all n-th roots of unity, then we have $\xi = \zeta^{\ell}$, where $(n, \ell) = 1$ (otherwise, ξ is not a primitive n-th root of unity). Let p be a prime factor of ℓ. It suffices to show that ζ^p is a root of $\varphi_n(x)$, because we may replace ζ by ζ^p and continue our arguments.

Note that p is a factor of ℓ, and $(p, n) = 1, p \nmid n$. Let

$$x^n - 1 = \varphi_n(x)h(x), \quad h(x) \in \mathbf{Z}[x]$$

If ζ^p is not a root of $\varphi_n(x)$, then it must be a root of $h(x)$. Therefore, $h(\zeta^p) = 0$, and ζ is a root of $h(x^p)$. Since $\varphi_n(x)$ is the minimal polynomial of ζ over $\mathbf{Q}$, then we must have $\varphi_n(x) \mid h(x^p)$. We shall modulo the above equation by p, namely, let $\sigma : \mathbf{Z}[x] \to (\mathbf{Z}/p)[x]$ be the canonical map, then we have

$$x^n - 1 = \sigma(\varphi_n(x))\sigma(h(x)) = \overline{\varphi}_n(x)\overline{h}(x)$$
$$\overline{\varphi}_n(x) \mid \overline{h}(x^p) = \overline{h}(x)^p$$

The last equality follows from the arithmetic rules of a finite field. The last line of the above means that in some algebraic closure, $\overline{\Omega}$, of $\mathbf{Z}_p$, $\overline{\varphi}_n(x)$ and $\overline{h}(x)$ will have a common root. However, $x^n - 1$ will not have a multiple root: the derivative of $x^n - 1$, $nx^{n-1} \neq 0 \in \mathbf{Z}_p[x]$, has no common factor with $x^n - 1$, note that $n \not\equiv 0$ (Mod p). Henceforth, we deduce a contradiction, and conclude ζ^p is a root of $\varphi_n(x)$.

(3) This follows from (2) and the definition of $\varphi(n)$. ∎

Let us continue our study of the roots of unity. Let us call the minimal polynomial of a primitive n-th root of unity $\varphi_n(x)$ the *n-th cyclotomic polynomial*. Note that a n-th root of unity, η, must be a primitive m-th root of unity for a unique $m \mid n$, and a root of $\varphi_m(x)$. Therefore, we have

$$x^n - 1 = \prod(x - \eta) = \prod_{m|n} \varphi_m(x)$$

Note that *deg* $\varphi_m(x) = \varphi(m)$ the Euler number. We have the following formula by comparing the degrees on both sides of the above equation,

$$n = \sum_{m|n} \varphi(m)$$

∎

Example 10. *Let p be a prime number > 2. Can we construct with straight-edge and compass a regular p^2-gon? The answer is no. Note that if $p = 3$, since we can construct a regular triangle, and can not construct a regular 9-gon, then 60° can not be trisected. Therefore, we answer the problem of trisecting an angle.*

Note that the construction of any regular p^2-gon implies the construction of a regular p^2-gon inscribing the unity circle. Hence it suffices to show that the primitive p^2-th roots can not be constructed. Let ζ be any primitive p^2-th root of unity. Its minimal polynomial over **Q** *is $\varphi_n(x)$. We observe that the degree of the p^2-th cyclotomic polynomial $\varphi_n(x)$ is $\varphi(p^2) = p(p-1)$. If ζ can be constructed, then there must be a chain of quadratic extensions $\mathbf{K}_i$ of* **Q** *such that*

$$\mathbf{Q} \overset{2}{\subset} \mathbf{K}_1 \overset{2}{\subset} \cdots \overset{2}{\subset} \mathbf{K}_n, \quad \zeta \in \mathbf{K}_n$$

Then we have the following conclusion,

$$\begin{aligned} 2^n &= [\mathbf{K}_n : \mathbf{Q}] = [\mathbf{K}_n : \mathbf{Q}[\zeta]]\ [\mathbf{Q}[\zeta] : \mathbf{Q}] \\ &= \ell p(p-1) \end{aligned}$$

Note that p is odd, and the above equation is impossible. Henceforth, a regular p^2-gon can not be constructed.

Similar to the trisecting angle problem, we can show that for any prime $p > 2$, there is a constructible angle which can not be equally divided into p parts. We start with the whole angle 2π. If a regular p-gon can not be constructed, then the whole angle 2π can not be equally divided into p parts. If a regular p-gon can be constructed, then by the previous arguments, a regular p^2-gon can not be constructed. Therefore, the inner angle at any vertex of the regular p-gon can not be divided equally in p parts.

Let us consider the problem of construction of regular n-gon. Let

$$n = 2^m \prod_{p_i \ odd} p_i$$

If $m \mid n$, then a suitable subset of the set of vertices of a regular n-gon will be the set of vertices of a regular m-gon. Therefore, all odd primes must be distinct Fermat numbers. This is a necessary condition for th e constructibility of a regular n-gon. Later on (cf Examples 16. & 17.), we will show that this necessary condition is sufficient. ∎

Exercises

(1) Find the orders of all elements in $\mathbf{K}^*$, where $\mathbf{K}$ is a field of 16 elements.

(2) Let $\mathbf{K}$ be a finite field of p^n elements where p is a prime number. Show that there are $\varphi(p^n - 1)/n$ primitive polynomials.

(3) Let $\mathbf{K}$ be a finite field of p^n elements where p is a prime number. Show that a polynomial $f(x) \in \mathbf{Z}_p[x]$ is primitive if and only if $f(x)$ satisfies the following two conditions,

(i) $f(x) \mid x^{2^n-1} - 1$.

(ii) $f(x) \nmid x^{(2^n-1)/q} - 1$ for all q prime and $q \mid 2^n - 1$.

(4) Let $\mathbf{K}$ be a finite field of 8 elements. Factor $x^{16} - x$ in $\mathbf{K}[x]$.

(5) Let $\mathbf{K}$ be a finite field of p^n elements where p is a prime number. Find the number of monic quadratic irreducible polynomials in $\mathbf{K}[x]$.

(6) Let $\mathbf{K}$ be a finite field of p^n elements where p is a prime number. Let Ω be an algebraic closure of $\mathbf{K}$. Find the number of quadratic extensions of $\mathbf{K}$ in Ω.

(7) Prove that the ideal (3) is a maximal ideal in the ring of Gaussian integers $\mathbf{Z}[i]$.

(8) Prove that a finite integral domain is a field.

(9) Show that the quotient rings of the ring of Gaussian integers $\mathbf{Z}[i]$ over the ideals $(1+i), (2+i)$ are fields. Find their characteristics.

(10) Let $\mathbf{K}$ be a finite field of p^n elements where p is a prime number. Show that

(i) If $(r, p^n - 1) = 1$, then every elements in $\mathbf{K}$ has a r-th root. Especially, every element in $\mathbf{K}$ has a p-th root.

(ii) If $p^n - 1 = rs$, and $\alpha \in \mathbf{K}$, then there is a r-th root of $\alpha \Leftrightarrow \alpha^s = e$ the multiplicative identity.

(11) Let $\mathbf{K}$ be a finite field of p^n elements where p is a prime number. Let $\alpha \in \mathbf{K}$ be a root of an irreducible polynomial $f(x) \in \mathbf{Z}[x]$ of degree m. Show that the roots of $f(x)$ are precisely,

$$\alpha^p, \alpha^{p^2}, \cdots, \alpha^{p^m} = \alpha$$

(12) Let p be any prime, and n be any positive integer. Prove that there is an irreducible polynomial $f(x) \in \mathbf{Z}_p[x]$ of degree n.

(13) Let k be an odd integer. Show that $\mathbf{Q}[e^{2\pi i/k}] = \mathbf{Q}[e^{\pi i/k}]$.

(14) Prove that $e^{\pi i/4}$ is of degree 2 over $\mathbf{Q}[i]$.

(15) Let p be a prime number, and $\mathbf{K}= \mathbf{Q}[e^{2\pi i/p}]$. Prove that

$$\mathbf{K} \cup \mathsf{R} = \mathbf{Q}[e^{2\pi i/p} + e^{-2\pi i/p}]$$

§5 Separable Algebraic Extension

This section is mainly about fields of characteristic $p > 0$. Except Theorem 5.22., it may be skipped by the reader who is concerned with only characteristic 0 fields or only study the subfields of complex field $\mathbf{C}$. In this case, the term "separable" should be omitted in the statement of Theorem 5.22., and the discussions about finite fields should be omitted from the proof of Theorem 5.22..

In the general field theory, an irreducible polynomial in $\mathbf{K}[x]$ may have multiple roots in an algebraic closure Ω of $\mathbf{K}$. Note that α is a multiple root of a polynomial $f(x) \Leftrightarrow f(\alpha) = f'(\alpha) = 0$, where $f'(x)$ is the derivative of $f(x)$. Let us examine the following example,

Example 11. *(1) If the characteristic of a field* $\mathbf{K}$ *is* 0, *then any irreducible polynomial* $f(x) \in \mathbf{K}[x]$ *has only simple roots in an algebraic closure* Ω *of* $\mathbf{K}$: *let* α *be a root of* $f(x)$, *and*

$$f(x) = a_0 x^n + a_1 x^{n-1} + \cdots + a_{n-1}x + a_n, \qquad a_0 \neq 0$$

Then we have the following derivative $f'(x)$ *of* $f(x)$,

$$f'(x) = na_0 x^{n-1} + \cdots + a_{n-1}$$

Since the characteristic of $\mathbf{K}$ *is* 0, *then* $na_0 \neq 0$, *and* $f'(x) \neq 0, deg\ f'(x) = n-1$. *The polynomial* $f(x)$ *is assumed to be the minimal non-zero polynomial satisfied by* α. *Therefore,* $f'(\alpha) \neq 0$, *and* $f(x)$ *has no multiple root.*

(2) Let $\mathbf{F}$ *be a field of characteristic* $p > 0$. *Let* $y = z^p$, $\mathbf{F}(y) \subset \mathbf{F}(z)$ *be rational function fields in the variables* y, z *respectively. Clearly,* z *is algebraic over* $\mathbf{F}(y)$ *satisfying the following equation*

$$f(x) = x^p - y = 0$$

We shall use Theorem 3.12.. Note that $\mathbf{F}(y)$ *is the quotient field of* $\mathbf{F}[y]$ *which is a U.F.D., and* $x^p - y$ *is a primitive polynomial in* $\mathbf{F}[y][x] = \mathbf{F}[y, x]$. *Since it is linear in* y, *then it is easy to see that it is an irreducible primitive polynomial in* $\mathbf{F}[y][x]$. *Therefore, it follows from Theorem 3.12. that it is irreducible in* $\mathbf{F}(y)[x]$. *We conclude that it is the minimal polynomial of* z *over* $\mathbf{F}(y)$.

Let Ω *be an algebraic closure of* $\mathbf{F}(z)$. *Then it is easy to see that* Ω *is an algebraic closure of* $\mathbf{F}(y)$. *In it, we have*

$$f(x) = x^p - y = x^p - z^p = (x - z)^p$$

Therefore, this irreducible polynomial $x^p - y$ *has only one root* z *with multiplicity* p. *Let us examine the derivative of the polynomial* $f(x) = x^p - y$. *Note that* y *is considered as an element in the constant field* $\mathbf{K}(y)$. *We have*

$$f'(x) = px^{p-1} = 0$$

∎

Definition 5.8. *(1) Let* $\mathbf{K}$ *be a field, and* $0 \neq g(x) \in \mathbf{K}[x]$. *If* $g(x)$ *has no multiple root in an algebraic closure* Ω *of* $\mathbf{K}$, *then* $g(x)$ *is called a* ***separable polynomial.***

(2) Let $\mathbf{L}$ *be an extension of* $\mathbf{K}$, *and* $\alpha \in \mathbf{L}$. *If* α *is algebraic over* $\mathbf{K}$ *and the minimal polynomial of it over* $\mathbf{K}$ *is separable, then* α *is called a* ***separable algebraic element*** *over* $\mathbf{K}$.

∎

Discussion

(1) If the characteristic of $\mathbf{K}$ is 0, then every algebraic element is separable algebraic element.

∎

We have the following simple criteria for separable algebraic elements.

Lemma 1. *Let* $\mathbf{L}$ *be an extension of* $\mathbf{K}$. *Then we have* α *is separable algebraic over* $\mathbf{K} \Leftrightarrow$ α *is a root of some separable polynomial. Especially, if* α *is separable algebraic over a subfield* $\mathbf{R}$ *of* $\mathbf{K}$, *then* α *is separable algebraic over* $\mathbf{K}$.

Proof: ($\Rightarrow$) The minimal polynomial is separable.

($\Leftarrow$) Let $f(x)$ be the minimal polynomial of α, which is a root of a separable polynomial $g(x)$, over $\mathbf{K}$. Then we have

$$f(x) \mid g(x)$$

If $g(X)$ has no multiple root in an algebraic closure Ω of $\mathbf{K}$, then $f(x)$ has no multiple root in Ω.

For the last part of the lemma, let $h(x)$ be the minimal polynomial of α over $\mathbf{R}$, then $h(x)$ is separable, and our statement follows.

∎

Lemma 2. *Let* $\mathbf{L}$ *be an extension of* $\mathbf{K}$. *Then we have* α *is separable algebraic over* $\mathbf{K} \Leftrightarrow$ *the derivative* $f'(x)$ *of the minimal polynomial* $f(x)$ *of* α *over* $\mathbf{K}$ *is not zero.*

Proof: ($\Rightarrow$) If $f'(x) = 0$, then it follows from Theorem 3.20. that α is a multiple root of $f(x)$. Impossible.

($\Leftarrow$) Let Ω be an algebraic closure of $\mathbf{K}$. Let α' be any root of $f(x)$ in Ω. Then $f(x)$ is the minimal polynomial of α'. Since $f'(x) \neq 0$, $deg\ f'(x) < deg\ f(x)$, then α' can not be a root of a non-zero polynomial $f'(x)$ of smaller degree, and $f'(\alpha') \neq 0$. Therefore, it follows from Theorem 3.20. that α' is not a multiple root of $f(x)$. Therefore, $f(x)$ is a separable polynomial.

∎

Definition 5.9. *Let* $\mathbf{K}$ *be a field. If any algebraic element* α *in any extension* $\mathbf{L}$ *is separable algebraic over* $\mathbf{K}$, *then* $\mathbf{K}$ *is called a* ***perfect field.***

∎

Theorem 5.16. *If a field* $\mathbf{K}$ *is either with characteristic zero or finite, then* $\mathbf{K}$ *is perfect.*

Proof: It is easy to see a characteristic 0 field is perfect. Let us assume that $\mathbf{K}$ is a finite field of p^n elements. Then it follows from Theorem 5.14. that all elements of $\mathbf{K}$ are roots of the following polynomial

$$f(x) = x^{p^n} - x$$

We have

$$f'(x) = p^n x^{p^n - 1} - 1 = -1 \neq 0$$

It is easy to see that $f(x)$ and $f'(x) = -1$ has no common root. Therefore, $f(x)$ is a separable polynomial, and our theorem follows from Lemma 1.

∎

We have the following definition,

Definition 5.10. *Let* **L** *be an extension of* **K** *with characteristic* $p > 0$, *and* $\alpha \in$ **L**. *If there is a non-negative integer* ℓ *such that* $\alpha^{p^{\ell}} \in$ **K**, *then* α *is called a* ***purely inseparable algebraic element*** *over* **K**. ∎

Discussion

(1) The number ℓ may be 0 in the above definition. Note that an element α is in **K** if and only if the minimal polynomial of it over **K** is linear. It follows that every element in **K** is separable and purely inseparable algebraic over **K**.

(2) If α is purely inseparable over **K**, then the minimal equation of α over **K** is of the form $x^{p^{\ell'}} - \alpha^{p^{\ell'}}$: we may assume the number ℓ is the minimal one with $\alpha^{p^{\ell}} \in$ **K**. We claim $x^{p^{\ell}} - \alpha^{p^{\ell}}$ is the minimal polynomial of α over **K**. Suppose not. Let $f(x)$ be the minimal polynomial of α over **K** with $deg\ f(x) = mp^s < p^{\ell}$. Since

$$f(x) \mid x^{p^{\ell}} - \alpha^{p^{\ell}} = (x - \alpha)^{p^{\ell}} \in \mathbf{K}[\alpha][x]$$

Therefore, we have

$$f(x) = (x - \alpha)^{mp^s} \in \mathbf{K}[\alpha][x]$$

The constant term, α^{mp^s}, of $f(x)$ is in **K**. We must have

$$\alpha^{p^s} = (\alpha^{p^{\ell}})^i (\alpha^{mp^s})^j, \qquad \text{for some suitable } i, j.$$

Therefore, $\alpha^{p^s} \in$ **K**. Note that $p^s < p^{\ell}$, it contradicts our assumption on ℓ. ∎

Lemma 3. *Let* **L** *be an extension of a field* **K** *of characteristic* $p > 0$, *and* $\alpha \in$ **L**. *Then* $\alpha \in$ **K** *if and only if* α *is separable and purely inseparable algebraic over* **K**.

Proof. ($\Rightarrow$) See the preceding discussion.

($\Leftarrow$) If α is separable and purely inseparable algebraic over **K**, let the minimal polynomial of it be $f(x)$, then we have

$$f(x) \mid x^{p^{\ell}} - \alpha^{p^{\ell}} = (x - \alpha)^{p^{\ell}} \in \mathbf{K}[\alpha][x]$$

and $f(x)$ has only one root. Since $f(x)$ is separable, $f(x)$ has no multiple root, then $f(x)$ must be linear. We conclude that $\alpha \in$ **K**. ∎

We want to show that in an algebraic closure, Ω, of a field **K**, all separable algebraic elements form a field. Hence, the set of separable algebraic elements over **K** will be closed with respect to the four arithmetical operations. For our long sequence of deductions, we need the following definition and theorems.

Definition 5.11. *Let* **L** *be an extension of a field* **K**. *We have the following definitions,*

(1) *If all elements in* **L** *are separable algebraic over* **K**, *then* **L** *is called a* ***separable algebraic extension*** *of* **K**.
(2) *If all elements in* **L** *are purely inseparable algebraic over* **K**, *then* **L** *is called a* ***purely inseparable algebraic extension*** *of* **K**.

∎

We have the following criterion of finite separable algebraic extensions,

Theorem 5.17. *Let* **K** *be a field of characteristic* $p > 0$, *and* **L** *a finite extension of* **K**. *Let*

$$\mathbf{KL}^p = \{\sum_{i=1}^{n} k_i\ell_i^p : k_i \in \mathbf{K}, \ell \in \mathbf{L}.\}$$

Namely, $\mathbf{KL}^p$ *is the vector space spanned by* $\mathbf{L}^p$ *over* **K**. *Then we have,*

(1) $\mathbf{KL}^p$ *is a field and a finite extension over* **K**.
(2) **L** *is a separable algebraic extension of* $\mathbf{K} \Leftrightarrow \mathbf{KL}^p = \mathbf{L}$.

Proof: (1) Let $\alpha, \beta \in \mathbf{KL}^p \subset \mathbf{L}$. We claim $\alpha \pm \beta, \alpha\beta, \beta^{-1} \in \mathbf{KL}^p$, where $\beta \neq 0$. Since they are all in $\mathbf{K}[\alpha, \beta] \subset \mathbf{KL}^p$. Therefore, $\mathbf{KL}^p$ is a field. Furthermore, $\mathbf{KL}^p \subset \mathbf{L}$, it is clear that $\mathbf{KL}^p$ is a finite extension of **K**.

(2)($\Rightarrow$) We assume that **L** is separable algebraic over **K**. Let $\alpha \in \mathbf{L}$. We want to prove that α is separable and purely inseparable algebraic over $\mathbf{KL}^p$. Note then it follows from Lemma 3 that $\alpha \in \mathbf{KL}^p$.

Since **K** is a subfield of $\mathbf{KL}^p$, then it follows from Lemma 1 that α is separable algebraic over $\mathbf{KL}^p$. Furthermore, $\alpha^p \in \mathbf{KL}^p$. Hence α is purely inseparable algebraic over $\mathbf{KL}^p$. We are done.

(2)($\Leftarrow$) Let $\{\ell_1, \ell_2, \cdots, \ell_m\}$ be a basis of **L** as a vector space over **K**. Clearly, $\{\ell_1^p, \cdots, \ell_m^p\}$ is a generating set of $\mathbf{KL}^p$ over **K**. Therefore, we have

$$\mathbf{KL}^p = \mathbf{L} \Leftrightarrow \{\ell_1^p, \cdots, \ell_m^p\} \text{ is a basis of } \mathbf{L}.$$

Let α be any element in **L** with algebraic degree n over **K**. Then the set $\{1, \alpha, \alpha^2, \cdots, \alpha^{n-1}\}$ is linearly independent over **K**. We may extend it to a basis of **L** as a vector space over **K**. It follows from the preceding arguments that $\{1, \alpha^p, \alpha^2p, \cdots, \alpha^{(n-1)p}\}$ is linearly independent.

Let $f(x)$ be the minimal polynomial of α over **K** with the following expression,

$$f(x) = \sum_{i=0}^{n} k_{n-i}x^i$$

We claim that $f(x)$ is a separable polynomial. Note then it follows from the definition that α is separable over $\mathbf{K}$, and we are done. Suppose not. Then $f(x)$ is not a separable polynomial, and the derivative $f'(x)$ must be identical zero, namely,

$$\begin{aligned} f'(x) = \sum_{i=1}^{n} ik_{n-i}x^{i-1} = 0 &\Rightarrow ik_{n-i} = 0 \ \forall\, i \\ &\Rightarrow p \mid i \text{ if } k_{n-i} \neq 0 \\ &\Rightarrow f(x) \in \mathbf{K}[x^p] \end{aligned}$$

Therefore, we have a non-trivial linear relation among $\{1, \alpha^p, \alpha^{2p}, \cdots, \alpha^{(n-1)p}\}$ as follows,

$$f(\alpha) = \sum_{i=0}^{n/p} k_{n-ip}\alpha^{ip}$$

A contradiction. We conclude that $f(x)$ is a separable polynomial, and α is separable over $\mathbf{K}$. ∎

We will use the preceding criterion of separability for the following theorems,

Theorem 5.18. *Let* $\mathbf{L}$ *be an extension of* $\mathbf{K}$, *and* $\alpha \in \mathbf{L}$.

(1) *If* α *is separable algebraic over* $\mathbf{K}$, *then* $\mathbf{K}[\alpha]$ *is a separable algebraic extension of* $\mathbf{K}$.
(2) *If* α *is purely inseparable algebraic over* $\mathbf{K}$, *then* $\mathbf{K}[\alpha]$ *is a purely inseparable algebraic extension of* $\mathbf{K}$.

Proof: (1) According to the preceding theorem, it suffices to prove $\mathbf{K}(\mathbf{K}[\alpha])^p = \mathbf{K}[\alpha]$. Evidently, it suffices to prove α is separable and purely inseparable algebraic over $\mathbf{K}(\mathbf{K}[\alpha])^p$.

Since $\mathbf{K}$ is a subfield of $\mathbf{K}(\mathbf{K}[\alpha])^p$, then it follows from Lemma 1 that α is separable algebraic over $\mathbf{K}(\mathbf{K}[\alpha]\ )^p$. Furthermore, $\alpha^p \in \mathbf{K}(\mathbf{K}[\alpha])^p$. Hence α is purely inseparable algebraic over $\mathbf{K}(\mathbf{K}[\alpha])^p$. We are done.

(2) If $\alpha^{p^\ell} \in \mathbf{K}$, let $h(\alpha)$ be any element in $\mathbf{K}[\alpha]$. Then we have

$$h(\alpha)^{p^\ell} = h^*(\alpha^{p^\ell}) \in \mathbf{K}$$

∎

Theorem 5.19. *Let* $\mathbf{L}$ *be an extension of* $\mathbf{K}$ *of characteristic* $p > 0$, $\alpha \in \mathbf{L}$. *Suppose that* α *is algebraic over* $\mathbf{K}$ *with the minimal polynomial* $f(x)$ *of degree* n. *Let* ℓ *be a non-negative integer such that*

$$f(x) \in \mathbf{K}[x^{p^\ell}], \qquad f(x) \notin \mathbf{K}[x^{p^{\ell+1}}]$$

Let

$$\beta = \alpha^{p^\ell}$$

Then we have

(1) $\mathbf{K} \subset \mathbf{K}[\beta] \subset \mathbf{K}[\alpha]$.
(2) *β is separable algebraic over $\mathbf{K}$ of degree n/p^ℓ.*
(3) *α is purely inseparable algebraic over $\mathbf{K}[\beta]$ of degree p^ℓ.*

Proof: The above (1) is trivial. Let $f(x)$ be the minimal polynomial of α over $\mathbf{K}$, and

$$g(x^{p^\ell}) = f(x)$$

Then we have

$$g(x) \notin \mathbf{K}[x^p], \quad g(\beta) = 0, \quad deg\ g(x) = \frac{n}{p^\ell}$$

Note that α is a root of $x^{p^\ell} - \beta \in \mathbf{K}[\beta][x]$, and $g(\beta) = f(\alpha) = 0$. Therefore, we

$$[\mathbf{K}[\alpha] : \mathbf{K}[\beta]] \le p^\ell, \qquad [\mathbf{K}[\beta] : \mathbf{K}] \le \frac{n}{p^\ell}.$$

On the other hand, it follows from Theorem 5.6. that

$$n = [\mathbf{K}[\alpha] : \mathbf{K}] = [\mathbf{K}[\alpha] : \mathbf{K}[\beta]]\ [\mathbf{K}[\beta] : \mathbf{K}]$$

Therefore, we conclude

$$[\mathbf{K}[\alpha] : \mathbf{K}[\beta]] = p^\ell, \qquad [\mathbf{K}[\beta] : \mathbf{K}] = \frac{n}{p^\ell}.$$

The conclusions (2) & (3) follow. ∎

We have the following theorem similar to Theorem 5.6..

Theorem 5.20. *Let $\mathbf{L}$ be a finite separable algebraic extension of $\mathbf{S}$, and $\mathbf{S}$ a finite separable algebraic extension of $\mathbf{K}$. Then $\mathbf{L}$ is a finite separable algebraic extension of $\mathbf{K}$.*

Proof: It follows from Theorem 5.6. that $\mathbf{L}$ is a finite extension of $\mathbf{K}$. It is easy to verify the following equation,

$$\mathbf{KL}^p = \mathbf{K}(\mathbf{SL})^p = \mathbf{KS}^p\mathbf{L}^p = \mathbf{SL}^p = \mathbf{L}.$$

Therefore, $\mathbf{L}$ is a finite separable algebraic extension of $\mathbf{K}$. ∎

Similar to algebraic closure, we have the concept of separable algebraic closure,

Theorem 5.21. *Let $\mathbf{L}$ be an extension of a field $\mathbf{K}$ of characteristic $p > 0$. Let the separable algebraic closure of $\mathbf{K}$, $\mathbf{K}^S_L$, be defined as $\mathbf{K}^S_L = \{\alpha : \alpha \in \mathbf{L}, \alpha$ is separable algebraic over $\mathbf{K}.\}$. Then we have*

(1) *$\mathbf{L} \supset \mathbf{K}^S_L \supset \mathbf{K}$ are field extensions.*
(2) *$\mathbf{L}$ is a purely inseparable algebraic extension of $\mathbf{K}^S_L$, and $\mathbf{K}^S_L$ is a separable algebraic extension of $\mathbf{K}$.*
(3) *If $\mathbf{L} \supset \mathbf{R}$ which is a finite extension of $\mathbf{K}^S_L$, then $[\mathbf{R}: \mathbf{K}^S_L] = p^\ell$ for some suitable ℓ.*

Proof: (1) It suffices to show that $\mathbf{K}_L^S$ is a field. Let $\alpha, \beta \in \mathbf{K}_L^S$. Then we have $\mathbf{K}_L^S \supset \mathbf{K}[\alpha, \beta] \supset \mathbf{K}[\alpha] \supset \mathbf{K}$. It follows from Theorem 5.20. that $\mathbf{K}[\alpha, \beta]$ is a finite separable algebraic extension of $\mathbf{K}$. Therefore, $\mathbf{K}_L^S$ is a field.

(2) It follows from the definition that $\mathbf{K}_L^S$ is a separable algebraic extension of $\mathbf{K}$. Let α be any element in $\mathbf{L}$. Let us use the notations of Theorem 5.19.. Then $\beta \in \mathbf{K}_L^S$, and α is purely inseparable over $\mathbf{K}_L^S$.

(3) Let the following be a chain of field extensions,

$$\mathbf{K}_L^S \subset \mathbf{K}_L^S[\gamma_1] \subset \cdots \subset \mathbf{K}_L^S[\gamma_1, \cdots, \gamma_s] \subset \mathbf{K}_L^S[\gamma_1, \cdots, \gamma_{s+1}] = \mathbf{R}.$$

Since γ_i is purely inseparable algebraic over $\mathbf{K}$, then it is purely inseparable algebraic over $\mathbf{K}_L^S[\gamma_1, \cdots, \gamma_{i-1}]$. Therefore, its minimal polynomial over the said field is of the form $x^{p^s} - c$. Henceforth, the field degrees of the fields in the above chain over the preceding one is of the form p^s. All statements follow. ∎

The following theorem clarify the structure of a finite separable algebraic extension. Note that the discussions about finite field should be omitted if the reader is concerned with only the characteristic 0 fields,

Theorem 5.22. *An extension field* $\mathbf{L}$ *of* $\mathbf{K}$ *is called a* ***simple extension*** *if* $\mathbf{L}=\mathbf{K}[\alpha]$ *for some suitable* α, *in this situation, the element* α *is called a* ***primitive element*** *of* $\mathbf{L}$ *over* $\mathbf{K}$. *If* $\mathbf{L}$ *is a finite separable algebraic extension of a field* $\mathbf{K}$, *then* $\mathbf{L}$ *is a simple extension of* $\mathbf{K}$.

Proof: Since $\mathbf{L}$ is a finite extension of $\mathbf{K}$, the we must have,

$$\mathbf{L} = \mathbf{K}[\alpha_1, \cdots, \alpha_n]$$

If we can show that $\mathbf{k}[\alpha, \beta] = \mathbf{K}[\gamma]$, then in finitely many steps, the theorem can be proved.

We have two cases, (i) $\mathbf{K}$ is a finite field, (ii) $\mathbf{K}$ is an infinite field.

Case (i). If $\mathbf{K}$ is a finite field, then a finite extension $\mathbf{K}[\alpha, \beta]$ is also a finite field. It follows from Theorem 5.14. that the multiplicative group $\mathbf{K}[\alpha, \beta]^*$ is a cyclic group. Let γ be a group generator, then it is easy to see that $\mathbf{K}[\alpha, \beta] = \mathbf{K}[\gamma]$. We are done.

Case (ii). We assume that $\mathbf{K}$ is an infinite field. Let Ω be an algebraic closure of $\mathbf{K}[\alpha, \beta]$. Let $f(x)$ be the minimal polynomial of α over $\mathbf{K}$, and $g(x)$ be the minimal polynomial of β over $\mathbf{K}$. Let the decompositions of $f(x), g(x)$ in $\Omega[x]$ be as follows,

$$f(x) = \prod_{i=1}^{n} (x - \alpha_i), \quad \alpha_1 = \alpha$$

$$g(x) = \prod_{j=1}^{m} (x - \beta_j), \quad \beta_1 = \beta$$

Since we assume that α, β are separable algebraic over $\mathbf{K}$, then $f(x), g(x)$ have no multiple roots. Since $\mathbf{K}$ is infinite, then we may find $c \in \mathbf{K}$ with

$$c \neq -\frac{\alpha_1 - \alpha_i}{\beta_1 - \beta_j}, \qquad i \geq 2, j \geq 2$$
$$\alpha_1 + c\beta_1 \neq \alpha_i + c\beta_j, \qquad i \geq 2, j \geq 2$$

Let $\gamma = \alpha_1 + c\beta_1 = \alpha + c\beta$. We claim $\mathbf{K}[\alpha, \beta] = \mathbf{K}[\gamma]$.

Clearly, we have

$$\mathbf{K}[\alpha, \beta] \supset \mathbf{K}[\gamma]$$

Since we have $\alpha = \gamma - c\beta$, then it suffices to prove $\beta \in \mathbf{K}[\gamma]$.

Let $h(x)$ be the minimal polynomial of β over $\mathbf{K}[\gamma]$. We want to show $h(x)$ is linear. Note then we will be done. Since β satisfies two polynomial equations $g(x) = 0, f(\gamma - cx) = 0$, then $h(x)$ must be a common divisor of $g(x), f(\gamma - cx)$. Let us consider the common divisors of $g(x)$ and $f(\gamma - cx)$ in $\Omega[x]$. It follows from the selection of c that

$$\gamma - c\beta_j = \alpha_1 + c\beta_1 - c\beta_j \neq \alpha_i, \qquad i \geq 1, j \geq 2$$

Therefore, $f(\gamma - c\beta_j) \neq 0$ for all $j \geq 2$, and $(x - \beta_j)$ is not a factor of $f(\gamma - cx)$. We conclude that the only common factor is $(x - \beta_1)$. Henceforth, $h(x) = x - \beta_1 = x - \beta$, and $\beta \in \mathbf{K}[\gamma]$.

∎

Example 12. *The condition on the separability of* $\mathbf{L}$ *over* $\mathbf{K}$ *is crucial in Theorem 5.22.. Let us consider an example with* $\mathbf{L}$ *not separable over* $\mathbf{K}$.

Let x, y *be two variables, and consider the field extension* $\mathbb{Z}_p(x^p, y^p) \subset \mathbb{Z}_p(x, y)$. *It is easy to see that*

$$[\mathbb{Z}_p(x, y) : \mathbb{Z}_p(x^p, y^p)] = p^2$$

Let $\gamma \in \mathbb{Z}_p(x, y)$ *be any element. Then it is easy to see*

$$\gamma^p \in \mathbb{Z}_p(x^p, y^p)$$

Therefore, we have

$$[\mathbb{Z}_(x^p, y^p)[\gamma] : \mathbb{Z}_p(x^p, y^p)] \leq p$$
$$\mathbb{Z}_p(x, y) = \mathbb{Z}_p(x^p, y^p)[x, y] \neq \mathbb{Z}_p(x^p, y^p)[\gamma]$$

Henceforth, $\mathbb{Z}_p(x^p, y^p)[x, y]$ *is not a simple extension of* $\mathbb{Z}_p(x^p, y^p)$.

∎

Exercises

(1) Prove that an algebraic extension of a perfect field is perfect.
(2) Let **L** be an extension of a field **K** of characteristic $p > 0$, and $\alpha \in$ **L**. Show that if α is separable over **K**$[\alpha^p]$, then $\alpha \in$ **K**$[\alpha^p]$.
(3) Let **K** be a field of characteristic $p > 0$. Show that **K** is perfect if and only if every element in **K** has a p-th root in **K**.
(4) Let **L** be an extension of a field **K** of characteristic $p > 0$, and $\alpha \in$ **L**. Show that α is separable algebraic over **K** if and only if **K**$[\alpha]$=**K**$[\alpha^{p^n}]$ for all positive integer n.
(5) Let Ω be an algebraic closure of a field **K**. Let $f(x)$ be an irreducible polynomial in **K**$[x]$. Show that all roots of $f(x)$ in Ω have the same multiplicities.
(6) Let **K** be a field of characteristic $p > 0$. Let n be a positive integer with $p \mid n$. Show that there does not exist n distinct roots of the equation $x^n - 1 = 0$.
(7) Let **L**$\supset$ **R***supset* **K** be field extensions. Suppose that **L** is purely inseparable over **K**. Show that **R** is purely inseparable over **K**.
(8) Let **L**$\supset$ **R***supset* **K** be field extensions. Suppose that **L** is purely inseparable over **R**, and **R** is purely inseparable over **K**. Show that **L** is purely inseparable over **K**.
(9) Let **L** be an extension of a field **K** of characteristic $p > 0$, and $\alpha, \beta \in$ **L**. Suppose that α is separable algebraic over **K**, and β is purely inseparable algebraic over **K**. Show that **K**$[\alpha, \beta]$=**K**$[\alpha + \beta]$. Moreover, if $\alpha \neq 0$ and $\beta \neq 0$, then **K**$[\alpha, \beta]$ =**K**$[\alpha\beta]$.
(10) Let **L** be a finite extension of degree n of a field **K** with characteristic $p > 0$. Suppose that $p \nmid n$. Show that **L** is a separable algebraic extension of **K**.
(11) Let **L**=**K**$[\alpha, \beta]$ be an extension of a field **K** of characteristic $p > 0$, $\alpha^p, \beta^p \in$ **K**, and [**L**: **K**]=p^2. Show that **L** is not a simple extension of **K**.
(12) Let **L** be a finite extension of a field **K** of characteristic $p > 0$. Show that **L** is a simple extension of **K** if and only if there are finitely many fields between **L** and **K**.

§6 Galois Theory

The center of Galois Theory is to show a deep connection between the set of roots, $\{\alpha_i\}$, of a polynomial $f(x) = \prod_i (x - \alpha_i)$ and the permutation group **G** of all roots $\{\alpha_i\}$.

We have the following definitions.

Definition 5.12. *Let Ω be an algebraic closure of a field* **K**, *and $f(x) \in$* **K**$[x]$ *be a non-constant polynomial. Let $f(x)$ be factored out into product of linear polynomials in $\Omega[x]$ as follows,*

$$f(x) = a_0 \prod_{i=1}^{n} (x - \alpha_i)$$

Then $\mathbf{K}[\alpha_1, \alpha_2, \cdots, \alpha_n]$ *is called the splitting field of* $f(x)$ *over* $\mathbf{K}$, *or simply the splitting field of* $f(x)$. ∎

Definition 5.13. *Let* $\mathbf{L}$ *be an algebraic extension of a field* $\mathbf{K}$. $\mathbf{L}$ *is called a normal extension of* $\mathbf{K}$ *if for any irreducible polynomial* $f(x) \in \mathbf{K}[x]$, *as long as* $f(x)$ *has one root in* $\mathbf{L}$, *then* $f(x)$ *can be factored out into a product of linear polynomials in* $\mathbf{L}[x]$. ∎

Example 13. *(1) The algebraic closure* Ω *of* $\mathbf{K}$ *is a normal extension of* $\mathbf{K}$.

(2) Let $\mathbf{L}$ *be a finite extension of a finite field* $\mathbf{K}$. *Then* $\mathbf{L}$ *is a normal extension of* $\mathbf{K}$: *let the cardinality of* $\mathbf{L}$ *be* p^ℓ *where p is the characteristic of* $\mathbf{K}$, *and* $f(x)$ *be an irreducible polynomial in* $\mathbf{K}[x]$ *with one root* $\alpha \in \mathbf{L}$. *Note that* α *is a root of the following*

$$x^{p^\ell} - x = 0 \in \mathbf{K}[x]$$

Therefore, we have

$$f(x) \mid x^{p^\ell} - x = \prod_{\beta_j \in L} (x - \beta_j) \in \mathbf{L}[x]$$

Henceforth, $f(x)$ *can be factored out into a product of linear polynomials in* $\mathbf{L}[x]$. *We conclude that* $\mathbf{L}$ *is a normal extension of* $\mathbf{K}$.

(3) Let $\mathbf{K}=\mathbf{Q}$, *and* $\mathbf{L}=\mathbf{Q}[\sqrt[3]{2}]$. *Then* $\mathbf{Q}[\sqrt[3]{2}]$ *is not a normal extension of* $\mathbf{Q}$: *let* $f(x) = x^3 - 2$. *Then* $f(x)$ *is irreducible in* $\mathbf{Q}[x]$ *by Eisenstein Criterion, and* $f(x)$ *has a root* $\sqrt[3]{2}$ *in* $\mathbf{Q}[\sqrt[3]{2}]$. *Note that* $\mathbf{Q}[\sqrt[3]{2}] \subset \mathbf{R}$ *the real field. In the complex field* $\mathbf{C}$, $f(x)$ *has two other complex roots as follows,*

$$\sqrt[3]{2}e^{\frac{2\pi i}{3}}, \quad \sqrt[3]{2}e^{\frac{4\pi i}{3}}$$

It is easy to see that they are non-reals, and not in $\mathbf{Q}[\sqrt[3]{2}]$. *Therefore,* $f(x)$ *can not be factored into a product of linear polynomials in* $\mathbf{Q}[\sqrt[3]{2}]$. ∎

We have the following theorem,

Theorem 5.23. *A field* $\mathbf{L}$ *is a finite normal extension of a field* $\mathbf{K} \Leftrightarrow \mathbf{L}$ *is the splitting field of a non-constant polynomial* $f(x) \in \mathbf{K}[x]$ *where* $\mathbf{K}$ *is a field.*

Proof: ($\Rightarrow$) Since $\mathbf{L}$ is a finite extension of $\mathbf{K}$, then $\mathbf{L}=\mathbf{K}[\alpha_1, \cdots, \alpha_n]$ for some suitable α_i. Let $f_i(x)$ be the minimal polynomial of α_i over $\mathbf{K}$. Since $f_i(x)$ has one root $\alpha_i \in \mathbf{L}$, and $\mathbf{L}$ is a normal extension of $\mathbf{K}$, then $f_i(x)$ can be factored out as a product of linear polynomials. It is easy to see that $\mathbf{L}$ is the splitting field of the following polynomial

$$f(x) = \prod_i f_i(x)$$

($\Leftarrow$) Let **L** be the splitting field of $f(x)$ where

(1) $$f(x) = a_0 \prod_{i=1}^{n}(x-\alpha_i) = a_0(x^n - \theta_1 x^{n-1} + \cdots + (-1)^i\theta_i x^{n-i} + \cdots + (-1)^n\theta_n) \in \mathbf{K}[x]$$

Note that $\theta_i \in \mathbf{K}$ is the i-th elementary symmetric polynomial of $\alpha_1, \cdots, \alpha_n$.

Let $g(x) \in \mathbf{K}[x]$ be an irreducible polynomial with one root β in $\mathbf{L}=\mathbf{K}[\alpha_1, \cdots, \alpha_n]$. Then we have the following expression for β,

(2) $$\beta = \sum_{finite} a_{i_1\cdots i_n}\alpha_1^{i_1}\cdots\alpha_n^{i_n}$$

Let $\mathbf{S}_n$ be the permutation group on the n digits $\{1, 2, \cdots, n\}$. For any $\sigma \in \mathbf{S}_n$, let $\sigma(\beta)$ be defined as

(3) $$\sigma(\beta) = \sum_{finite} a_{i_1\cdots i_n}\alpha_{\sigma(1)}^{i_1}\cdots\alpha_{\sigma(n)}^{i_n} \in \mathbf{L}$$

Let $h(x)$ be defined as

(4) $$h(x) = \prod_{\sigma\in S_n}(x - \sigma(\beta)) = x^{n!} + \cdots + b_j x^{n!-j} + \cdots + b_{n!}$$

Obviously, the above Eq (4) is symmetric with respect to all permutations in $\mathbf{S}_n$. Therefore, all coefficients b_j are symmetric polynomials in $\alpha_1, \cdots, \alpha_n$. It follows from Theorem 3.16. that all coefficients b_j are polynomials of the elementary symmetric polynomials $\theta_1, \cdots, \theta_n$. Henceforth, we have

$$b_j \in \mathbf{K}[\theta_1, \cdots, \theta_n] = \mathbf{K}, \quad h(x) \in \mathbf{K}[x]$$

Note that β is a root of $g(x)$ and $h(x)$, and $f(x)$ is irreducible. We have

$$g(x) \mid h(x)$$

Because $h(x)$ can be factored out into a product of linear polynomials in $\mathbf{L}[x]$, then $g(x)$ can be factored out into a product of linear polynomials in $\mathbf{L}[x]$. We conclude that **L** is a finite normal extension of **K**. ∎

Theorem 5.24. *Let* **L** *be a finite extension of a field* **K**, *and* Ω *be an algebraic closure of* **L**. *Then we have* **L** *is a normal extension of* **K** $\Leftrightarrow$ *any* **K**-*injection of* **L** *into* Ω *is a* **K**-*automorphism of* **L**.

Proof: ($\Rightarrow$) Let $\mathbf{L}= \mathbf{K}[\alpha_1, \cdots, \alpha_n]$, and the minimal polynomial of α_i over **K** be $f_i(x)$. Since **L** is a normal extension of **K**, then $f_i(x)$ can be factored out into a product of linear polynomials. Let σ be any **K**-injection of **L** into Ω. Then we have

$$0 = \sigma(f_i(\alpha_i)) = f_i(\sigma(\alpha_i))$$

Therefore, $\sigma(\alpha_i)$ is a root of $f_i(x)$, and is in **L**. It follows at once that σ is a **K**-automorphism of **L**.

($\Leftarrow$) Let **L**= $\mathbf{K}[\alpha_1, \cdots, \alpha_n]$, and the minimal polynomial of α_i over **K** be $f_i(x)$. Let $f(x) = \prod_i f_i(x)$. We claim that $f(x)$ can be factored out into a product of linear polynomials in $\mathbf{L}[x]$. Note then **L** is the splitting field of $f(x)$ over **K**, and is a normal extension of **K**.

Suppose that one of $f_i(x)$'s can not be factored out completely, say $f_1(x)$. Therefore, inside Ω, $f_1(x)$ has a root $\overline{\alpha}_1 \notin \mathbf{L}$. It follows from Theorem 5.10. that there is a **K**-injection $\sigma_1 : \mathbf{K}[\alpha_1] \to \mathbf{K}[\overline{\alpha}_1]$ such that

$$\sigma_1(\alpha_1) = \overline{\alpha}_1$$

Let $g_2(x)$ be the minimal polynomial of α_2 over $\mathbf{K}[\alpha_1]$, and $\overline{\alpha}_2$ be a root of $\sigma_1(g(x))$ in Ω. It follows from Theorem 5.10. again that there is a **K**-injection $\sigma_2 : \mathbf{K}[\alpha_1, \alpha_2] \to \mathbf{K}[\overline{\alpha}_1, \overline{\alpha}_2]$ such that

$$\sigma_2(\alpha_1) = \overline{\alpha}_1, \quad \sigma_2(\alpha_2) = \overline{\alpha}_2$$

Steps by steps, we show that there are elements $\overline{\alpha}_1, \cdots, \overline{\alpha}_n \in \Omega$, and a **K**-injection $\sigma_n : \mathbf{K}[\alpha_1, \cdots, \alpha_n] \to \mathbf{K}[\overline{\alpha}_1, \cdots, \overline{\alpha}_n]$ such that

$$\sigma_n(\alpha_i) = \overline{\alpha}_i, \quad \forall\, i = 1, 2, \cdots, n$$

Since $\overline{\alpha}_1 \notin \mathbf{L}$, then σ_n is not a **K**-automorphism of **L**. This contradicts our assumption. ∎

Corollary. *Let* **K** *be a field,* $f(x) \in \mathbf{K}[x]$ *be irreducible, and* **L** *be the splitting field of* $f(x)$ *over* **K** *in an algebraic closure* Ω *of* **K**. *Let* α_1, α_2 *be any two roots of* $f(x)$. *Then there is a* **K**-*automorphism* σ *of* **L** *such that* $\sigma(\alpha_1) = \alpha_2$.

Proof. Note that **L** is a normal extension of **K**. We have

$$\mathbf{K}[\alpha_1] \approx \mathbf{K}[x]/(f(x)) \approx \mathbf{K}[\alpha_2]$$

It follows from Theorem 5.10. that the above **K**-isomorphism can be extended to an injection from **L** to Ω. Our preceding theorem shows that the injection is a **K**-automorphism of **L**.

∎

The following is the central concept of this section,

Definition 5.14. *A field* **L** *is called a* ***Galois extension*** *of a field* **K** *if* **L** *is a finite normal separable extension of* **K**.

∎

The central thesis of Galois Theory is the relations between the permutation groups and the Galois extensions. For these purposes, we have the following definition,

Definition 5.15. *(1) Let* **S** *be an algebraic extension of a field* **R**. *Let* $G(S/R)$ *be the group of* **R**-*isomorphisms of* **S**. *If* **S** *is a Galois extension of* **R**, *then* $G(S/R)$ *is called the Galois group of* **S** *over* **R**.

(2) Let H be an automorphism group of a field **S**. *Let $F(H)$ be the set of invariants of H, namely $F(H) = \{s : s \in \mathbf{S}, \sigma(s) = s \ \forall\, \sigma \in H\}$.* ∎

We have the following lemma,

Lemma. *The set $F(H)$ is a subfield of* **S**, *the invariant field of H, or the fixed field of H.*

Proof: We always have

$$\sigma(0) = 0, \sigma(1) = 1, \qquad \forall\, \sigma \in H$$

Therefore, $0, 1 \in F(H)$. Moreover, let $\alpha, \beta \in F(H)$. Then we have

$$\sigma(\alpha \pm \beta) = \alpha \pm \beta, \quad \sigma(\alpha\beta) = \alpha\beta, \quad \forall\, \sigma \in H$$

Therefore, $\alpha \pm \beta, \alpha\beta \in F(H)$. If $\beta \neq 0$, then we have

$$1 = \sigma(1) = \sigma(\beta\beta^{-1}) = \beta\sigma(\beta^{-1}), \quad \forall\, \sigma \in H$$

That is

$$\sigma(\beta^{-1}) = \beta^{-1}, \quad \forall\, \sigma \in H$$

Henceforth, $\beta^{-1} \in F(H)$. We conclude that $F(H)$ is a field. ∎

We have the following important theorem which establishes a bijective map from the set of all intermediate fields between a Galois extension **L** of **K** to the set of all subgroups of $G(L/K)$.

Theorem 5.25. (The Fundamental Theorem of Galois Theory). *Let* **L** *be a Galois extension of* **K**. *Then we have*

(1) *Let* **S** *be an intermediate field between* **L** *and* **K**, *namely* **L**$\subset$ **S**$\subset$ **K**. *Then*
   (i) **L** *is a Galois extension of* **S**. *$G(L/S)$ is a subgroup of $G(L/K)$.*
   (ii) *The order of the group $G(L/S)$, $o(G(L/S)) = [\mathbf{L} : \mathbf{S}]$.*
   (iii) *The invariant field of $G(L/S)$, $F(G(L/S)) =$***S**.

(2) *Let H be a subgroup of $G(L/K)$. Then*
   (iv) *The invariant field $F(H)$ is an intermediate field between* **L** *and* **K**.
   (v) *The order of H, $o(H) = [\mathbf{L} : F(H)]$.*
   (vi) *The Galois group $G(L/F(H)) = H$.*

Therefore, we have bijective maps in the following diagram

$$\{\text{ all intermediate fields } \mathbf{S} \ \} \overset{G(L/\cdot)}{\underset{F}{\longleftrightarrow}} \{\text{ all subgroups } H \ \}$$

Moreover, the normal subgroups H of $G(L/K)$ correspond exactly to the normal extensions **S** *of* **K**. *In this case, we have the following canonical isomorphism*

$$G(S/K) \approx G(L/K)/G(L/S)$$

Proof: (1) (i) Since **L** is a finite separable extension of **K**, then it follows that **L** is a finite separable extension of **S**. Furthermore, **L** is a splitting field of a polynomial $f(x) \in \mathbf{K}[x] \subset \mathbf{S}[x]$. Therefore, **L** is the splitting field of the same polynomial $f(x)$ over **S**, and **L** is normal over **S**. We conclude that **L** is a Galois extension of **S**.

(ii) It follows from Theorem 5.22. that **L** is a simple extension of **S**, **L**=**S**$[\alpha]$ for some suitable α. Let $f(x)$ be the minimal polynomial of α over **S**. Since $f(x)$ has one root α in **L**, and **L** is a normal extension of **S**, then $f(x)$ can be factored out as follows in $\mathbf{L}[x]$,

$$f(x) = \prod_{i=1}^{n}(x - \alpha_i), \quad \alpha_1 = \alpha$$

Furthermore, α is separable algebraic over **S**. Then the minimal polynomial is separable, and all α_i are distinct.

We claim $o(G(L/S)) \geq [\mathbf{L} : \mathbf{S}] = n$. It follows from Theorem 5.10. that the identity map id of **S** can be extended to an isomorphism $\sigma_i : \mathbf{S}[\alpha_1] \to \mathbf{S}[\alpha_i]$ with

$$\sigma(\alpha_1) = \alpha_i$$

Clearly, we have

$$n = [\mathbf{S}[\alpha_1] : \mathbf{S}] = [\mathbf{S}[\alpha_i] : \mathbf{S}], \quad \mathbf{S}[\alpha_i] \subset \mathbf{S}[\alpha_1]$$

Therefore, we conclude that $\mathbf{S}[\alpha_i]$=$\mathbf{S}[\alpha_1]$, and the isomorphisms σ_i are all distinct **S**-automorphisms of **L**. Henceforth, we have $o(G(L/S)) \geq [\mathbf{L} : \mathbf{S}] = n$.

We claim $o(G(L/S)) = [\mathbf{L} : \mathbf{S}] = n$. Let $\sigma \in G(L/S)$. Then we have

$$0 = \sigma(f(\alpha_1)) = f(\sigma(\alpha_1))$$

Therefore, $\sigma(\alpha_1)$ is a root of $f(x)$, and

$$\sigma(\alpha_1) = \alpha_i, \quad \text{for some suitable } i.$$

Then it is easy to see that $\sigma = \sigma_i$, and $\{\sigma_1, \cdots, \sigma_n\} = G(L/S)$. We have the equality.

(iii) Note that we have naturally $F(G(L/S)) \supset \mathbf{S}$. Let $\beta \in \mathbf{L}\backslash\mathbf{S}$, we claim that there is $\sigma \in G(L/S)$ such that $\sigma(\beta) \neq \beta$, in other words, then $\beta \notin F(G(L/S))$. Clearly, this implies our statement.

Let the minimal polynomial of β over **S** be $g(x)$ which is separable. Let the decomposition of $g(x)$ in $\mathbf{L}[x]$ be

$$g(x) = \prod_{i=1}^{m}(x - \beta_i), \quad m > 1, \beta_1 = \beta$$

Let $\mathbf{R}_1$ =$\mathbf{S}[\beta_1]$, $\mathbf{R}_2$ =$\mathbf{S}[\beta_2]$. It follows from Theorem 5.10. that there is a **S**-isomorphism $\gamma : \mathbf{R}_1 \to \mathbf{R}_2$ such that

$$\gamma(\beta_1) = \beta_2 \neq \beta_1$$

Note that L=K[α] =$\mathbf{R}_1[\alpha]$ =$\mathbf{R}_2[\alpha]$. Let the minimal polynomial of α over $\mathbf{R}_1$ be $h(x)$. Let Ω be an algebraic closure of L. Note that Ω is an algebraic closure of $\mathbf{R}_2$. Let $\overline{\alpha}$ be a root of $\gamma(h(x))$. It follows from Theorem 5.10. that the K-isomorphism $\gamma : \mathbf{R}_1 \to \mathbf{R}_2$ can be extended to a K-isomorphism $\gamma' : \mathbf{R}_1[\alpha] \to \mathbf{R}_2[\overline{\alpha}]$. Recall that $\mathbf{R}_1[\alpha] = \mathbf{L}$, and L is a normal extension of S. It follows from Theorem 5.24. that $\gamma' \in G(L/S)$, and $\gamma'(\beta) \neq \beta$.

(2) (iv) It follows from Lemma that the invariant field $F(H)$=S is an intermediate field between L and K.

(v) Let S=$F(H)$. It is trivial to see that $H \subset G(L/S)$. It follows from (ii) that

(1) $$o(H) \leq o(G(L/S)) = o(G(L/F(H)) = [L:S] = [L:F(H)]$$

Let L=S[α], and $H = \{\gamma_1, \cdots, \gamma_m\}$. Let

$$q(x) = \prod_{i=1}^{m}(x - \gamma_i(\alpha)) = x^m + c_1x^{m-1} + \cdots + c_m$$

Clearly, we have

$$\gamma_j(q(x)) = \prod_{i=1}^{m}(x - \gamma_j\gamma_i(\alpha)) = q(x), \quad j = 1, \cdots, m$$

Therefore, $\gamma_j(c_i) = c_i$ for all j. In other words, $c_i \in$ S, and $q(x) \in$ S[x]. Since α is a generator of L over S, and a root of $q(x)$ which is of degree m. then we have

(2) $$[L:F(H)] = [L:S] \leq m \leq o(H).$$

Combining the above two inequalities, we have

$$[L:F(H)] = o(H).$$

(vi) It follows from the inequalities (1) & (2) of (v) that

$$o(H) = o(G(L/F(H)))$$

Since H is a subgroup of $G(L/F(H))$, then they must be equal.

Let us consider the last part of our theorem. Our discussions will consist of two parts: (vii) assume that S is a normal extension of K, we claim $G(L/S)$ is a normal subgroup of $G(L/K)$ and there is a canonical map π such that

$$\pi : G(S/K) \approx G(L/K)/G(L/S)$$

(viii) assume that H is a normal subgroup of $G(L/K)$, we claim $F(H)$ is a normal extension of K. Note then our theorem is finished.

(vii) Assume that S is a normal extension of K. Let $\sigma \in G(L/K)$, and Ω be an algebraic closure of L. Then σ will induce a K-injection of S into Ω. It follows from Theorem 5.24. that the restriction of σ to S, $\overline{\sigma}$, is in $G(S/K)$. Let τ be the restriction map, namely

$$\tau : G(L/K) \to G(S/K), \text{ with } \tau(\sigma) = \overline{\sigma}$$

It is easy to see that τ is a homomorphism, and $ker(\tau) = G(L/S)$. To prove (vii), it suffices to show τ is surjective.

Let $\overline{\rho}$ be any element in $G(S/K)$. Let L=S$[\alpha]$ for some suitable α, and $h(x)$ be the minimal polynomial of α over S. Let $\overline{\alpha}$ be any root of $\overline{\rho}(h(x))$ in Ω. It follows from Theorem 5.10. that $\overline{\rho}$ can be extended to a K-injection, ρ, from S$[\alpha]$ (=L) to S$[\overline{\alpha}] \subset \Omega$. Since L is a normal extension of K, then it follows from Theorem 5.24. that ρ is in $G(L/K)$. It is trivial to see that the restriction of ρ to S is $\overline{\rho}$, namely

$$\tau(\rho) = \overline{\rho}$$

(viii) Assume that H is a normal subgroup of $G(L/K)$. Let S=$F(H)$, $m = o(H) = [L : F(H)] = [L : S], \ell = [G(L/K) : H]$, and $G(L/K) = H\gamma_1 \cup H\gamma_2 \cup \cdots \cup H\gamma_\ell$ for some suitable $\gamma_1, \gamma_2, \cdots, \gamma_\ell$. Note that we have

$$m\ell = [L : K] = o(G(L/K))$$
$$\ell = [S : K]$$

Since S is a finite separable extension of K, then there is β such that S=K$[\beta]$. We have

$$\sigma\gamma_i(\beta) = \gamma_i\sigma'(\beta) = \gamma_i(\beta), \quad \forall\, \sigma \in H, i = 1, 2, \cdots, \ell$$

where σ' is some suitable element in H. We have at once,

$$\gamma_i(\beta) \in \mathbf{S}, \quad i = 1, 2, \cdots, \ell$$

Let $h(x)$ be defined as

$$h(x) = \prod_{i=1}^{\ell}(x - \gamma_i(\beta))$$

Then it is easy to see that $h(x)$ is invariant under the group actions of $G(L/K)$. Therefore, $h(x) \in \mathbf{K}[x]$. Since one of $\gamma_i's$ may be taken to be the identity, then β is a root of $h(x)$ and S is the splitting field of $h(x)$, Therefore, $F(H)$ is a normal extension of K if H is a normal subgroup of $G(L/K)$. ∎

Corollary. *Let* L *be a Galois extension of a field* K. *Then we have* $o(G(L/K)) = [L : K]$. ∎

Definition 5.16. *Let* **L** *be a Galois extension of* **K**. *If the Galois group $G(L/K)$ is cyclic, or abelian, or solvable, then* **L** *is called a cyclic extension, or abelian extension, or solvable extension of* **K**. ∎

Example 14. *Let us compute a concrete example. Let* **L** *be the splitting field of $x^3 - 2$ in the complex field* **C**. *We have*

$$x^2 - 2 = (x - \sqrt[3]{2})(x - \sqrt[3]{2}e^{\frac{2\pi i}{3}})(x - \sqrt[3]{2}e^{\frac{4\pi i}{3}})$$

$\mathbf{L}=\mathbf{Q}[\sqrt[3]{2}, \sqrt[3]{2}e^{\frac{2\pi i}{3}}, \sqrt[3]{2}e^{\frac{4\pi i}{3}}] = \mathbf{Q}[\sqrt[3]{2}, e^{\frac{2\pi i}{3}}]$. *We want to compute the Galois group $G(L/\mathbf{Q})$.*

Let $\zeta = e^{\frac{2\pi i}{3}}$, $\mathbf{R}=\mathbf{Q}[\sqrt[3]{2}]$, *and* $\mathbf{S}= \mathbf{Q}[\zeta]$. *The minimal polynomial of ζ over* **Q** *is the third cyclotomic polynomial as follows (cf Example 5.),*

$$\frac{x^3 - 1}{x - 1} = x^2 + x + 1 = (x - \zeta)(x - \zeta^2)$$

Note that **S** *is a Galois extension of* **Q**, *and there are two* **Q***-automorphisms as follows,*

$$\gamma_1(\zeta) = \zeta$$
$$\gamma_2(\zeta) = \zeta^2$$

Furthermore, note that

$$[\mathbf{R} : \mathbf{Q}] = 3, \qquad [\mathbf{S} : \mathbf{Q}] = 2$$
$$[\mathbf{L} : \mathbf{S}] \leq 3$$
$$2, 3 \mid [\mathbf{L} : \mathbf{Q}] = [\mathbf{L} : \mathbf{S}]\ [\mathbf{S} : \mathbf{Q}] \leq 6$$

Therefore, we have

$$[\mathbf{L} : \mathbf{Q}] = 6, \quad \text{and } [\mathbf{L} : \mathbf{S}] = 3$$

It follows from Theorem 5.10. that γ_1, γ_2 have three extensions as follows,

$$\gamma_1 \begin{cases} \sigma_1 : \sigma_1(\zeta) = \zeta,\ \sigma_1(\sqrt[3]{2}) = \sqrt[3]{2} \\ \sigma_2 : \sigma_2(\zeta) = \zeta,\ \sigma_2(\sqrt[3]{2}) = \sqrt[3]{2}\zeta \\ \sigma_3 : \sigma_3(\zeta) = \zeta,\ \sigma_3(\sqrt[3]{2}) = \sqrt[3]{2}\zeta^2 \end{cases}$$
$$\gamma_2 \begin{cases} \sigma_4 : \sigma_4(\zeta) = \zeta^2,\ \sigma_4(\sqrt[3]{2}) = \sqrt[3]{2} \\ \sigma_5 : \sigma_2(\zeta) = \zeta^2,\ \sigma_5(\sqrt[3]{2}) = \sqrt[3]{2}\zeta \\ \sigma_6 : \sigma_6(\zeta) = \zeta^2,\ \sigma_6(\sqrt[3]{2}) = \sqrt[3]{2}\zeta^2 \end{cases}$$

The group $G(L/\mathbf{Q})$ of 6 automorphisms, σ_i, above can be represented as the permutation group, S_3, on three digits as follows. Let

$$\alpha_1 = \sqrt[3]{2}, \quad \alpha_2 = \sqrt[3]{2}\zeta, \quad \alpha_3 = \sqrt[3]{2}\zeta^2$$

Then we have that following correspondence,

$$\sigma_1 \to (1)(2)(3), \quad \sigma_2 \to (1,2,3), \quad \sigma_3 \to (1,3,2)$$
$$\sigma_4 \to (1)(2,3), \quad \sigma_5 \to (1,2)(3), \quad \sigma_6 \to (1,3)(2)$$

In general, it is easy to see that if **L** *is the splitting field of a polynomial $f(x)$ of degree n, then the Galois group $G(L/K)$ is the permutation group of the roots of $f(x)$, and therefore, $G(L/K)$ can be considered as a subgroup of the permutation group S_n on n digits.* ∎

Example 15. *Let us consider a finite extension* **L** *of a finite field* **K**. *Let the cardinality of* **K** *be p^n, and field degree of* **L** *over* **K** *be m. Then the cardinality of* **L** *is p^{nm}. Since* **K** *is finite, then* **K** *is perfect, and* **L** *is a separable algebraic extension of* **K**. *Moreover,* **L** *is the splitting field of the following equation in an algebraic closure Ω of* **L**,

$$x^{p^{nm}} - x = 0$$

Therefore, **L** *is a normal extension, and hence a Galois extension of* **K**.

Let us compute the Galois group $G(L/K)$. Let ρ be the Frobinus map,

$$\rho(\alpha) = \alpha^p$$

It follows from Theorem 5.15. that ρ^n is a **K***-automorphism of* **L** *of order m. Furthermore, it follows from Corollary of Theorem 5.25. that*

$$o(G(L/K)) = [\mathbf{L}:\mathbf{K}] = m$$

Therefore, ρ^n generated $G(L/K)$, and

$$G(L/K) = \{id, \rho^n, \rho^{2n}, \cdots, \rho^{(m-1)n}\}$$

We conclude that every finite extension of a finite field is a cyclic extension. ∎

Example 16. (The constructibility of a regular n-gon). *(1) Let us consider the case of a prime number p. Example 8. shows a necessary condition for a regular p-gon to be constructible is that p must be a Fermat number. We shall show that the preceding necessary condition is sufficient.*

Recall that we work inside the complex field **C**. *Let p be a Fermat number, namely,*

$$p = 2^{2^q} + 1$$

for some suitable q. Let ζ be a root of the following p-th cyclotomic polynomial,

$$\varphi_p(x) = \frac{x^p - 1}{x - 1} = x^{p-1} + x^{p-2} + \cdots + x + 1 = \prod_{i=0}^{p-1}(x - \zeta^i)$$

It is easy to see that $\mathbf{Q}[\zeta]$ *is the splitting field of* $\varphi(x)$ *in* $\mathbf{C}$. *Since the characteristic of* $\mathbf{C}$ *is* 0, *then* $\mathbf{Q}[\zeta]$ *is a Galois extension of* $\mathbf{Q}$. *It follows from Corollary of Theorem 5.25. that*

$$2^{2^q} = [\mathbf{Q}[\zeta] : \mathbf{Q}] = o(G(\mathbf{Q}[\zeta]/\mathbf{Q}))$$

It follows from §6 of Chapter II that the Galois group $G(\mathbf{Q}[\zeta]/\mathbf{Q})$ *is a 2-group, and there is a chain of normal subgroups* G_i *(cf Theorem 2.14.) such that*

$$\begin{aligned} &G_{2^{2^q}} = G(\mathbf{Q}[\zeta]/\mathbf{Q}) \\ &G_{2^{2^q}} \triangleright \cdots \triangleright G_i \triangleright G_{i-1} \triangleright \cdots \triangleright G_0 = \{e\} \\ &[G_i : G_{i-1}] = 2 \end{aligned}$$

The Fundamental Theorem of Galois Theory implies that there is a chain of field extensions as follows,

$$\begin{aligned} &\mathbf{K}_{2^{2^q}} = \mathbf{Q} \\ &\mathbf{K}_{2^{2^q}} \subset \cdots \subset \mathbf{K}_i \subset \mathbf{K}_{i-1} \subset \mathbf{K}_0 = \mathbf{Q}[\zeta] \\ &[\mathbf{K}_{i-1} : \mathbf{K}_i] = 2 \end{aligned}$$

It follows from the necessary and sufficient condition of constructibility of Example 6. that a regular p-gon can be constructed if p is a Fermat number.

(2) Let us consider the general case. A necessary condition (cf Example 10.) for a regular n-gon to be constructible, where $n > 2$, *is that n must be of the following form,*

$$n = p_1 p_2 \cdots p_s 2^m$$

where $p_1, p_2, \cdots, p_s$ *must be distinct prime and Fermat numbers. We shall show that this is sufficient.*

Under the preceding condition on the number n, we claim that the angle $2\pi/n$ *can be constructed. Note that then regular n-gon can be constructed.*

We use ***partial fractions.*** *We have*

$$\frac{2\pi}{n} = 2\pi\frac{1}{n} = 2\pi\left(\sum \frac{a_i}{p_i} + \frac{b}{2^m}\right) = \sum \frac{2\pi a_i}{p_i} + \frac{2\pi b}{2^m}$$

where a_i, b *are integers.*

Since p_i *are Fermat numbers, then the angle* $(2\pi)/p_i$ *can be constructed. Since We know how to bi-sect an angle, then* $(2\pi)/2^m$ *can be constructed. It is easy to conclude from the above formula that* $(2\pi)/n$ *can be constructed.*

We have established that a regular n-gon can be constructed if and only if n is of the following form,

$$n = p_1 p_2 \cdots p_s 2^m$$

where $p_1, p_2, \cdots, p_s$ are distinct prime and Fermat numbers.

Example 17. *The ancient Greek could construct regular a n-gon for $n = 3, 4, 5, 6, 8, 12,$. Gauss discovered the constructibility of regular 17-th gon in his teenage years and it motivated him to start a career of mathematician.[4] Let us use Example 16. to produce this result.*

Note that $17 = 2^{2^2} + 1$ is a Fermat number. Let $\zeta = e^{(2\pi i)/17}$, and

$$\varphi_{17}(x) = x^{16} + x^{15} + \cdots + x + 1 = \prod_{i=1}^{16}(x - \zeta^i)$$

Let $\mathbf{L}=\mathbf{Q}[\zeta]$, and $\sigma \in G(L/\mathbf{Q})$ be defined as

$$\sigma(\zeta) = \zeta^3$$

Then it is easy to check

$$\begin{aligned}\sigma : \zeta &\mapsto \zeta^3 \mapsto \zeta^9 \mapsto \zeta^{10} \mapsto \zeta^{13} \mapsto \zeta^5 \mapsto \zeta^{15} \mapsto \zeta^{11} \mapsto \zeta^{16} \\ &\mapsto \zeta^{14} \mapsto \zeta^8 \mapsto \zeta^7 \mapsto \zeta^4 \mapsto \zeta^{12} \mapsto \zeta^2 \mapsto \zeta^6 \mapsto \zeta\end{aligned}$$

and order of σ is 16. We have the following chain of subgroups,

$$G(L/\mathbf{Q}) = \{\sigma\} \triangleright \{\sigma^2\} \triangleright \{\sigma^4\} \triangleright \{\sigma^8\} \triangleright \{e\}.$$

Let

$$\begin{aligned}\alpha_8 &= \zeta + \zeta^{16} \\ \alpha_4 &= \zeta + \zeta^{13} + \zeta^{16} + \zeta^4 \\ \alpha_2 &= \zeta + \zeta^9 + \zeta^{13} + \zeta^{15} + \zeta^{16} + \zeta^8 + \zeta^4 + \zeta^2 \\ \alpha &= \sum_{i=1}^{16} \zeta^i = -1\end{aligned}$$

Clearly, we have

$$\sigma^8(\alpha_8) = \alpha_8, \quad \sigma^4(\alpha_4) = \alpha_4, \quad \sigma^2(\alpha_2) = \alpha_2, \quad \sigma(\alpha) = \alpha$$

We have the following quadratic extensions,

$$\mathbf{Q} \subset \mathbf{Q}[\alpha_2] \subset \mathbf{Q}[\alpha_2, \alpha_4] \subset \mathbf{Q}[\alpha_2, \alpha_4, \alpha_8] \subset \mathbf{Q}[\zeta]$$

---

[4] Gauss ordered a regular 17-th gon to be carved on his tombstone

It suffices to write down the quadratic equations step by step. For the first extension, we have the following equation,

$$(x-\alpha_2)(x-(\alpha-\alpha_2)) = x^2 - \alpha x + 4\alpha = x^2 + x - 4$$

For the second extension, we have the following equation,

$$(x-\alpha_4)(x-(\alpha_2-\alpha_4)) = x^2 - \alpha_2 x + \alpha = x^2 - \alpha_2 x - 1$$

For the third extension, we have the following equation,

$$(x-\alpha_8)(x-(\alpha_4-\alpha_8)) = x^2 - \alpha_4 x + \frac{1}{2}(\alpha_4^2 + \alpha_4 - \alpha_2 - 4)$$

For the last extension, we have the following equation

$$(x-\zeta)(x-(\alpha_8-\zeta)) = x^2 - \alpha_8 x + 1$$

Therefore, zeta can be constructed. ∎

Example 18. *Let us continue our discussion of Example 9.. Let us use its notations. Given ζ a primitive n-th root of unity, then ζ^ℓ is a primitive n-th root of unity if and only if $(n,\ell)=1$. Therefore, we have*

$$\varphi_n(x) = \prod_{(\ell,n)=1} (x-\zeta^\ell), \qquad 0 \le \ell < n$$

Clearly, $G(\mathbf{Q}[\zeta]/\mathbf{Q})$ consists of the isomorphisms σ_ℓ for $(\ell,n)=1, 0 \le \ell < n$ defined as follows

$$\sigma_\ell(\zeta) = \zeta^\ell$$

Let $\mathbf{Z}_n^*$ be the multiplicative group of $\mathbf{Z}_n = \{[m] : (n,m) = 1\}$. Then there is a map $\tau : G(\mathbf{Q}[\zeta]/\mathbf{Q}) \to \mathbf{Z}_n^*$ with $\tau(\sigma_\ell) = [\ell]$. It is easy to see that τ is an isomorphism. We identify $G(\mathbf{Q}[\zeta]/\mathbf{Q})$ with $\mathbf{Z}_n^*$. ∎

The following definition and theorem are useful in understanding the Galois extensions.

Definition 5.17. *Let* **R** *and* **S** *be subfields of a field* **L**. *Then the* ***composition***, **R·S**, *of* **R** *and* **S** *is the smallest subfield of* **L** *which contains both* **R** *and* **S**. ∎

Lemma. *Let* **L** *and* **S** *be subfields of a field* **R**. *Suppose that* **R** *is either algebraic over* **L** *or* **S**. *Then the* ***composition*** **L·S**$= \{\sum_{finite} a_i b_i : a_i \in \mathbf{L},\ b_i \in \mathbf{S}\}$.

Proof: Suppose that **R** is algebraic over **L**. Then it is easy to see that $\mathbf{L}\cdot\mathbf{S} = \mathbf{L}[\mathbf{S}] = \{\sum_{finite} a_i b_i : a_i \in \mathbf{L},\ b_i \in \mathbf{S}\}$. ∎

Theorem 5.26. *Let* **L** *be a Galois extension of a field* **K**, *and* Ω *be an algebraic closure of* **L**. *Let* **S** *be an extension of* **K**. *Then the composition,* **L·S**, *is a Galois extension of* **S**, *and the Galois group* $G((L\cdot S)/S)$ *is a subgroup of* $G(L/K)$.

Proof: Since **L** is a Galois extension of **K**, then **L** =**K**$[\alpha]$, for some suitable α, and **L** is the splitting field of the minimal polynomial $f(x)$ over **K** in Ω. Clearly, **L·S** is the splitting field of $f(x)$ over **S** in Ω. It is trivial to see that **L·S** is a finite separable extension of **S**. Therefore, **L·S** is a Galois extension of **S**.

Let us define a map $\tau : G((L\cdot S)/S) \to G(L/K)$ as

$$\tau(\sigma(\ell)) = \sigma(\ell), \quad \forall \ell \in \mathbf{L}, \sigma \in G((L\cdot S)/S)$$

Then it follows at once that $\tau(\sigma)$ is a **K**-isomorphism from **L** to Ω. Since **L** is normal over **K**, then it follows from Theorem 5.24. that $\tau(\sigma) \in G(L/K)$. We shall show that τ is injective.

Suppose that $\tau(\sigma) = id$. Then we have

$$\tau(\sigma(\alpha)) = \alpha$$

Note that **L·S**=**S**$[\alpha]$. We conclude that

$$\sigma = id$$

∎

Example 19. *Let* Ω *be an algebraic closure of a field* **S** *of characteristic* 0. *Let* **R** *be the splitting field of the following equation over* **S**,

$$x^n - 1 = 0$$

We claim that the Galois group $G(R/S)$ *is a subgroup of the multiplicative group* $\mathbf{Z}^* = \{[m] : (m,n) = 1\}$, *and therefore,* **R** *is an abelian extension of* **S**.

Clearly, the set G *of all roots of* $x^n - 1$ *form a finite multiplicative group. It follows from Theorem 5.13. that* G *is a cyclic group. Let* ζ *be a generator of the group* G. *Then we have* **R**=**S**$[\zeta]$.

Since **S** *is of characteristic* 0, *then it contains the prime field* **Q**=**K**. *Let* **L**=**Q**$[\zeta]$. *Then* **L** *is a Galois extension of* **K**, *and* **L**· **S**=**S**$[\zeta]$=**R**. *It then follows from our preceding Theorem 5.26. that* $G(R/S) = G(L\cdot S/S)$ *is a subgroup of* $G(L/K) = Z_n^*$ *(cf Example 18.).* ∎

Exercises

(1) Let **L** be the splitting field of $x^{10} - 1$ over **Q** in **C**. Find [**L** : **Q**]. Is **L** a cyclic extension of **Q**?
(2) Let **L** be the splitting field of $x^8 - 1$ over **Q** in **C**. Find all intermediate fields, and find a primitive element α for each intermediate field.
(3) Let **L** be the splitting field of $x^3 + x^2 + 1$ over **Q** in **C**. Find [**L** : **Q**].
(4) Let **L** be the splitting field of $x^9 + x^3 + 1$ over **Q** in **C**. Find [**L** : **Q**].
(5) Let ζ_n be a primitive n-th root of unity. Show that

$$\mathbf{Q}[\zeta_n + \zeta_m] = \mathbf{Q}[\zeta_n, \zeta_m].$$

(6) Let ζ_p be a primitive p-th root of unity where $p > 2$ is a prime number. Show that

$$\sqrt{(-1)^{(p-1)/2}p} \in \mathbf{Q}[\zeta_p]$$

(7) Find the minimal polynomial of $\sqrt[3]{2} + \sqrt{2}$ over **Q**, and the Galois group of the splitting field of the polynomial over **Q**.
(8) Find the Galois groups of the splitting fields of $x^4 - 5$ in **C** over (i) **Q**, (ii) $\mathbf{Q}[\sqrt{5}]$, (iii) $\mathbf{Q}[\sqrt{5}i]$.
(9) Find the Galois groups of the splitting fields of $x^3 - 10$ in **C** over (i) **Q**, (ii) $\mathbf{Q}[\sqrt{2}]$, (iii) $\mathbf{Q}[\sqrt{-3}]$.
(10) Find all subfields of $\mathbf{Q}[\omega, \sqrt[3]{2}]$, where

$$\omega = \frac{-1 - \sqrt{3}i}{2}$$

Which subfields are normal extensions of **Q**?
(11) Let $\mathbf{L}=\mathbf{Q}[e^{(2\pi i/7)}]$. Find all subgroups of the Galois group $G(L/\mathbf{Q})$. Find the invariant fields of all subgroups.
(12) Let α be a root of $x^3 + x^2 - 2x - 1$ in **C**. Show that $\mathbf{Q}[\alpha]$ is a normal extension of **Q**, and find the Galois group.
(13) Prove that the splitting field of a polynomial $f(x)$ of degree n over **K** in an algebraic closure Ω is generated by any $n - 1$ roots of $f(x)$ over **K**.
(14) Let **L** be the splitting field of the following polynomial over a field **K** in an algebraically closed field Ω,

$$f(x) = (x - u_1)^{n_1} \cdots (x - u_s)^{n_s}, \quad u_i \geq 1$$

Let

$$g(x) = (x - u_1) \cdots (x - u_s) = x^s + v_1 x^{s-1} + \cdots + v_s$$

and S=K$[v_1, v_2, \cdots, v_s]$. Show that

(i) S is a subfield of L.

(ii) L is the splitting field of $g(x)$ over S.

(iii) L is a Galois extension of S.

(15) Let L=K$[\alpha]$ be a Galois extension of a field K with Galois group $G(L/K) = G$. Let H be subgroup of G. Let $f(x)$ be the following polynomial,

$$f(x) = \prod_{\sigma \in H} (x - \sigma(\alpha))$$

Show that the fixed field of H is generated over K by all coefficients of $f(x)$.

(16) Let Ω be an algebraic closure of a field K of characteristic $p > 0$. Let $\alpha \in$ K such that there is no $\beta \in$ K with $\beta^p - \beta = \alpha$. Let L be the splitting field of $x^p - x - \alpha$ in Ω. Find the Galois group $G(L/K)$.

(17) Let K be a field of characteristic $p > 0$. Let $f(x) = x^p - x - a \in$ K$[x]$. Show that if $f(x)$ is reducible inK$[x]$, then $f(x)$ can be factored to a product of linear polynomials in K$[x]$.

(18) Let L be a finite extension of a field K. Prove that L is a normal extension of K if and only if any two irreducible factors $g(x), h(x) \in$ L$[x]$ of an irreducible polynomial $f(x) \in$ K$[x]$ must of the same degrees, namely $deg\ g(x) = deg\ h(x)$.

(19) Let x, y be variables, and $a, b, c, d \in$ **Z** with

$$n = |det \begin{pmatrix} a & c \\ b & d \end{pmatrix}| = |ad - bc| \neq 0$$

Let L=C(x, y), and K=C$(x^a y^b, x^c y^d)$. Show that

(i) L is a finite extension of K, and [L: K]=n.

(ii) L is a Galois extension of K, and find $G(L/K)$.

(20) Let L be a cyclic extension of K of algebraic degree n. Show that for every positive integer $d \mid n$, there is a unique intermediate field of degree d over K.

(21) Let Ω be an algebraic closed field, L be a subfield which is an extension of K, and $\alpha \in \Omega$ with K$[\alpha]$ a Galois extension of K. Prove that the Galois group $G(K[\alpha]/K)$ = the Galois group $G(L[\alpha]/L)$ if and only if K$[\alpha] \cap$L =K.

§7 Solve Equation by Radicals

The method of solving a quadratic equation was known to the ancient Babylonian, Egyptian, Chinese, Indian and Greek. For the following standard equation,

$$ax^2 + bx + c = 0$$

the general solutions can be written as

$$x = \frac{-b \pm \sqrt{b^2 - 4ac}}{2a}$$

It is natural to seek the solutions of the cubic or higher equations. In the 16-th century, Cardano's formulas for the cubic and the quartic equation were known[5] (cf Examples 20. & 21.). The solutions of quadratic, cubic and quartic equations are general formulas involving radicals of the coefficients of the given equation. Since then a search was on for formula solutions involving only radicals of coefficients of equations of degree 5 or higher.

The insolvability of the quintic equation was due to Abel and Ruffini[6]. In Galois Theory, we have a complete and systematic treatment of solving equations by radicals. The climax of this section is Theorem 5.31. (Galois Theorem) which determines the solvability of an equation by radicals using group-theoretic criterion.

For the simplicity of our treatment, we assume that the field $\mathbf{K}$ is of **characteristic** 0, and fix an algebraic closure Ω of $\mathbf{K}$. All fields are subfields of Ω, and all elements are inside Ω. Note that the separability is free under our assumption, hence a field $\mathbf{L}$ is a Galois extension of a field $\mathbf{K}$ if and only if $\mathbf{L}$ is the splitting field of a polynomial $f(x)$ over $\mathbf{K}$. We introduce the following definitions.

Definition 5.18. *If* $\mathbf{L}$ *is the splitting field over* $\mathbf{K}$ *of the following equation, then* $\mathbf{L}$ *is called a **radical extension** of* $\mathbf{K}$,

$$x^n - a = 0, \quad n \text{ positive integer, } a \in \mathbf{K}.$$

∎

Definition 5.19. *If there exists a chain of extensions as follows,*

$$\mathbf{K} = \mathbf{K}_1 \subset \mathbf{K}_2 \subset \cdots \subset \mathbf{K}_i \subset \mathbf{K}_{i+1} \subset \cdots \subset \mathbf{K}_n$$

such that the splitting field $\mathbf{L}$ *of* $f(x)$ *over* $\mathbf{K}$ *is contained in* $\mathbf{K}_n$, *then we say that* $f(x)$ *can be solved by radicals.* ∎

Discussion

(1) The above Definition 5.19. gives a precise definition for the classical notion of a formula solution involving only the radicals of the coefficients of the equation. ∎

---

[5]Chinese Mathematician Wang Hsiao-Thung solved some numerical cubic equations in the 7-th century, the general solution of cubic equation was discovered by del Ferro, and later by Tartaglia, the solution of the quartic equation was discovered by Ferrari. The results of Tartaglia and Ferrari were published by Cardano in 1545, and they are known as Cardano's formulas.

[6]Abel solved the problem in 1821, Paolo Ruffini gave an imperfect solution in 1799. The theorem of insolvability of quintic equation is known as Abel-Ruffini Theorem.

Before we prove Theorem 5.31., we shall discuss two simple cases.

Theorem 5.27. *If* L *is a radical extension of* K, *then the Galois group* $G(L/K)$ *is solvable.*

Proof: Let L be the splitting field of the following equation,

$$(1) \qquad x^n - a = 0$$

(1) Let us consider the simple case that $a = 1$, namely we have the following equation,

$$x^n - 1 = 0$$

Recall Example 19., let ζ be a root of the above equation, and ζ is a generator of the cyclic group of all roots. Then it follows from Example 19. that L=K[ζ], and $G(K[\zeta]/K)$ is abelian, hence solvable.

(2) In general, let us assume that $a \neq 1$. Let α be a root of Eq (1). Then we have L=K $[\alpha, \alpha\zeta, \alpha\zeta^2, \cdots, \alpha\zeta^{n-1}]$=K $[\alpha, \zeta]$. Let S =K[ζ]. Then we have the following chain of Galois extensions.

$$\mathrm{K} \subset \mathrm{S} \subset \mathrm{L}$$

Corresponding to the above chain, we have a chain of normal subgroups as follows,

$$G(L/K) \triangleright G(L/S) \triangleright \{e\}$$

To show that $G(L/K)$ is solvable is to show that the following two quotient groups are abelian.

$$G(L/K)/G(L/S)(\approx G(S/K)), \quad G(L/S)/\{e\}(\approx G(L/S))$$

It follows from (1) that the first group is abelian. Let us show that the second group is abelian.

Let $\sigma \in G(L/S)$. Then $\sigma(\alpha)$ must be a root of Eq (1). Therefore, we have,

$$\sigma(\alpha) = \alpha\zeta^{j(\sigma)}, \quad 0 \leq j(\sigma) \leq n-1$$

Note that the action of σ is totally determined by $j(\sigma)$. Let us define a map $\tau : G(L/S) \to \mathbf{Z}_n$ the additive group of $\mathbf{Z}$ modulo $n\mathbf{Z}$ with

$$\tau(\sigma) = [j(\sigma)]$$

It is easy to see that

$$\sigma_1\sigma_2(\alpha) = \sigma_1(\alpha\zeta^{j(\sigma_2)}) = \sigma_1(\alpha)\zeta^{j(\sigma_2)} = \alpha\zeta^{j(\sigma_1)+j(\sigma_2)}$$

and τ is a group homomorphism. It is easy to see τ is injective. Therefore, $G(L/S)$ can be considered as a subgroup of $\mathbf{Z}_n$, and is an abelian group. ∎

Theorem 5.28. *Let* **L** *be the splitting field of a polynomial $f(x)$ over* **K**. *If $G(L/K)$ is a cyclic group, then $f(x)$ can be solved by radicals.*

Proof: We shall construct a chain of radical extensions,

$$\mathbf{K} \subset \mathbf{K}_1 \subset \mathbf{K}_2$$
$$\mathbf{L} \subset \mathbf{K}_2$$

(1) Let us consider the following equation,

$$x^n - 1 = 0$$

Recall Example 19., let ζ be a root of the above equation, and ζ is a generator of the cyclic group of all roots. Then it follows from Example 19. that

$$x^n - 1 = \prod_{i=0}^{n-1}(x - \zeta^i)$$

Let $\mathbf{K}_1 = \mathbf{K}[\zeta]$.

(2) We shall construct $\mathbf{K}_2$. Let $\mathbf{L}=\mathbf{K}[\alpha]$ for some suitable α. Let $\mathbf{K}_2= \mathbf{K}[\zeta,\alpha]$. Clearly, we have $\mathbf{L}=\mathbf{K}[\alpha] \subset \mathbf{K}[\zeta,\alpha] = \mathbf{K}_2$. To prove our theorem, it suffices to show that $\mathbf{K}_2$ is a radical extension of $\mathbf{K}_1$.

(3) It follows from Theorem 5.26. that the Galois group $G(K_2/K_1)$ is a subgroup of the Galois group $G(L/K)$ which is a cyclic group, therefore, $G(K_2/K_1)$ is a cyclic group. Let $G(K_2/K_1) = \{1, \sigma, \cdots, \sigma^{m-1}\}$ where $m = o(G(K_2/K_1))$ and $m \mid n$.

(4) We shall employ a method, the diagonalization of a cyclic linear action σ, which is commonly used in linear algebra.

Let $\eta = \zeta^{n/m}$. We have $\eta^m = 1$ and

$$x^m - 1 = \prod_{i=0}^{m-1}(x - \eta^i)$$

Note that σ is a linear transformation from the $\mathbf{K}_1$-vector space $\mathbf{K}_2$ to itself. The minimal polynomial, which is the last invariant factor, $c_\ell(x)$, of σ is the following,

$$x^m - 1 = \prod_{j=1}^{m}(x - \eta^j) = 0$$

Further note that $\mathbf{K}_2$ is an m-dimensional vector space. Therefore, the above is the characteristic polynomial of σ. Since there are primitive m-th roots of unity in $\mathbf{K}_1$, then there are characteristic vectors, $\beta_j \neq 0$, of the characteristic values η^j. Therefore, we have

$$\sigma^j(\beta_1) = \eta^j\beta_1, \quad \forall\, 0 \le j \le m-1$$
$$\sigma(\beta_1^m) = (\eta\beta_1)^m = \eta^m\beta^m = \beta^m$$

Note then we have $\beta_1^m \in \mathbf{K}_1$, and β_1 satisfies the following equation over $\mathbf{K}_1$,

$$x^m - \beta_1^m = \prod_{j=1}^{m}(x - \eta^j\beta_1)$$

Let $\mathbf{L}=\mathbf{K}_1[\beta_1]$. Then $\mathbf{L}$ is a radical extension of $\mathbf{K}_1$.

(5) We claim $\mathbf{L}=\mathbf{K}_2$. Note then our theorem is established. It is clear that $\beta_1 \in \mathbf{K}_2$ and $\mathbf{L} \subset \mathbf{K}_2$. On the other hand, none of the σ^j's are the identity map on $\mathbf{L}$, therefore, we have

$$m \leq o(G(L/K_1)) = [\mathbf{L}: \mathbf{K}_1] \leq [\mathbf{K}_2 : \mathbf{K}_1] = m$$

We conclude that $\mathbf{L}=\mathbf{K}_2$. ∎

The preceding Theorem 5.27. & Theorem 5.28. are the special cases of Theorem 5.31 (Galois Theorem). To prove Theorem 5.31., we need the following lemma.

Lemma 1. *For any given chain of radical extensions as follows,*

$$\mathbf{K} = \mathbf{K}_1 \subset \mathbf{K}_2 \subset \cdots \subset \mathbf{K}_n$$

there exists another chain of radical extensions as follows,

$$\mathbf{K} = \mathbf{L}_1 \subset \mathbf{L}_2 \subset \cdots \subset \mathbf{L}_m$$

such that $\mathbf{K}_n \subset \mathbf{L}_m$, and $\mathbf{L}_m$ is a Galois extension of $\mathbf{K}$.

Proof: We shall use induction on the length n of the first chain. If $n = 1$, then we let $m = 1$. Suppose that $n > 1$, and we have the following chain of radical extensions,

$$\mathbf{K} = \mathbf{L}_1 \subset \mathbf{L}_2 \subset \cdots \subset \mathbf{L}_{m'}$$

such that $\mathbf{K}_{n-1} \subset \mathbf{L}_{m'}$, and $\mathbf{L}_{m'}$ is a Galois extension of $\mathbf{K}$. Let $\mathbf{L}_{m'} = \mathbf{K}[\alpha]$ for some suitable α.

Since $\mathbf{K}_n$ is a radical extension of $\mathbf{K}_{n-1}$, then let $\mathbf{K}_n$ be the splitting field of the following equation over $\mathbf{K}_{n-1}$,

$$x^\ell - a = 0, \quad a \in \mathbf{K}_{n-1} \subset \mathbf{L}_{m'}$$

Let $f(x)$ be the product of all conjugates of the above equation over $\mathbf{K}$, namely

$$f(x) = \prod_{\sigma \in G(L_{m'}/K)} (x^\ell - \sigma(a))$$

Clearly, we have

(i) $\sigma(a) \in \mathbf{L}_{m'}$.
(ii) $f(x) \in \mathbf{K}[x]$.

Let us start with $\mathbf{L}_{m'}$ and form radical extensions by joining the roots of the factors $x^{\ell} - \sigma(a)$ of $f(x)$ one by one. Then we have the following chain of radical extensions,

$$\mathbf{K} = \mathbf{L}_1 \subset \mathbf{L}_2 \subset \cdots \subset \mathbf{L}_{m'} \subset \cdots \subset \mathbf{L}_m$$

Clearly, we have $\mathbf{K}_n \subset \mathbf{L}_m$. It suffices to show that $\mathbf{L}_m$ is a Galois extension of $\mathbf{K}$.

Let $\mathbf{L}_{m'} = \mathbf{K}[\alpha]$ and the minimal polynomial of α over $\mathbf{K}$ be $g(x)$. Then it is clear that $\mathbf{L}_m$ is the splitting field of $f(x)g(x)$ over $\mathbf{K}$. Therefore, $\mathbf{L}_m$ is a Galois extension of $\mathbf{K}$, and our lemma is established. ∎

Now we shall prove the first half of Theorem 5.31..

Theorem 5.29. *Let* $\mathbf{L}$ *be the splitting field of* $f(x)$ *over* $\mathbf{K}$. *If* $f(x)$ *can be solved by radicals, then the Galois group* $G(L/K)$ *is solvable.*

Proof: (1) Note that $\mathbf{L}$ is a Galois extension of $\mathbf{K}$. It follows from Definition 5.19. that there exists a chain of radical extensions as follows,

$$\mathbf{K} = \mathbf{K}_1 \subset \mathbf{K}_2 \subset \cdots \subset \mathbf{K}_i \subset \mathbf{K}_{i+1} \subset \cdots \subset \mathbf{K}_n$$

such that $\mathbf{L} \subset \mathbf{K}_n$. It follows from Lemma 1. that there exists another chain of radical extensions as follows,

$$\mathbf{K} = \mathbf{L}_1 \subset \mathbf{L}_2 \subset \cdots \subset \mathbf{L}_m$$

such that $\mathbf{K}_n \subset \mathbf{L}_m$, and $\mathbf{L}_m$ is a Galois extension of $\mathbf{K}$. Therefore, we have $\mathbf{L} \subset \mathbf{L}_m$. It follows from the Fundamental Theorem of Galois Theory that

$$G(L_m/K) \triangleright G(L_m/L) \triangleright \{e\}$$
$$G(L/K) \approx G(L_m/K)/G(L_m/L)$$

It follows from Theorem 2.22. that it suffices to prove $G(L_m/K)$ is a solvable group. Note that the quotient group of a solvable group is solvable.

(2) It follows from the Fundamental Theorem of Galois Theory that we have the following chain of groups,

$$G(L_m/K) \supset G(L_m/L_2) \supset \cdots \supset G(L_m/L_i) \supset G(L_m/L_{i+1}) \supset \cdots \supset G(L_m/L_m) = \{e\}$$

Since $\mathbf{L}_{i+1}$ is a radical extension of $\mathbf{L}_i$, then $\mathbf{L}_{i+1}$ is a Galois extension of $\mathbf{L}_i$. Therefore, we have

$$G(L_m/K) \triangleright G(L_m/L_2) \triangleright \cdots \triangleright G(L_m/L_i) \triangleright G(L_m/L_{i+1}) \triangleright \cdots \triangleright G(L_m/L_m) = \{e\}$$
$$G(L_{i+1}/L_i) \approx G(L_m/L_i)/G(L_m/L_{i+1})$$

It follows from Theorem 5.27. that the quotient groups $G(L_{i+1}/L_i)$ are all solvable. It is easy to refine the above normal series such that all quotient groups are commutative. Therefore, $G(L_m/K)$ is solvable. ∎

To prove the other half of Theorem 5.31., we need the following lemma.

Lemma 2. *Let* $\mathbf{L}\supset \mathbf{S}\supset \mathbf{K}$ *be field extensions. Suppose that there are a chain of radical extensions of* $\mathbf{K}$ *and a chain of radical extensions of* $\mathbf{S}$ *as follows,*

$$\mathbf{K} = \mathbf{S}_1 \subset \mathbf{S}_2 \subset \cdots \subset \mathbf{S}_n$$
$$\mathbf{S} = \mathbf{L}_1 \subset \mathbf{L}_2 \subset \cdots \subset \mathbf{L}_m$$

such that $\mathbf{S}\subset \mathbf{S}_n$, *and* $\mathbf{L}\subset \mathbf{L}_m$. *Then there exists a chain of radical extensions as follows,*

$$\mathbf{K} = \mathbf{K}_1 \subset \mathbf{K}_2 \subset \cdots \subset \mathbf{K}_\ell$$

such that $\mathbf{L}\subset \mathbf{K}_\ell$.

Proof: Let $\mathbf{L}_{i+1}$ be the splitting field of the following equation over $\mathbf{L}_i$,

$$f_i(x) = x^{m_i} - a_i, \quad a_i \in \mathbf{L}_i$$

We shall construct the chain of radical extensions $\mathbf{K}_j$. Let $\mathbf{K}_j = \mathbf{S}_j$ for $j = 1, 2, \cdots, n$, and $\mathbf{K}_{n+1}$ be the splitting field of $f_1(x)$ over $\mathbf{K}_n$, $\cdots$, $\mathbf{K}_{n+j}$ be the splitting field of $f_j(x)$ over $\mathbf{K}_{n+j-1}$, $\cdots$, $\mathbf{K}_{n+m-1}$ be the splitting field of $f_{m-1}(x)$ over $\mathbf{K}_{n+m-2}$. It is easy to see that we have the following chain of radical extensions,

$$\mathbf{K} = \mathbf{K}_1 \subset \mathbf{K}_2 \subset \cdots \subset \mathbf{K}_{n+m-1}$$

such that $\mathbf{L}\subset \mathbf{K}_{n+m-1}$. ∎

Theorem 5.30. *Let* $\mathbf{L}$ *be the splitting field of a polynomial* $f(x) \in \mathbf{K}[x]$. *If* $G(L/K)$ *is solvable, then* $f(x)$ *can be solved by radicals.*

Proof: Since $G(L/K)$ is solvable, then there is a normal series of subgroups such that the quotient groups are commutative. It is easy to refine the normal series to the following such that the quotient groups are cyclic,

$$G(L/K) \triangleright H_{n-1} \triangleright H_{n-2} \triangleright \cdots \triangleright H_0 = \{e\}$$

It follows from the Fundamental Theorem of Galois Theory that there corresponds a chain of cyclic extensions as follows,

$$\mathbf{K} \subset \mathbf{K}_1 \subset \mathbf{K}_2 \subset \cdots \subset \mathbf{L}$$

Let $\mathbf{K}_{i+1}$ be the splitting field of $f_{i+1}(x)$ over $\mathbf{K}_i$. It follows from Theorem 5.28. and Definition 5.19. that there are the following chains of radical extensions with $\mathbf{S}_{jn_j} \supset \mathbf{K}_{j+1}$,

$$\mathbf{K} = \mathbf{S}_{01} \subset \mathbf{S}_{02} \subset \cdots \subset \mathbf{S}_{0n_0}$$
$$\mathbf{K}_1 = \mathbf{S}_{11} \subset \mathbf{S}_{12} \subset \cdots \subset \mathbf{S}_{1n_1}$$
$$\mathbf{K}_2 = \mathbf{S}_{21} \subset \mathbf{S}_{22} \subset \cdots \subset \mathbf{S}_{2n_2}$$
$$\cdots\cdots\cdots$$
$$\mathbf{L} = \mathbf{S}_{n1} \subset \mathbf{S}_{n2} \subset \cdots \subset \mathbf{S}_{nn_n}$$

Note that $\mathbf{K}_1 \subset \mathbf{S}_{0n_0}$, and $\mathbf{S}_{1n_1} \subset \mathbf{S}_{1n_1}$. It follows from Lemma 2. that there is a chain of radical extensions of $\mathbf{K}$ as follows,

$$\mathbf{K} = \mathbf{S'}_{01} \subset \mathbf{S'}_{02} \subset \cdots \subset \mathbf{S'}_{0n'_0}$$

such that $\mathbf{K}_2 \subset \mathbf{S}_{1n_1} \subset \mathbf{S'}_{0n'_0}$. Let us consider the above chain and the preceding chain starting with $\mathbf{K}_2$. Using Lemma 2 again, we may combine them into one. In finite steps, we construct a chain of radical extensions,

$$\mathbf{K} = \mathbf{K}_1 \subset \mathbf{K}_2 \subset \cdots \subset \mathbf{K}_\ell$$

such that $\mathbf{L} \subset \mathbf{K}_\ell$. Therefore, $f(x)$ can be solved by radicals. ∎

We have our main theorem of this section.

Theorem 5.31. (Galois Theorem). *Let* $\mathbf{L}$ *be the splitting field of* $f(x) \in \mathbf{K}[x]$. *Then* $f(x)$ *can be solved by radicals if and only if the Galois group* $G(L/K)$ *is solvable.*

Proof: (⇒) Theorem 5.29.. (⇐) Theorem 5.30.. ∎

Example 20. *We shall deduce the Cardano Formula of the cubic equations.*

(1) (i) We shall consider the generic case with the indeterminate coefficients. Let $\mathbf{F}$ *be a field containing a primitive third root of unity* ω *with* $\omega \neq 1$, *and* $\omega^3 = 1$. *Let* x_1, x_2, x_3 *be variables with the elementary symmetric polynomials* $\Theta_1, \Theta_2, \Theta_3$ *as follows,*

$$\begin{aligned}
\Theta_1 &= x_1 + x_2 + x_3 \\
\Theta_2 &= x_1x_2 + x_2x_3 + x_3x_1 \\
\Theta_3 &= x_1x_2x_3
\end{aligned}$$

Let $\mathbf{K}$=$\mathbf{F}[\Theta_1, \Theta_2, \Theta_3]$. *Let* $\mathbf{L}$ *be the splitting field of the following polynomial* $f(x)$ *over* $\mathbf{K}$,

$$f(x) = x^3 - \Theta_1 x^2 + \Theta_2 x - \Theta_3 = (x - x_1)(x - x_2)(x - x_3)$$

It is easy to see that $\mathbf{L}$=$\mathbf{F}[x_1, x_2, x_3]$, *and the Galois group* $G(L/K) = S_3$ *the symmetric group on three digits. We have the following normal series in* S_3,

$$S_3 \triangleright A_3 \triangleright \{e\}$$

Note that the quotient groups of the above series are cyclic. Therefore, it follows from Theorem 5.31. that the equation $f(x)$ *can be solved by radicals over* $\mathbf{K}$. *Let us carry out the solving process.*

(ii) Corresponding to the above normal series of S_3, *there is a chain of field extensions,*

$$\begin{aligned}
&\mathbf{K} \subset \mathbf{S} \subset \mathbf{L} \\
&[\mathbf{S} : \mathbf{K}] = o(S_3/A_3) = 2 \\
&[\mathbf{L} : \mathbf{S}] = o(A_3) = 3
\end{aligned}$$

Let the discriminant Δ *be defined as*

$$\Delta = (x_2 - x_1)(x_3 - x_2)(x_3 - x_2)$$

Then it is easy to see that $\Delta \in \mathbf{S}\backslash\mathbf{K}$, *and* $\mathbf{S} = \mathbf{K}[\Delta]$. *The minimal polynomial* $g(x)$ *of* Δ *over* $\mathbf{K}$ *is the following,*

$$g(x) = x^2 - \Delta^2 = x^2 + 4\Theta_1^3\Theta_3 - \Theta_1^2\Theta_2^2 - 18\Theta_1\Theta_2\Theta_3 + 4\Theta_2^3 + 27\Theta_3^2$$

It is easy to see

$$\Delta = \sqrt{-4\Theta_1^3\Theta_3 + \Theta_1^2\Theta_2^2 + 18\Theta_1\Theta_2\Theta_3 - 4\Theta_2^3 - 27\Theta_3^2}$$

Let $\sigma = (1,2,3) \in A_3$, *and*

$$\beta_1 = x_1 + \omega^2\sigma(x_1) + \omega\sigma^2(x_1) = x_1 + \omega^2 x_2 + \omega x_3$$
$$\beta_2 = x_1 + \omega\sigma(x_1) + \omega^2\sigma^2(x_1) = x_1 + \omega x_2 + \omega^2 x_3$$

Then we have

$$\sigma(\beta_1) = \omega\beta_1, \qquad \sigma^2(\beta_1) = \omega^2\beta_1$$
$$\sigma(\beta_2) = \omega^2\beta_2, \qquad \sigma^2(\beta_2) = \omega\beta_2$$
$$\sigma(\beta_1\beta_2) = \beta_1\beta_2, \qquad \sigma(\beta_i^3) = \beta_i^3, \quad i = 1,2$$

It follows that $\mathbf{K}[\beta_1] = \mathbf{K}[\beta_2] = \mathbf{K}[\Delta] = \mathbf{S}$, *and* $\beta_1^3, \beta_2^3, \beta_1\beta_2 \in \mathbf{S}$.

(iii) Note that

$$\omega^2 + \omega + 1 = 0, \quad \omega^2 + \frac{1}{2} = -(\omega + \frac{1}{2})$$
$$\omega = -\frac{1 - \sqrt{-3}}{2}$$

We write down β_1^3 *explicitly as*

$$\beta_1^3 = \Theta_1^3 - \frac{9}{2}(\Theta_1\Theta_2 - 3\Theta_3) + 3(\omega + \frac{1}{2})\Delta$$

Therefore, we have the following expression for β_1 *and* $\beta_1\beta_2$,

$$\beta_1 = \sqrt[3]{\Theta_1^3 - \frac{9}{2}(\Theta_1\Theta_2 - 3\Theta_3) + 3(\omega + \frac{1}{2})\Delta}\ \omega^j, \quad j = 0,1,2$$
$$\beta_1\beta_2 = \Theta_1^2 - 3\Theta_2$$

Therefore, once the value of β_1 *is selected, then the value of* β_2 *is fixed. Recall that we have the following system of three equations,*

$$\begin{cases} \Theta_1 = x_1 + x_2 + x_3 \\ \beta_1 = x_1 + \omega^2 x_2 + \omega x_3 \\ \beta_2 = x_1 + \omega x_2 + \omega^2 x_3 \end{cases}$$

The solutions of the above system are the following

$$\begin{cases} x_1 = \frac{1}{3}(\Theta_1 + \beta_1 + \beta_2) \\ x_2 = \frac{1}{3}(\Theta_1 + \omega^2\beta_1 + \omega\beta_2) \\ x_3 = \frac{1}{3}(\Theta_1 + \omega\beta_1 + \omega^2\beta_2) \end{cases}$$

(2) Let us consider the special cases of the constant coefficients. We are given the following equation,

$$x^3 - \theta_1 x^2 + \theta_2 x - \theta_3 = 0$$

over a field **K**. *We shall 'kill' the coefficient θ_1 by a changing variable of the form $x \to x + \theta_1/3$. We may assume that $\theta_1 = 0$. In the previous discussions, we make the following substitutions: $\Theta_1 \mapsto 0, \Theta_2 \mapsto \theta_2, \Theta_3 \mapsto \theta_3$. Then we have the following simplified situation,*

$$\Delta \mapsto \overline{\Delta} = \sqrt{-\theta_2^3 - 27\theta_3^2}$$

$$\beta_1^3 \mapsto \overline{\beta}_1^3 = \frac{27}{2}\theta_3 + \frac{3}{2}\sqrt{-3}\,\overline{\Delta}$$

$$\beta_2 \mapsto \overline{\beta}_2^3 = \frac{27}{2}\theta_3 - \frac{3}{2}\sqrt{-3}\,\overline{\Delta}$$

$$x_1 = \sqrt[3]{\frac{\theta_3}{2} + \sqrt{\frac{\theta_3^2}{4} + \frac{\theta_2^3}{27}}} + \sqrt[3]{\frac{\theta_3}{2} - \sqrt{\frac{\theta_3^2}{4} + \frac{\theta_2^3}{27}}}$$

$$x_2 = \sqrt[3]{\frac{\theta_3}{2} + \sqrt{\frac{\theta_3^2}{4} + \frac{\theta_2^3}{27}}\omega^2} + \sqrt[3]{\frac{\theta_3}{2} - \sqrt{\frac{\theta_3^2}{4} + \frac{\theta_2^3}{27}}\omega}$$

$$x_1 = \sqrt[3]{\frac{\theta_3}{2} + \sqrt{\frac{\theta_3^2}{4} + \frac{\theta_2^3}{27}}\omega} + \sqrt[3]{\frac{\theta_3}{2} - \sqrt{\frac{\theta_3^2}{4} + \frac{\theta_2^3}{27}}\omega^2}$$

The above are the so-called Cardano Formula of the cubic equations. ∎

Example 21. *We shall deduce the Cardano's Formula of the quartic equations.*

(1) (i) We shall consider that generic case of the indeterminate coefficients. Let **F** *be a field containing a primitive third root of unity ω with $\omega \neq 1$, and $\omega^3 = 1$, and the primitive 4-th root of unity i with $i^4 = 1, i^2 \neq 1$. Let x_1, x_2, x_3, x_4 be variables with the elementary symmetric polynomials $\Theta_1, \Theta_2, \Theta_3, \Theta_4$ as follows,*

$$\begin{aligned} &\Theta_1 = x_1 + x_2 + x_3 + x_4 \\ &\Theta_2 = x_1x_2 + x_2x_3 + x_3x_1 + x_1x_4 + x_2x_4 + x_3x_4 \\ &\Theta_3 = x_1x_2x_3 + x_1x_2x_4 + x_1x_3x_4 + x_2x_3 + x_4 \\ &\Theta_4 = x_1x_2x_3x_4 \end{aligned} \tag{1}$$

Let $\mathbf{K}=\mathbf{F}[\Theta_1,\Theta_2,\Theta_3,\Theta_4]$. *Let* $\mathbf{L}$ *be the splitting field of the following polynomial* $f(x)$ *over* $\mathbf{K}$,

$$f(x)=x^4-\Theta_1x^3+\Theta_2x^2-\Theta_3x+\Theta_4=(x-x_1)(x-x_2)(x-x_3)(x-x_4) \tag{2}$$

It is easy to see that $\mathbf{L}=\mathbf{F}[x_1,x_2,x_3,x_4]$, *and the Galois group* $G(L/K)=S_4$ *the symmetric group on four digits. We have the following normal series in* S_3,

$$S_4 \rhd A_3 \rhd K_4 \rhd N \rhd \{e\}$$

where $K_4=\{e,(12)(34),(13)(24),(14)(23)\}$ *is the Klein four-group, and* $N=\{e,(12)(34)\}$. *Note that the quotient groups of the above series are cyclic. Therefore, it follows from Theorem 5.31. that the equation* $f(x)$ *can be solved by radicals over* $\mathbf{K}$. *Let us carry out the solving process.*

(ii) Corresponding to the above normal series of S_4, *there is a chain of field extensions,*

$$\begin{aligned}&\mathbf{K}\subset\mathbf{R}_1\subset\mathbf{R}_2\subset\mathbf{R}_3\subset\mathbf{L}\\&[\mathbf{R}_2:\mathbf{K}]=o(R_4/K_4)=6\\&[\mathbf{L}:\mathbf{R}_2]=o(K_4)=4\end{aligned}$$

(iii) Let y_1,y_2,y_3 *be defined as follows,*

$$\begin{aligned}y_1&=x_1x_2+x_3x_4\\y_2&=x_1x_3+x_2x_4\\y_3&=x_1x_4+x_2x_3\end{aligned} \tag{3}$$

It is easy to see that y_1,y_2,y_3 *are invariant under the actions of* K_4. *Therefore,* $y_1,y_2,y_3\in\mathbf{R}_2$. *We claim that* $\mathbf{K}[y_1,y_2,y_3]=\mathbf{R}_2$. *Note that we already have* $\mathbf{K}[y_1,y_2,y_3]\subset\mathbf{R}_2$.

Note that y_1,y_2,y_3 *are roots of the following equation,*

$$\begin{aligned}g(x)&=(x-y_1)(x-y_2)(x-y_3)\\&=x^3-(y_1+y_2+y_3)x^2+(y_1y_2+y_1y_3+y_2y_3)x-y_1y_2y_3\\&=x^3-\Theta_2+(\Theta_1\Theta_3-4\Theta_4)x-\Theta_4(\Theta_1^2-4\Theta_2)-\Theta_3^2\end{aligned} \tag{4}$$

Therefore, $g(x)\in\mathbf{K}[x]$, *and* $\mathbf{K}[y_1,y_2,y_3]$ *is the splitting field of* $g(x)$ *over* $\mathbf{K}$. *We shall compute the Galois group. The restrictions of the elements of* S_4 *to* $\mathbf{K}[y_1,y_2,y_3]$ *form the permutation group* S_3 *whose order is 6. Therefore, we have*

$$[\mathbf{K}[y_1,y_2,y_3]:\mathbf{K}]\geq 6$$

It follows that $\mathbf{K}[y_1,y_2,y_3]=\mathbf{R}_2$.

(iv) Using the method of Example 19., we may solve for y_1, y_2, y_3 from the expression of $g(x)$. Recall Θ_1, and let $\beta_1, \beta_2, \beta_3$ be defined as follows,

$$(5)\quad \begin{cases} \Theta_1 = x_1 + x_2 + x_3 + x_4 \\ \beta_1 = x_1 + x_2 - x_3 - x_4 \\ \beta_2 = x_1 - x_2 + x_3 - x_4 \\ \beta_3 = x_1 - x_2 - x_3 + x_4 \end{cases}$$

It is easy to see that for any $\sigma \in K_4$, we always have $\sigma(\beta_i) = \pm\beta_i$, and $\sigma(\beta_1\beta_2\beta_3) = \beta_1\beta_2\beta_3$. In fact, we have

$$(6)\quad \begin{aligned} \beta_1^2 &= \Theta_1^2 - 4\Theta_2 + 4y_1 \\ \beta_2^2 &= \Theta_1^2 - 4\Theta_2 + 4y_2 \\ \beta_1^2 &= \Theta_1^2 - 4\Theta_2 + 4y_3 \\ \beta_1\beta_2\beta_3 &= 8\Theta_3 - 4\Theta_1\Theta_2 + \Theta_1^3 \end{aligned}$$

Solving the system of equations (6), we have

$$(7)\quad \begin{cases} x_1 = \frac{1}{4}(\Theta_1 + \beta_1 + \beta_2 + \beta_3) \\ x_2 = \frac{1}{4}(\Theta_1 + \beta_1 - \beta_2 - \beta_3) \\ x_3 = \frac{1}{4}(\Theta_1 - \beta_1 + \beta_2 - \beta_3) \\ x_4 = \frac{1}{4}(\Theta_1 - \beta_1 - \beta_2 + \beta_3) \end{cases}$$

(2) Let us consider the special cases of the constant coefficients. We are given the following equation,

$$x^4 - \theta_1 x^3 + \theta_2 x^2 - \theta_3 x + \theta_4 = 0$$

over a field **K**. We shall 'kill' the coefficient θ_1 by a changing variable of the form $x \to x + \theta_1/4$. We may assume that $\theta_1 = 0$. In the previous discussions, we make the following substitutions: $\Theta_1 \mapsto 0, \Theta_2 \mapsto \theta_2, \Theta_3 \mapsto \theta_3, \Theta_4 \mapsto \theta_4$. Then we have the following simplified situation: the defining equation for y_1, y_2, y_3 is

$$\overline{g}(x) = x^3 - \theta_2 x^2 - 4\theta_4 x + 4\theta_2\theta_4 - \theta_3^2$$

and $\beta_1, \beta_2, \beta_3$ are given by

$$\begin{aligned} \beta_1^2 &= -4\theta_2 + 4y_1 \\ \beta_2^2 &= -4\theta_2 + 4y_2 \\ \beta_1^2 &= -4\theta_2 + 4y_3 \\ \beta_1\beta_2\beta_3 &= 8\theta_3 \end{aligned}$$

The solutions are given by

$$\begin{cases} x_1 = \frac{1}{4}(\beta_1 + \beta_2 + \beta_3) \\ x_2 = \frac{1}{4}(\beta_1 - \beta_2 - \beta_3) \\ x_3 = \frac{1}{4}(-\beta_1 + \beta_2 - \beta_3) \\ x_4 = \frac{1}{4}(-\beta_1 - \beta_2 + \beta_3) \end{cases}$$

∎

We shall show that there are quintic equations which can not be solved by radicals. In general, the Galois group of the splitting field $\mathbf{L}$ of any quintic equation $f(x)$ over the rational field $\mathbf{Q}$ is most likely to be S_5 which is non-solvable by Theorem 2.27.. Instead of this general argument, we shall give a concrete example $f(x)$ to show that in some case, the Galois group of its splitting field over the rational field $\mathbf{Q}$ is S_5. Then it follows from Theorem 5.31. that the polynomial $f(x)$ can not be solved by radicals.

We need the following lemma.

Lemma 3. *Let p be a prime number, and G a subgroup of S_p. If the following two conditions are satisfied, then $G = S_p$,*

(1) *G is transitive, namely, for any $i, j \in \{1, 2, \cdots, p\}$, there exists $\sigma \in G$ such that $\sigma(i) = j$.*
(2) *There is a transpose $(k, \ell) \in G$.*

Proof: (1) We shall define an equivalence relation in the set $\{1, 2, \cdots, p\}$ of p elements as follows,

$$i \sim j \Leftrightarrow i = j \text{ or } (i, j) \in G$$

We have

(i) Reflexion: $i \sim i$.
(ii) Symmetry: $i \sim j \Leftrightarrow j \sim i$.
(iii) Transition: $i \sim j,\ j \sim k \Rightarrow (j, k)(i, j)(j, k) = (i, k) \in G \Rightarrow i \sim k$.

(2) Let $E(i) = \{j : i \sim j\}$ the equivalence class of i. Recall that for any $\sigma \in S_p$, the corresponding inner automorphism τ_σ is defined as

$$\tau_\sigma((i, j)) = \sigma(i, j)\sigma^{-1} = (\sigma(i), \sigma(j))$$

Therefore, it is easy to see that the inner automorphism τ_σ will induce an injection from $E(i)$ to $E(\sigma(i))$. Replacing σ by σ^{-1}, we see at once that the said injection is a bijection. Therefore, all equivalence sets $E(i)$ have the same cardinalities. Since the total cardinality is a prime number p. Every $E(i)$ is either a singleton or the whole set. We already assume that there is a transpose $(i, j) \in G$, hence every $E(i)$ can not be a singleton. Therefore, there is only one equivalence set, and all transposes are in G. It follows from Theorem 2.23. that $G = S_p$.

∎

Example 22. (Insolvability of Some Quintic Equations). *Let* L *be the splitting field of the following polynomial* $f(x)$ *over the rational field* $\mathbf{Q}$,

$$f(x) = 2x^5 - 10x + 5$$

It follows from Eisenstein Criterion that the above polynomial is irreducible. Let $G = G(L/\mathbf{Q})$. *It follows from Corollary of Theorem 5.24. that* G *is transitive on the five roots of* $f(x)$. *We claim that there is a transpose in* G. *Note then it follows from Lemma 3. that* $G = S_5$ *is a non-solvable group. Therefore, it follows from Theorem 5.31. that the polynomial* $f(x)$ *can not be solved by radicals.*

We shall use Calculus to study the graph of the polynomial $f(x)$. *We shall show that* $f(x)$ *has three real roots and two non-real roots.*

Let us consider the first and the second derivatives of $f(x)$. *We have*

$$f'(x) = 10x^4 - 10 = 10(x^4 - 1)$$
$$f''(x) = 40x^3$$

Note that $f'(x) = 0$ *has two real roots* ± 1, $f''(1) > 0, f''(-1) < 0, f(1) = -3$, $f(-1) = 13$. *We have a local minimal point at* $x = 1$, *and a local maximal point at* $x = -1$. *Further note that*

$$lim_{x \to \infty} f(x) = \infty$$
$$lim_{x \to -\infty} f(x) = -\infty$$

The graph of $f(x)$ *is as follows,*

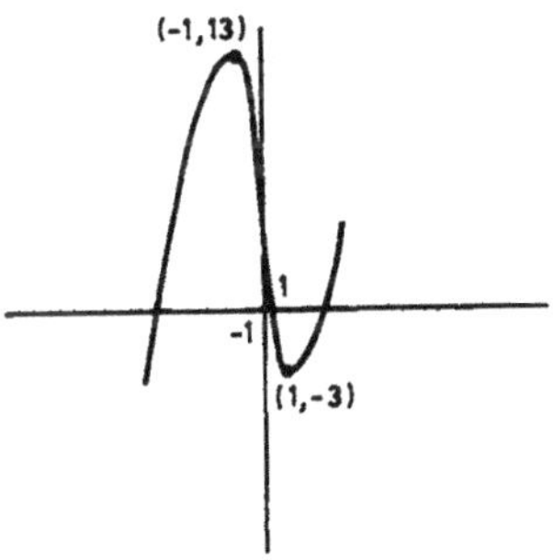

Therefore, $f(x)$ has precisely three real roots. Let $a+bi$ be another complex root of $f(x)$ where $b \neq 0$. Then we have

$$0 = \overline{f(a+bi)} = f(\overline{a+bi}) = f(a-bi)$$

and $a-bi$ is the fifth root of $f(x)$. We conclude that $f(x)$ has three real roots $\alpha_1, \alpha_2, \alpha_3$, and two non-real roots $\alpha_4 = a+bi$, $\alpha_5 = a-bi$.

The complex conjugation, which is an automorphism of the complex field $\mathbf{C}$, is a $\mathbf{Q}$-automorphism of $\mathbf{L}$. Note that the complex conjugation which will fix all three real roots, $\alpha_1, \alpha_2, \alpha_3$, and interchange the two non-real roots, α_4, α_5, is an element in $G = G(L/\mathbf{Q})$. Therefore, G has a transpose as we claimed. We have shown that G is isomorphic to S_5. Therefore, the given equation can not solved by radical. ∎

Exercises

(1) Let q be a prime number which is different than the characteristic of a field $\mathbf{K}$, and $a \in \mathbf{K}$. Show that $x^q - a$ is either irreducible over $\mathbf{K}$, or has a root in $\mathbf{K}$.

(2) Let q be a prime number, and $a \in \mathbf{Q}$. Suppose that $f(x) = x^q - a$ is irreducible in $\mathbf{Q}[x]$. Let $\mathbf{L}$ be the splitting field of $f(x)$ over $\mathbf{Q}$ in $\mathbf{C}$. Show that the Galois group $G(L/\mathbf{Q})$ is isomorphic to the group of maps of $\mathbf{Z}_q$ of the following form,

$$\sigma_{[k],[\ell]}([i]) = [ki + \ell], \qquad [k] \neq 0$$

(3) Let $\theta_1, \theta_2, \cdots, \theta_n$ be n variables. Prove that the polynomial

$$f(x) = x^n - \theta_1 x^{n-1} + \cdots + (-1)^i \theta_i x^{n-i} + \cdots + (-1)^n \theta_n$$

is a separable polynomial over $\mathbf{K}(\theta_1, \cdots, \theta_n)$. Let $\mathbf{L}$ be the splitting field of $f(x)$ in some algebraic closure of $\mathbf{K}(\theta_1, \cdots, \theta_n)$. Show that the Galois group is S_n.

(4) Let $\mathbf{L}$ be a Galois extension of a field $\mathbf{K}$ of characteristic 0. Assume that every intermediate field is a Galois extension of $\mathbf{K}$. Show that $\mathbf{L}$ is a radical extension of $\mathbf{K}$.

(5) Is there a Galois extension $\mathbf{L}$ over $\mathbf{Q}$ of degree 3?

8 Field Polynomial and Field Discriminant

We shall use linear algebra to study field extensions. Let **K** be a field, and **L** a finite field extension of **K** with basis $\{w_1, \cdots, w_n\}$. For any $a \in \mathbf{L}$, there corresponds a linear transformation A defined by the multiplication of a as follows,

$$A(\alpha) = a\alpha$$

Let

$$aw_i = \sum_j a_{ij} w_j$$

Then the matrix form, (a_{ij}), of A can be used to give the following definition,

Definition 5.20. *The characteristic polynomial $\chi_A(x) = det\ (xI - A)$ of A is called the field polynomial $F(x)$ of a.* ∎

Discussion

(1) The field polynomial $F(x)$ is independent of the vector space basis $\{w_1, \cdots, w_n\}$ selected for **L** over **K**.

(2) The field polynomial $F(x)$ depends on the element a, and the fields **L**, **K**. Note that we have $deg\ F(x) = [\mathbf{L} : \mathbf{K}] = n$. The field polynomial of 1, when 1 is considered as an element in **K**, is $x - 1$, and when 1 is considered as an element in **L**, is $(x-1)^n$ respectively.

(3) Clearly we have

$$aI \begin{pmatrix} w_1 \\ w_2 \\ \cdot \\ \cdot \\ w_n \end{pmatrix} = A \begin{pmatrix} w_1 \\ w_2 \\ \cdot \\ \cdot \\ w_n \end{pmatrix}$$

Namely,

$$(aI - A) \begin{pmatrix} w_1 \\ w_2 \\ \cdot \\ \cdot \\ w_n \end{pmatrix} = \begin{pmatrix} 0 \\ 0 \\ \cdot \\ \cdot \\ 0 \end{pmatrix}$$

Therefore, we have

$$det\ (aI - A) = 0$$

Henceforth, $F(a) = 0$, and a is a root of its field polynomial. ∎

We will prove the following theorem.

Theorem 5.32. *Let* **L** *be a finite extension of a field* **S** *which is a finite extension of* **K** *and* $a \in \mathbf{S}$. *Let the field polynomial of* a *as an element of* **L** *be* $F(x)$, *the field polynomial of* a *as an element of* **S** *be* $G(x)$, *and the minimal polynomial of* a *over* **K** *be* $f(x)$. *Then we have*

(1) $\mathbf{L}=\mathbf{K}[a] \Leftrightarrow F(x) = f(x)$.
(2) *Let* $m = [\mathbf{L} : \mathbf{S}]$. *Then* $F(x) = G(x)^m$.

Proof: (1) It follows from the above Discussion (3) that a is a root of $F(x)$. Therefore, we have

$$f(x) \mid F(x)$$

Since their degrees are equal, and both are monic polynomials, then they must be equal.

(2) Let $[\mathbf{S} : \mathbf{K}]=s$. Let $\{u_1, u_2, \cdots, u_s\}$ be a basis for **S** over **K**, and $\{v_1, v_2, \cdots, v_m\}$ be a basis for **L** over **S**. Let $\mathbf{L}_i$ be the following subspace,

$$\mathbf{L}_i = \oplus_{j=1}^{s} u_j v_i \mathbf{K}$$

Then we have

$$\mathbf{L} = \oplus_{i=1}^{m} \mathbf{L}_i$$

Each $\mathbf{L}_i$ is an invariant subspace of A with the characteristic polynomial of the restriction of A to $\mathbf{L}_i$ equaling $G(x)$. Therefore, we have $F(x) = G(x)^m$. ∎

Similar to the definitions of trace and determinant of a matrix, we have the following definition.

Definition 5.21. *We define the trace of* a, $Tr_{L/K}(a)$, *as trace*$(A) = \sum a_{ii}$. *We define the norm of* a, $N_{L/K}(a)$, *as det* (A). *In other words, in the following field polynomial* $F(x)$ *of* a, *we have*

$$F(x) = x^n - Tr_{L/K}(a)x^{n-1} + \cdots + (-1)^n N_{L/K}(a)$$

∎

We have the following trivial theorem.

Theorem 5.33. *Let* **L** *be a finite extension of* **K**. *Then we have*

(1) $Tr_{L/K} : \mathbf{L} \rightarrow \mathbf{K}$ *is a* **K**-*linear transformation of* **L** *to* **L**.
(2) $N_{L/K} : \mathbf{L} \rightarrow \mathbf{K}$ *is a multiplicative map, namely* $N_{L/K}(ab) = N_{L/K}(a)N_{L/K}(b)$.
(3) *Let* $[\mathbf{L} : \mathbf{K}]=n$, *and* $a \in \mathbf{K}$. *Then we have* $Tr_{L/K}(a) = na, N_{L/K}(a) = a^n$.

Proof: They are trivial. ∎

Theorem 5.34. *Let* **L** *be a finite extension of a field* **S** *which is a finite extension of* **K**, *and* $a \in \mathbf{S}$. *Let* $m = [\mathbf{L} : \mathbf{K}]$. *Then we have*

(1) $Tr_{L/K}(a) = mTr_{S/K}(a)$.
(2) $N_{L/K}(a) = N_{S/K}(a)^m$.

Proof: They are trivial. ∎

Example 23. *Let $d \in \mathbf{Q}$ such that d is not a square. Let* $\mathbf{L}=\mathbf{Q}[\sqrt{d}]$, *and* $\mathbf{K}=\mathbf{Q}$. *Let* $\{w_1, w_2\} = \{1, \sqrt{d}\}$, *and $a = \alpha + \beta\sqrt{d} \in \mathbf{L}$. Then we have*

$$(\alpha + \beta\sqrt{d})1 = \alpha 1 + \beta\sqrt{d}$$
$$(\alpha + \beta\sqrt{d})\sqrt{d} = \beta d1 + \alpha\sqrt{d}$$

We have

$$A = \begin{pmatrix} \alpha & \beta \\ \beta d & \alpha \end{pmatrix}$$

The characteristic polynomial of A is as follows,

$$det \begin{pmatrix} x-\alpha & -\beta \\ -\beta d & x-\alpha \end{pmatrix} = x^2 - 2\alpha x + (\alpha^2 - \beta^2 d)$$

Therefore, we have

$$Tr_{L/K}(\alpha + \beta\sqrt{d}) = 2\alpha = (\alpha + \beta\sqrt{d}) + (\alpha - \beta\sqrt{d})$$
$$N_{L/K}(\alpha + \beta\sqrt{d}) = \alpha^2 - \beta^2 d = (\alpha + \beta\sqrt{d})(\alpha - \beta\sqrt{d})$$

∎

We have the following definition.

Definition 5.22. *Let $\{w_1, w_2, \cdots, w_n\}$ be a vector space basis of* $\mathbf{L}$ *over* $\mathbf{K}$. *Then the* ***discriminant*** *of the basis $\{w_1, w_2, \cdots, w_n\}$, $Dis\{w_1, \cdots, w_n\}$, is defined as*

$$Dis\{w_1, \cdots, w_n\} = det\ (Tr_{L/K}(w_i w_j))$$

∎

Discussion

(1) Let $\{w'_1, \cdots, w'_n\}$ be another basis of $\mathbf{L}$ over $\mathbf{K}$. Let

$$w'_i = \sum b_{ik} w_k$$

Then we have

$$\begin{aligned} Dis\{w'_1, \cdots, w'_n\} &= det\ (Tr_{L/K}(w'_i w'_j)) \\ &= det\ (Tr_{L/K}(\sum_{k,\ell} b_{ik} b_{j\ell} Tr_{L/K}(w_k w_\ell))) \\ &= det\ (Tr_{L/K}(w_k w_\ell))(det\ B)^2 \\ &= Dis\{w_1, \cdots, w_n\}(det\ B)^2 \end{aligned}$$

Since $det\ B \neq 0$, then we have

$$Dis\{w_1, \cdots, w_n\} = 0 \Leftrightarrow Dis\{w'_1, \cdots, w'_n\} = 0$$

∎

Theorem 5.35. *Let* **L** *be a finite extension of* **K**, *and* $\{w_1, \cdots, w_n\}$ *a basis of* **L** *over* **K**. *Then we have*

$$Dis\{w_1, \cdots, w_n\} \neq 0 \Leftrightarrow \mathbf{L} \text{ is a separable algebraic extension of } \mathbf{K}.$$

Proof: ($\Rightarrow$) Let $\mathbf{K}_L^S$ be a separable closure of **K** in **L**. If $\mathbf{K}_L^S$=**L**, then **L** is separable algebraic over **K**, and we are done. Otherwise, $\mathbf{K}_L^S \neq$**L**, **L** is purely inseparable over $\mathbf{K}_L^S$, and and

$$[\mathbf{L} : \mathbf{K}_L^S] = p^r, \qquad r \geq 1$$

Let α be any element in **L**. We claim that $Tr_{L/K}(\alpha) = 0$. Note then $Tr_{L/K}(w_i w_j) = 0$, and $Dis\{w_1, \cdots, w_n\} = 0$.

There are two cases, (i) $\alpha \in \mathbf{K}_L^S$, (ii) $\alpha \notin \mathbf{K}_L^S$.

Case (i). Let the field polynomial of α as an element in $\mathbf{K}_L^S$ be $G(x)$ as follows,

$$G(x) = x^m + a_1 x^{m-1} + \cdots + a_m$$

It follows from Theorem 5.32. that

$$F(x) = G(x)^{p^r} = x^{mp^r} + 0x^{mp^r - 1} + \cdots$$

Therefore, $Tr_{L/K}(\alpha) = 0$.

Case (ii). We shall use Theorem 5.19.. Then there is an integer $\ell \geq 1$ such that the minimal polynomial $f(x)$ of α over **K** is in $\mathbf{K}[x^{p^\ell}]$. It follows from Theorem 5.32. that the field polynomial of α as an element in **L** is of the following form,

$$F(x) = f(x)^s \in \mathbf{K}[x^{p^\ell}]$$

Therefore, $Tr_{L/K}(\alpha) = 0$.

($\Leftarrow$) Since **L** is a finite separable extension of **K**. then **L** is a simple extension of **K**, and **L**=**K**$[\alpha]$ for some suitable α. Let us take $\{1, \alpha, \alpha^2, \cdots, \alpha^{n-1}\}$ as a basis of **L** over **K**. Let Ω be an algebraic closure of **L**, and $f(x)$ be the minimal polynomial of α over **K**. Note that the field polynomial of α, $F(x)$, is $f(x)$. Let $f(x)$ be split completely in Ω as follows,

$$f(x) = \prod_{i=1}^{n} (x - \alpha_i), \qquad \alpha_1 = \alpha, \alpha_i \neq \alpha_j$$

Then we have

$$Tr_{L/K}(\alpha) = \sum \alpha_i$$

Let us compute $Tr_{L/K}(\alpha^j)$. We claim

$$Tr_{L/K}(\alpha^j) = \sum \alpha_i^j$$

Let the splitting field of $f(x)$ in Ω be **S** which is a Galois extension of **K** with Galois group $G(S/K)$. In the collection $\{\alpha_1^j, \alpha_2^j, \cdots, \alpha_n^j\}$, some elements may be identical. It is easy to see that each element appears with the same multiplicities. Picking all distinct elements from it to form a set $\{\beta_1, \beta_2, \cdots, \beta_m\}$. Then we have that $m \mid n$, and the polynomial $g(x)$ defined as

$$g(x) = \prod_{i=1}^{m}(x - \beta_i)$$

is the minimal polynomial of α^j over **K**. Then we have

$$Tr_{K[\alpha^j]/K}(\alpha^j) = \sum \beta_i$$

$$Tr_{L/K}(\alpha^j) = \frac{n}{m}\sum \beta_i = \sum_{i=1}^{n} \alpha_i^j$$

Therefore, we have the following computations,

$$Dis\{1, \alpha, \cdots, \alpha^{n-1}\} = \det \begin{pmatrix} n & \sum \alpha_i & \cdot & \cdot & \sum \alpha_i^{n-1} \\ \sum \alpha_i & \sum \alpha_i^2 & \cdot & \cdot & \sum \alpha_i^n \\ \cdot & \cdot & \cdot & \cdot & \cdot \\ \cdot & \cdot & \cdot & \cdot & \cdot \\ \sum \alpha_i^n & \sum \alpha_i^{n+1} & \cdot & \cdot & \sum \alpha_i^{2n-1} \end{pmatrix}$$

$$= \det \begin{pmatrix} 1 & 1 & \cdot & \cdot & 1 \\ \alpha_1 & \alpha_2 & \cdot & \cdot & \alpha_n \\ \cdot & \cdot & \cdot & \cdot & \cdot \\ \cdot & \cdot & \cdot & \cdot & \cdot \\ \alpha_1^{n-1} & \alpha_2^{n-1} & \cdot & \cdot & \alpha_n^{n-1} \end{pmatrix} \begin{pmatrix} 1 & \alpha_1 & \cdot & \cdot & \alpha_1^{n-1} \\ 1 & \alpha_2 & \cdot & \cdot & \alpha_2^{n-1} \\ \cdot & \cdot & \cdot & \cdot & \cdot \\ \cdot & \cdot & \cdot & \cdot & \cdot \\ 1 & \alpha_n & \cdot & \cdot & \alpha_n^{n-1} \end{pmatrix}$$

$$= \det \begin{pmatrix} 1 & 1 & \cdot & \cdot & 1 \\ \alpha_1 & \alpha_2 & \cdot & \cdot & \alpha_n \\ \cdot & \cdot & \cdot & \cdot & \cdot \\ \cdot & \cdot & \cdot & \cdot & \cdot \\ \alpha_1^{n-1} & \alpha_2^{n-1} & \cdot & \cdot & \alpha_n^{n-1} \end{pmatrix}^2 = (\textstyle\prod_{i>j}(\alpha_i - \alpha_j))^2 \neq 0$$

∎

Exercises

(1) Let α be a root of $x^3 - 2$ in **C**, and $L = \mathbf{Q}[\alpha]$. Let $\beta = -2\alpha^2 + \alpha - 1$. Find $Tr_{L/\mathbf{Q}}(\beta)$, $N_{L/\mathbf{Q}}(\beta)$, and $Dis\{1, \alpha, \alpha^2\}$.

(2) Let α be a root of $x^3 - 2x + 2$ in C, and L= Q[α]. Let $\beta = \alpha^2 + \alpha - 1$. Find $Tr_{L/\mathbf{Q}}(\beta)$, $N_{L/\mathbf{Q}}(\beta)$, and $Dis\{1, \alpha, \alpha^2\}$.
(3) Let D be a square-free integer, and L=Q[$\sqrt{D}$]. Let $\beta \in$ L. Find $Tr_{L/\mathbf{Q}}(\beta)$, $N_{L/\mathbf{Q}}(\beta)$, and $Dis\{1, \sqrt{D}\}$.
(4) Let L=K[α] be a finite extension over K of degree n. Let Ω be an algebraic closure of L. Let the minimal polynomial $f(x)$ of α over K be split completely in Ω as

$$f(x) = \prod_{i=1}^{n}(x - \alpha_i), \qquad \alpha = \alpha_1$$

Let β be any element in L with the following expression,

$$\beta = \beta_1 = a_0 + a_1\alpha_1 + \cdots + a_{n-1}\alpha_{n-1}, \qquad a_j \in \mathbf{K}$$

Let

$$\beta_i = a_0 + a_1\alpha_i + \cdots + a_{n-1}\alpha_{n-1}$$

Prove that

$$Tr_{L/K}(\beta) = \sum \beta_i, \qquad N_{L/K}(\beta) = \prod \beta_i$$

(5) Let L=K[α] be a finite extension over K of degree n. Let $\beta \in$ L. Let the minimal polynomial $g(x)$ of β over K be as follows,

$$g(x) = x^m + b_1x^{m-1} + \cdots + b_m$$

Let $\ell = n/m$. Prove that

$$Tr_{L/K}(\beta) = -\ell\beta_1, \qquad N_{L/K}(\beta) = (-1)^n\beta_m^\ell$$

(6) Let L be a finite extension of Q. Is $Tr_{L/\mathbf{Q}}$ surjective? Is $N_{L/\mathbf{Q}}$ surjective?
(7) (Hilbert's Theorem 90.) Let L be a finite cyclic extension of K with Galois group $G(L/K) = \{\sigma\}$. For any element $\beta \in$ L, $N_{L/K}(\beta) = 1 \Leftrightarrow$ there exists $0 \neq \alpha \in$ L such that $\beta = \alpha/\sigma(\alpha)$.

§9 Lüroth's Theorem

In §1, we observe that there are transcendental numbers in C. Therefore, C is not an algebraic extension of Q. For any field K, there are functions which are not algebraic over the rational function field K$(x_1, x_2, \cdots, x_n)$. Those functions happen naturally. For instance, e^{ix} is not algebraic over C(x). We have the following definition.

Definition 5.23. *(1) If a field* **L** *is not an algebraic extension of a subfield* **K**, *then* **L** *is called a **transcendental extension** of* **K**.

(2) If finitely many elements $y_1, y_2, \ldots, y_n \in \mathbf{L}$ *satisfy a non-zero polynomial* $f(x_1, x_2, \cdots, x_n) = 0 \in \mathbf{K}[x_1, \cdots, x_n]$ *the polynomial ring of n variables over* **K**, *then* $y_1, \cdots, y_n$ *are said to be **algebraically dependent**, otherwise **algebraically independent**.*

(3) A maximal algebraically independent set of **L** *is called a **transcendental basis** of* **L**.

(4) If **L**=**K**$(\{y_i\})$ *for some transcendental basis, then* **L** *is said to be a **pure transcendental extension** of* **K**. ∎

The materials about transcendental extensions are very rich. We shall only present a simple and interesting theorem.

Theorem 5.36. (Lüroth's Theorem). *Let x be a variable, and* $\mathbf{K}(x) \supset \mathbf{L} \supsetneq \mathbf{K}$. *Then there is a transcendental element y such that* **L**=**K**(y).

Proof: (1) We claim that $\mathbf{K}(x)$ is algebraic over **L**. Let $\alpha(x) \in \mathbf{L} \backslash \mathbf{K}$ be a non-constant rational function. Then we have

$$\alpha(x) = \frac{r(x)}{s(x)}, \qquad r(x), s(x) \in \mathbf{K}[x], (r(x), s(x)) = 1$$

Clearly, x is a root of the following polynomial with z as the variable,

$$r(z) - \alpha(x)s(z) = 0 \in \mathbf{L}[z]$$

Therefore, **L**=**K**(x)=**L**$[x]$ is algebraic over **L**.

(2) Let the field degree of $\mathbf{K}(x)$ over **L** be m. Let the minimal polynomial $f(z)$ of x over **L** be as follows,

$$\begin{aligned} f(z) &= z^m + \alpha_1 z^{m-1} + \cdots + \alpha_m, \qquad \alpha_i \in \mathbf{L} \\ \alpha_i &= \frac{r_i(x)}{s_i(x)}, \quad r_i(x), s_i(x) \in \mathbf{K}[x] \end{aligned} \tag{1}$$

Note that not all α_i are constants, say $\alpha_k \notin \mathbf{K}$, otherwise x will be algebraic over **K**. Let us clear the denominators of Eq (1), and produce

$$\begin{aligned} &h(x, z) = h_0(x)z^m + h_1(x)z^{m-1} + \cdots + h_m(x) \\ &h_i(x) \in \mathbf{K}[x], (h_0(x), \cdots, h_m(x)) = 1, \alpha_i = \frac{h_i(x)}{h_0(x)} \end{aligned}$$

Recall that $\alpha_k \notin$**K**. Let $y = \alpha_k$. We claim the **L**= **K**(y). Note then our theorem will be proved.

(3) Clearly, it follows from $\alpha_k \in \mathbf{L}$ that $\mathbf{K}(y) \subset \mathbf{L}$. Let the the reduced form of y be as follows,

$$y = \frac{h_k(x)}{h_0(x)} = \frac{a(x)}{b(x)}$$
$$a(x), b(x) \in \mathbf{K}[x], (a(x), b(x)) = 1, max(deg\ a(x), deg\ b(x)) = n$$

Clearly, x satisfies the following polynomial equation

$$\begin{aligned} &a(x) - yb(x) = 0 \\ &A(x,z) = b(x)a(z) - a(x)b(z) = 0 \end{aligned} \tag{2}$$

It is easy to see that $[\mathbf{K}(x) : \mathbf{L}]$=$n$, and $A(x,z)$ is a primitive polynomial in z. Now there are two primitive polynomials $h(x,z), A(x,z) \in \mathbf{K}[x][z]$ satisfied by x, and $h(x,z)$ is irreducible. Therefore, we must have

$$h(x,z) \mid A(x,z), \qquad A(x,z) = B(x,z)h(x,z) \tag{3}$$

We shall prove that $B(x,z)$ is a constant in $\mathbf{K}$.

Comparing the x-*deg* on both sides of Eq (3), we have

$$deg_x\ h(x,z) = sup_i(deg_x\ h_i(x)) \geq sup(deg_x\ a(x), deg_x\ b(x)) = deg_x\ A(x,z)$$

We conclude that

$$deg_x\ B(x,z) = 0$$

Hence $B(x,z)$ is a polynomial in z only. If $B(x,z)$ is not a constant, then it follows from Eq (3) that $A(x,z)$ has a polynomial in z, namely $B(x,z)$, as a factor. Note that $A(x,z)$ is symmetric with respect to the interchange of x, z. Therefore, $A(x,z)$ will have a polynomial in x as a factor, contradicting the fact that $A(x,z)$ is primitive. We conclude that

$$A(x,z) = Bh(x,z), \qquad B \in \mathbf{K}$$

(4) We shall finish the proof of our theorem by the following observations:

$$\begin{aligned} deg_x\ A(x,z) &= deg_z\ A(x,z) \\ deg_x\ A(x,z) &= deg_x\ h(x,z) = n \\ deg_z\ A(x,z) &= deg_z\ h(x,z) = m \\ [\mathbf{K}(x) : \mathbf{K}(y)] &= n = m = [\mathbf{K}(x) : \mathbf{L}] \end{aligned}$$

Therefore, we conclude

$$\mathbf{L} = \mathbf{K}(y)$$

∎

Discussion

(1) Certainly, an attempt has been made to generalize Lüroth's Theorem to more variables. We have

Zariski-Castelnuovo's Theorem: Let $\mathbf{K}$ be an algebraically closed field of characteristic 0. Let $\mathbf{L}$ be an intermediate field between $\mathbf{K}(x, y)$ and $\mathbf{K}$, namely $\mathbf{K}(x, y) \supset \mathbf{L} \supset \mathbf{K}$. Then $\mathbf{L}$ equals either $\mathbf{K}(u, v)$ or $\mathbf{K}(u)$ or $\mathbf{K}$ where u, v are algebraically independent.

The assumption of characteristic 0 is essential, there are counter-examples if the characteristic is $p > 0$. For three or more variables, there are counter-examples due to Clemens-Griffith, Iskovskih-Manin, Artin-Mumford. ∎

Example 24. (Rational Curve). *(1) Let $\mathbf{K}$ be an algebraically closed field. An **algebraic affine plane curve** C is the set of all solutions of a polynomial equation as follows,*

$$f(x, y) = 0 \tag{1}$$

*If the above polynomial is irreducible, then we say the curve is an **irreducible curve**. An irreducible curve is said to be a **rational curve** if there is a parametrization as follows,*

$$\begin{cases} x = \alpha(t) \in \mathbf{K}(t) \\ y = \beta(t) \in \mathbf{K}(t) \end{cases} \tag{2}$$

where $\alpha(t), \beta(t)$ are not both constant in $\mathbf{K}$, and Eq (1) is satisfied by the substitution given by Eq (2).

*Let the **coordinate ring**, $\mathbf{K}[C]$, of an algebraic affine plane curve C be defined as the residue class ring $\mathbf{K}[x, y]/(f(x, y))$. Let the **function field**, $\mathbf{K}(C)$, of an irreducible curve be defined as the quotient field of its coordinate ring $\mathbf{K}[C]$.*

In the case of a rational curve C, we may define a map $\tau : \mathbf{K}(C) \to \mathbf{K}(t)$ as given by Eq (2), namely

$$\begin{cases} \tau(x) = \alpha(t) \in \mathbf{K}(t) \\ \tau(y) = \beta(t) \in \mathbf{K}(t) \end{cases} \tag{3}$$

It is easy to see that τ is injective. Therefore, it follows from Lüroth's Theorem that

$$\mathbf{K}[C] = \mathbf{K}(t') \tag{4}$$

On the other hand, if Eq (4) is true, then we can easily find $\alpha(t'), \beta(t')$ such that Eq (2) is satisfied. Henceforth, we may define a rational curve as an irreducible curve whose function field is isomorphic to $\mathbf{K}(t)$.

(2) We shall apply the above discussions to integration theory. Let $f(x, y)$ be a rational function in two variables. We want to compute the following integration,

$$\int f(\sin\theta, \cos\theta) d\theta$$

Let $x = \sin\ \theta$, $y = \cos\ \theta$. Then we have the following relation

$$x^2 + y^2 = 1 \tag{5}$$

It is easy to see that Eq (5) defines a rational curve with the following parametrization,

$$\begin{cases} x = \frac{2t}{t^2+1} \\ y = \frac{t^2-1}{t^2+1} \end{cases} \tag{6}$$

Further note that

$$\begin{aligned} d\,\theta &= \frac{1}{y} d\,x \\ d\,x &= \frac{2 - 2t^2}{(t^2+1)^2} \end{aligned} \tag{7}$$

Substituting Eqs (6) & (7) into our original integration, we get

$$\int f(\sin\ \theta, \cos\ \theta) d\,\theta = \int g(t) d\,t$$

for some rational function $g(t)$. Therefore, the integrations of trigonometric functions can be reduced to the integrations of rational functions. ∎

Example 25. *Not all irreducible curves are rational. Recall Example 9. of Chapter III. As proved there, a Fermat's curve over the complex field* C *defined by the following equation,*

$$x^n + y^n - 1 = 0, \qquad n \geq 3 \tag{1}$$

has no parametrization of the following form,

$$\begin{cases} x = \alpha(t) \in \mathbf{K}(t) \\ y = \beta(t) \in \mathbf{K}(t) \end{cases} \tag{2}$$

where $\alpha(t), \beta(t)$ are not both constant in $\mathbf{K}$. ∎

Exercises

(1) Prove that any two transcendence bases of **L** over **K** have the same cardinalities. This number is called the ***transcendence degree***, *Trdeg* (**L**/**K**), of **L** over **K**. (Hint: imitate the proof of vector space bases.)

(2) Let **L** be an extension of **S**, and **S** an extension of **K**. Prove that *Trdeg* (**L**/**S**) + *Trdeg* (**S**/**K**) = *Trdeg* (**L**/**K**).

(3) Show that the transcendence degree, *Trdeg* (**C**/**Q**), of **C** over **Q** is uncountable.

(4) Let $\mathbf{K}(x)$ be a transcendental extension of **K**, and $f(x) \in \mathbf{K}(x)$. Show that $\mathbf{K}(x) = \mathbf{K}(f(x))$ if and only if $f(x)$ is of the following form,

$$f(x) = \frac{ax+b}{cx+d}, \quad a, b, c, d \in \mathbf{K}, ad - bc \neq 0$$

(5) Let $\mathbf{K}(x)$ be a transcendental extension of **K**. Show that there is an algebraic extension **L** of $\mathbf{K}(x)$ of degree n such that **L** is a pure transcendental extension of **K**.

(6) Let x be a variable, and $f(x), g(x) \in \mathbf{K}[x]$ with $deg\ f(x) = n, deg\ g(x) = m$. Show that $\mathbf{K}(x)$ is of algebraic degrees n, m over the fields $\mathbf{K}(f(x))$, $\mathbf{K}(g(x))$ respectively.

(7) Let $\mathbf{L}=\mathbf{Q}(x)$ be a transcendental extension of **Q**. Let σ, τ be two **Q**-automorphisms of **L** defined by

$$\sigma(x) = -x, \qquad \tau(x) = x^{-1}$$

Let G be the group generated by σ, and τ. Find the fixed field **S** of G, and $[\mathbf{L} : \mathbf{S}]$.

(8) Show that $\mathbf{C}(x)$ is a Galois extension of $\mathbf{C}(x^n + x^{-n})$.

APPENDIX I

Set-Theoretic Notations

$\subset$: is a subset of
$\subseteq$: is a subset of or equal to
$\subsetneqq$: is a proper subset of
$\supset$: is a superset of
$\supseteq$: is a superset of or equal to
$\supsetneqq$: is a proper superset of
$\equiv$: is equivalence to
$\sim$: is similar to
$\approx$: is isomorphic to
$\in$: is an element of
$\ni$: contains as an element
$\notin$: is not an element of
$|$: is a divisor of
$\nmid$: is not a divisor of
$\cap$: intersection
$\cup$: union
$\setminus$: set-theoretic minus
$\oplus$: direct sum
$\rightarrow$: maps to (as set)
$\mapsto$: maps to (as element)
$\Rightarrow$: only if
$\Leftarrow$: if
$\Leftrightarrow$: if and only if

APPENDIX II

Peano's Axioms

For every number a, the *successor*, a^+, is defined. The Peano's axioms for the natural numbers, $\mathbf{Z}_+$, are as follows,

Axiom 1.: One, 1, is a natural number.
Axiom 2.: If a is a natural number, then the successor of a, a^+, is a natural number.
Axiom 3.: 1 is not the successor of any natural number.
Axiom 4.: Two natural numbers of which the successors are equal are themselves equal.
Axiom 5.: If a set $\mathbf{S}$ of natural numbers contains 1 and also the successor of every number in $\mathbf{S}$, then every natural number is in $\mathbf{S}$.

The fifth axiom is of cause the axiom of mathematical induction. We shall use Peano's axioms to define addition, $+$, and multiplication, $\cdot$, in the set of natural integers $\mathbf{Z}_+$.

Definition A.1. *We define the addition recursively as follows for a given a,*

(1) $a + 1 = a^+$.
(2) *If $a + b$ is defined, then we define $a + b^+ = (a + b)^+$.*

∎

It is not hard to see the addition is defined for all natural numbers b from Axiom 5. Let us show that the addition is commutative.

Lemma 1. *We always have $a^+ + b = a + b^+$.*

Proof: For given a, let $\mathbf{S}=\{b : a^+ + b = a + b^+.\}$. Then we have

$$a^+ + 1 = (a^+)^+ = (a+1)^+ = a + 1^+.$$

Therefore, $1 \in \mathbf{S}$. Suppose that $b \in \mathbf{S}$. Then we have

$$a^+ + b^+ = (a^+ + b)^+ = (a + b^+)^+ = a + (b^+)^+.$$

Therefore, $b^+ \in$ **S**. It follows from Axiom 5. that **S** is the set of all natural numbers.

Theorem A.1. *We always have $a + b = b + a$.*

Proof: (1) We claim that $a + 1 = 1 + a$ for all a. Let **S** be the set of natural numbers a such that $a + 1 = 1 + a$. Clearly, we have $1 + 1 = 1 + 1$. Then we have $1 \in$ **S**. Let $a \in$ **S** Then we have

$$a^+ + 1 = (a^+)^+ = (a + 1)^+ = (1 + a)^+ = 1 + a^+$$

Therefore, $a^+ \in$ **S**. It follows from Axiom 5. that **S** is the set of all natural numbers.

(2) Let us fix a, and let **S'**$=\{b : a + b = b + a\}$. It follows from (1) that $1 \in$ **S'**. Suppose that $b \in$ **S'**. Then we use the preceding lemma to deduce the following,

$$a + b^+ = (a + b)^+ = (b + a)^+ = b + a^+ = b^+ + a$$

Therefore, $b^+ \in$ **S'**. It follows from Axiom 5. that **S'** is the set of all natural numbers. ∎

We shall prove the associative law for the addition.

Theorem A.2. *We always have*

$$(a + b) + c = a + (b + c)$$

Proof: For given a, b, let **S**$=\{c : (a + b) + c = a + (b + c).\}$.

(1) We claim that $1 \in$ **S**. In fact, we have,

$$(a + b) + 1 = (a + b)^+ = a + b^+ = a + (b + 1)$$

Therefore, $1 \in$ **S**.

(2) Let $c \in$ **S**. Then we have

$$(a + b) + c^+ = ((a + b) + c)^+ = (a + (b + c))^+ = a + (b + c)^+ = a + (b + c^+)$$

Therefore, $c^+ \in$ **S**. It follows from Axiom 5. that **S** is the set of all natural numbers. ∎

We shall define the multiplication, $\cdot$, as follows,

Definition A.2. *We define the multiplication recursively as follows for a given a,*

(1) $a \cdot 1 = a$.

(2) *If $a \cdot b$ is defined, then we define $a \cdot b^+ = a \cdot b + a$.*

∎

Lemma 2. *We always have $a \cdot 1 = 1 \cdot a$.*

Proof: Let S=$\{a : a \cdot 1 = 1 \cdot a.\}$. Clearly, we have $1 \in$ **S**. Let $a \in$ **S**. Then we have

$$1 \cdot a^+ = 1 \cdot a + 1 = a \cdot 1 + 1 = a + 1 = a^+ = a^+ \cdot 1$$

Therefore, $a^+ \in$ **S**. It follows from Axiom 5. that **S** is the set of all natural numbers. ∎

Lemma 3. *We always have $b^+ \cdot a = b \cdot a + a$.*

Proof: For a fixed b, let S=$\{a : b^+ \cdot a = b \cdot a + a\}$. Then we have

$$b^+ \cdot 1 = b^+ = b + 1 = b \cdot 1 + 1$$

Therefore, $1 \in$ **S**. Let $a \in$ **S**. Then we have,

$$\begin{aligned} b^+ \cdot a^+ &= b^+ \cdot a + b^+ = (b \cdot a + a) + b^+ = b \cdot a + (a + b^+) \\ &= b \cdot a + (b + a^+) = (b \cdot a + b) + a^+ = b \cdot a^+ + a^+ \end{aligned}$$

Therefore, $a^+ \in$ **S**. It follows from Axiom 5. that **S** is the set of all natural numbers. ∎

We shall prove the commutative law for the multiplication.

Theorem A.3. *We always have $a \cdot b = b \cdot a$.*

Proof: Let us fix a, and let S=$\{b : a \cdot b = b \cdot a.\}$. Then it follows from the Lemma 2. that $1 \in$ **S**. Let $b \in$**S**. Then we have, by Lemma 3., the following

$$a \cdot b^+ = a \cdot b + a = b \cdot a + a = b^+ \cdot a.$$

Therefore, $b^+ \in$ **S**. It follows from Axiom 5. that **S** is the set of all natural numbers. ∎

We shall prove the distributive law.

Theorem A.4. *We always have $(a + b) \cdot c = a \cdot c + b \cdot c$.*

Proof: For fixed a, b, let S= $\{c : (a + b) \cdot c = a \cdot c + b \cdot c.\}$.

(1) We claim that $1 \in$ **S**. In fact, we have,

$$(a + b) \cdot 1 = a + b = a \cdot 1 + b \cdot 1$$

Therefore, $1 \in$ **S**.

(2) Let $c \in \mathbf{S}$. Then we have

$$\begin{aligned}(a+b)\cdot c^{+} &= (a+b)\cdot c+(a+b) = (a\cdot c+b\cdot c)+(a+b)\\ &= (a\cdot c+a)+(b\cdot c+b) = a\cdot c^{+}+b\cdot c^{+}.\end{aligned}$$

Therefore, $c^{+} \in \mathbf{S}$. It follows from Axiom 5. that $\mathbf{S}$ is the set of all natural numbers. ∎

We shall prove the associative law for the multiplication.

Theorem A.5. *We always have* $(a\cdot b)\cdot c = a\cdot(b\cdot c)$.

Proof: For fixed a, b, let $\mathbf{S}=\{c : (a\cdot b)\cdot c = a\cdot(b\cdot c).\}$.

(1) We claim that $1 \in \mathbf{S}$. In fact, we have,

$$(a\cdot b)\cdot 1 = a\cdot b = a\cdot(b\cdot 1)$$

Therefore, $1 \in \mathbf{S}$.

(2) Let $c \in \mathbf{S}$. Then we have

$$\begin{aligned}(a\cdot b)\cdot c^{+} &= (a\cdot b)\cdot c+(a\cdot b) = a\cdot(b\cdot c)+a\cdot b\\ &= a\cdot(b\cdot c+b) = a\cdot(b\cdot c^{+}).\end{aligned}$$

Therefore, $c^{+} \in \mathbf{S}$. It follows from Axiom 5. that $\mathbf{S}$ is the set of all natural numbers. ∎

From the natural numbers $\mathbf{Z}_{+}$, we may construct the integers $\mathbf{Z}$ and the rational numbers $\mathbf{Q}$. It is a easy matter to show that the arithmetic operations $+$, $\cdot$ obey the usual laws.

APPENDIX III

Homological Algebra

Example 1. *Homological algebra originates from topology. Let us consider the following triangle* $\triangle ABC$,

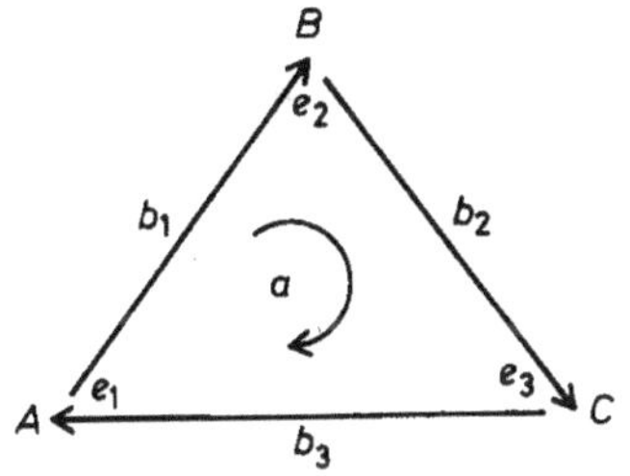

Let a *be the oriented triangle* $\triangle ABC$, b_1 *be the directed segment* $\overrightarrow{AB}$, b_2 *be the directed segment* $\overrightarrow{BC}$, b_3 *be the directed segment* $\overrightarrow{CA}$, *and the three vertices* A, B, C *be* e_1, e_2, e_3. *Let* d *be the* ***boundary operator***, *namely,*

$$\begin{aligned} d(a) &= b_1 + b_2 + b_3 \\ d(b_1) &= e_1 - e_2 \\ d(b_2) &= e_2 - e_3 \\ d(b_3) &= e_3 - e_1 \\ d(e_i) &= 0, \quad \forall\, i \end{aligned}$$

It is easy to verify that

$$d^2a = d(d(a)) = 0$$
$$d^2b_i = 0$$

Let us introduce an algebraic structure to the above situation. Let C_2 be $\mathbf{Z}a$= the free group generated by a, C_1 be $\mathbf{Z}b_1 \oplus \mathbf{Z}b_2 \oplus \mathbf{Z}b_3$= the free group generated by b_1, b_2, b_3, C_0 be $\mathbf{Z}e_1 \oplus \mathbf{Z}e_2 \oplus \mathbf{Z}e_3$= the free group generated by e_1, e_2, e_3. We may extend the map d naturally to homomorphisms in the following diagram,

$$C_2 \xrightarrow{d} C_1 \xrightarrow{d} C_0 \xrightarrow{d} 0$$

To clarify the situation, we will use different symbols for d in the following diagram.

$$C_2 \xrightarrow{d_2} C_1 \xrightarrow{d_1} C_0 \xrightarrow{d_0} 0$$

We have,

$$d_{i-1}d_i = 0$$

∎

Note that commutative groups are modules. We will define,

Definition A3.1. *(1) Let* **R** *be a ring. An* **R***-complex (C, d) is a collection of* **R***-modules C_i and* **R***-homomorphisms $d_i : C_i \to C_{i-1}$ with $d_{i-1}d_i = 0$ where $i \in \mathbf{Z}$,*

$$\cdots \xrightarrow{d_{i+1}} C_i \xrightarrow{d_i} C_{i-1} \xrightarrow{d_{i-1}} C_{i-1} \xrightarrow{d_{i-1}} \cdots$$

(2) Let $(C, d), (C', d')$ be **R***-complexes. A map $\alpha : (C, d) \to (C', d')$ is a collection (α_i) of maps $\alpha_i : C_i \to C'_i$ such that*

$$\alpha_{i-1}d_i = d'_i\alpha_i, \qquad \text{or simply} \qquad \alpha d = d'\alpha$$

The above equation may be indicated by saying the following diagram is commutative,

$$\begin{array}{ccc} C_i & \xrightarrow{d_i} & C_{i-1} \\ {\scriptstyle \alpha_i}\downarrow & & \downarrow{\scriptstyle \alpha_{i-1}} \\ C'_i & \xrightarrow{d'_i} & C'_{i-1} \end{array}$$

∎

Example 2. *Let* $\mathbf{R}$=$\mathbf{C}[x, y, z]$*. Let us consider the exterior product of the differentials as follows,*

$$dx \wedge dx = dy \wedge dy = dz \wedge dz = 0$$
$$dx \wedge dy = -dy \wedge dx$$
$$dy \wedge dz = -dz \wedge dy$$
$$dz \wedge dx = -dx \wedge dz$$

Let a complex (C, d) be defined as

$$C_0 = \mathbf{R}$$
$$C_{-1} = \mathbf{R}dx \oplus \mathbf{R}dy \oplus \mathbf{R}dz$$
$$C_{-2} = \mathbf{R}dy \wedge dz \oplus \mathbf{R}dz \wedge dx \oplus \mathbf{R}dx \wedge dy$$
$$C_{-3} = \mathbf{R}dx \wedge dy \wedge dz$$

$$d_0(f(x,y,z)) = \frac{\partial f}{\partial x}dx + \frac{\partial f}{\partial y}dy + \frac{\partial f}{\partial z}dz$$
$$d_{-1}(f_1dx + f_2dy + f_3dz) = (\frac{\partial f_3}{\partial y} - \frac{\partial f_2}{\partial z})dy \wedge dz$$
$$+ (\frac{\partial f_1}{\partial z} - \frac{\partial f_3}{\partial x})dz \wedge dx + (\frac{\partial f_2}{\partial x} - \frac{\partial f_1}{\partial y})dx \wedge dy$$
$$d_{-2}(f_1dy \wedge dz + f_2dz \wedge dx + f_3dx \wedge dy) = (\frac{\partial f_1}{\partial x} + \frac{\partial f_2}{\partial y} + \frac{\partial f_3}{\partial z})dx \wedge dy \wedge dz$$

and all other $C_i = 0, d_i = 0$. Then it is not hard to see that d_0 corresponds to the *gradient operator* ∇, d_{-1} corresponds to the *curl operator* $\nabla\times$, and d_{-2} corresponds to the *divergence operator* $\nabla\cdot$. Then the following two theorems in Advanced Calculus,

$$\nabla \times \nabla = 0, \quad \nabla \cdot \nabla\times = 0$$

simply mean,

$$d_{-1}d_0 = 0, \qquad d_{-2}d_{-1} = 0$$

Therefore, (C, d) is a complex. ∎

Definition A3.2. *Let (C, d) be an* **R***-complex. We use Z_i to denote the kernel of d_i, $ker(d_i)$. The elements in it will be called the **i-cycles** or simply the **cycles**. We use B_i to denote the image of d_{i+1}, $im(d_{i+1})$. The elements in it will be called the **i-boundaries** or simply the **boundaries**.* ∎

Discussion

(1) Since we always have $d_i d_{i+1} = 0$, then we have,

$$B_i \subset Z_i$$

∎

Definition A3.3. *(1) The quotient module Z_i/B_i is called the **i-th homology module**, $H_i = H_i(C)$, of the* **R***-complex (C, d).*

(2) If $H_i = 0$, namely the i-boundaries B_i= the i-cycles Z_i, then the complex is said to be exact at C_i.

(3) If $C_i = 0$ for all $i < 0$, then we say the complex is a positive complex or a chain complex. If $C_i = 0$ for all $i > 0$, then we say the complex is a negative complex or a cochain complex. Usually, we change the notations for the cochain complexes as follows,

$$C^i = C_{-i}, \quad d^i = d_{-i}, \quad B^i = B_{-i}, \quad Z^i = Z_{-i}, \quad H^i = H_{-i}$$

$$0 \longrightarrow C^0 \xrightarrow{d^0} C^1 \xrightarrow{d^1} C^2 \xrightarrow{d^2} \cdots$$

∎

Example 3. *Let us continue our discussion about Example 1.. Let us define $C_i = 0, d_i = 0$ for $i \neq 0, 1, 2$. Then we have,*

$$\begin{aligned}
Z_0 &= \mathbf{Z}e_1 \oplus \mathbf{Z}e_2 \oplus \mathbf{Z}e_3 \\
B_0 &= \mathbf{Z}(e_1 - e_2) + \mathbf{Z}(e_2 - e_3) + \mathbf{Z}(e_3 - e_1) \\
Z_1 &= B_1 = \mathbf{Z}(b_1 + b_2 + b_3) \\
Z_2 &= B_2 = 0
\end{aligned}$$

$$\begin{aligned}
H_0 &= Z_0/B_0 = \mathbf{Z} \\
H_1 &= 0 \\
H_2 &= 0 \\
H_i &= 0, \qquad \text{for all other } i
\end{aligned}$$

∎

Example 4. *Let us continue our discussion about Example 2..*

(1) Let us compute H^0. Note that

$$\begin{aligned}
&f(x, y, z) \in Z^0 \\
&\iff d^0(f(x, y, z)) = 0 \\
&\iff f(x, y, z) \in \mathbf{C}
\end{aligned}$$

Therefore, we have $Z^0 = \mathbf{C}$. Furthermore, we have $B^0 = 0$. We conclude

$$H^0 = \mathbf{C}$$

(2) Let us compute H^1. Recall the theorem from Advanced Calculus which states: 'the curl of a vector function is zero if and only if it is the gradient of a scalar function'. We have

$$\begin{aligned}
& f_1 dx + f_2 dy + f_3 dz \in Z_1 \\
& \Longleftrightarrow d^1(f_1 dx + f_2 dy + f_3 dz) = 0 \\
& \Longleftrightarrow \frac{\partial f_3}{\partial y} = \frac{\partial f_2}{\partial z}, \ \frac{\partial f_1}{\partial z} = \frac{\partial f_3}{\partial x}, \ \frac{\partial f_2}{\partial x} = \frac{\partial f_1}{\partial y} \\
& \Longleftrightarrow \text{there exists } g(x, y, z), \text{ such that } d^0(g(x, y, z)) = f_1 dx + f_2 dy + f_3 dz \\
& \Longleftrightarrow f_1 dx + f_2 dy + f_3 dz \in B^1
\end{aligned}$$

Therefore, we have

$$H^1 = 0$$

(3) Let us compute H^2. Recall the theorem from Advanced Calculus which states: 'the divergence of a vector function is zero if and only if it is the curl of a vector function'. We have

$$\begin{aligned}
& f_1 dy \wedge dz + f_2 dz \wedge dx + f_3 dx \wedge dy \in Z_2 \\
& \Longleftrightarrow d^2(f_1 dy \wedge dz + f_2 dz \wedge dx + f_3 dx \wedge dy) = 0 \\
& \Longleftrightarrow \frac{\partial f_1}{\partial x} + \frac{\partial f_2}{\partial y} + \frac{\partial f_3}{\partial z} = 0 \\
& \Longleftrightarrow \text{there exists } h_1 dx + h_2 dy + h_3 dz \in C^1, \text{ such that} \\
& \qquad d^1(h_1 dx + h_2 dy + h_3 dz) = f_1 dy \wedge dz + f_2 dz \wedge dx + f_3 dx \wedge dy \\
& \Longleftrightarrow f_1 dy \wedge dz + f_2 dz \wedge dx + f_3 dx \wedge dy \in B_2
\end{aligned}$$

Therefore, we have

$$H^2 = 0$$

(4) Let us compute H^3. It is easy to see that $Z^3 = C^3 = \mathbf{R} dx \wedge dy \wedge dz$. On the other hand, for any $f(x, y, z) dx \wedge dy \wedge dz \in C^3$, it is easy to find $g(x, y, z)$ such that,

$$\frac{\partial g}{\partial x} = f(x, y, z)$$

Therefore, we have,

$$\begin{aligned}
& d^2(g dy \wedge dz) = f dx \wedge dy \wedge dz \\
& B^3 = C^3 \\
& H^3 = 0
\end{aligned}$$

(5) It is easy to see that $H^i = 0$ for all other i. ∎

Example 5. *Let* $\mathbf{R}$=$\mathbb{C}((x))$ *the formal meromorphic function field in* x= *the quotient field of the formal power series ring* $\mathbb{C}[[x]]$. *Let a complex* (C, d) *be defined as*

$$\begin{aligned} C^0 &= \mathbf{R} = \mathbb{C}((x)) \\ C^1 &= \mathbf{R}dx = \mathbb{C}((x))dx \\ C^i &= 0, \quad \text{for all other } i. \\ d^0(f(x)) &= f'(x)dx \\ d^i &= 0, \quad \text{for all other } i. \end{aligned}$$

(1) Let us compute H^0. *Note that* $Z^0 = \mathbb{C}$ *and* $B^0 = 0$. *Therefore, we have* $H^0 = \mathbb{C}$.
(2) Let us compute H^1. *It is easy to see that* $Z^1 = C^1 = \mathbb{C}((x))$. *We have*

$$\begin{aligned} & f(x)dx = \sum_{i=-m}^{\infty} (a_i x^i)dx \in B^1 \\ \Longleftrightarrow\ & \text{there exists } g(x) \in \mathbb{C}((x)) \text{ such that} \\ & \quad g'(x) = f(x) \\ \Longleftrightarrow\ & a_{-1} = 0 \end{aligned}$$

Therefore, we conclude that

$$B^1 = \{\sum_{i=-m}^{\infty} a_i x^i dx : a_{-1} = 0\}$$

Let us define [illegible]*due map* $r : C^1 \to \mathbb{C}$ *as follows,*

$$r(\sum_{i=-m}^{\infty} a_i x^i dx) = a_{-1}$$

Then we see that $B^1 = ker(r)$ *and*

$$H^1 = Z^1/B^1 \approx \mathbb{C}$$

(3) It is easy to see that $H^i = 0$ *for all other* i. ∎

INDEX

www.ingramcontent.com/pod-product-compliance
Lightning Source LLC
Chambersburg PA
CBHW061620220326
41598CB00026BA/3829
9789810211967